Irina Rebrova
Re-Constructing Grassroots Holocaust Memory

SpatioTemporality/ RaumZeitlichkeit

Practices – Concepts – Media /
Praktiken – Konzepte – Medien

Edited by / Herausgegeben von
Sebastian Dorsch, Barbel Frischmann, Holt Meyer,
Susanne Rau, Sabine Schmolinsky,
Katharina Waldner

Editorial Board
Jean-Marc Besse (Centre national de la recherche scientifique de Paris),
Petr Bilek (Univerzita Karlova v Praze), Fraya Frehse (Universidade de São Paulo), Harry Maier (Vancouver School of Theology), Elisabeth Millán (De-Paul University, Chicago), Simona Slanicka (Universität Bern),
Jutta Vinzent (University of Birmingham), Guillermo Zermeño (Colegio de México)

Volume/Band 11

Irina Rebrova

Re-Constructing Grassroots Holocaust Memory

The Case of the North Caucasus

ISBN 978-3-11-068886-3
e-ISBN (PDF) 978-3-11-068899-3
e- ISBN (EPUB) 978-3-11-068904-4
ISSN 2365-3221

Library of Congress Control Number: 9783110688863

Bibliographic information published by the Deutsche Nationalbibliothek
The Deutsche Nationalbibliothek lists this publication in the Deutsche Nationalbibliografie;
detailed bibliographic data are available on the Internet at http://dnb.dnb.de.

© 2020 Walter de Gruyter GmbH, Berlin/Boston
Photograph on the book cover: The first obelisk at the killing site near the glass factory in
Mineralnye Vody, 1943, renovated in 2019 (author's photograph, 2019).
Typesetting: Integra Software Services Pvt. Ltd.
Printing and binding: CPI books GmbH, Leck

www.degruyter.com

Acknowledgements

In 2013 I moved to Germany to begin a new period of my life, both professional and private. Already holding a Russian doctorate [*kandidat nauk*] in History, I decided to conduct a PhD project in a European country. I was fortunate to join the amazing community of the Zentrum für Antisemitismusforschung (ZfA) at the Technische Universität Berlin, and to be supervised by its magnificent director Stefanie Schüler-Springorum. We had become acquainted several years before, during a conference on female experiences of World War II at the German Historical Institute in Warsaw. Over the past several years, this initial acquaintance has evolved into a very fruitful collaboration. After I had already begun my doctoral project, and after much persistence, I was happy to be taken on by my second supervisor, Dieter Pohl. It was important to me to have the insights of a specialist in Soviet Holocaust history, with both knowledge of Russian and experience conducting research in the North Caucasus. I am thankful to both supervisors for their invaluable comments and advice, their support and trust.

Many colleagues in Russia, Germany, Israel, the United States, Ukraine, and other countries shared their thoughts with me and gave me valuable feedback on my work in progress. All of them willingly answered my emails and messages, talked at length over the Internet, met in cafes to "discuss things over a cup of coffee," and commented on drafts of my chapters. I hope each one of you, my dear friends and colleagues, will recognise in these words my gratitude for your help during these past several years. I do not dare to name you all here, because the list would be too long and I would not want to leave anybody out. This is why, while thanking each of you personally, I will mention only a few names here. The incredible Galina Orlova brought me to the idea of applying Foucault's dispositif theory to Holocaust memory. This proved to be the best lens through which to reconstruct the life of Holocaust memory in a region not noted for its Jewish history, and to reveal the visible and hidden practices of remembrance of Holocaust victims throughout the Soviet and post-Soviet periods. Galina's ideas and interpretations inspired me to set about on a very ambitious task of observing the variety of networks through which Jews, as one of the Nazi victim groups in the North Caucasus, are remembered (or not). I had very fruitful discussions with Arkadi Zeltser, who commented on my chapter on monuments as he was writing his own book about Jewish Holocaust memorials in the Soviet Union. Leonid Terushkin provided me with thematic and pedagogical literature in Russian on teaching about the Holocaust, as well as new materials that he receives from Jews all over the world as Head of the Russian Research and Educational Holocaust Center Archive. Andrej Umansky, who wrote a PhD thesis on the history of the Holocaust in South Russia, became not only a friend but also a trusted colleague, who had the answer to all my historical doubts. I am pleased to acknowledge that within the very close circle of the researcher's world such like minds can come together without rivalry.

Research would not have been possible without travel to the North Caucasus. Between 2013 and 2016 I spent several months in each region, working in local archives, museums, and libraries; visiting monuments and memorial complexes devoted to "peaceful Soviet citizens" or Jews as victims of the Holocaust; conducting interviews with Holocaust survivors, members of local Jewish communities, history teachers at secondary schools, artists, local historians, and activists. In every single case, I met genuine interest in my topic and willingness to share with me their knowledge and sources. Collecting evidence for my research also took me to Moscow, then to Israel, Germany, and the United States. These acquaintanceships in the region and all over the world with people who have subsequently become good friends have been the most pleasant part of my project. Again, while not forgetting anyone – they can all be found in this book – I would like to single out a few: Irina Svetlichnaya, a history teacher at Ust-Labinsk's Gymnasium No. 5, provided me with teaching materials and shared her own experience in Holocaust education. Anatolii Karnaukh, a local historian from the village of Arzgir in Stavropol Krai, personifies the image of the altruistic researcher of the wartime and Holocaust history of his own village and region. Aleksandr Okhtov became my personal guide to the history of the rescuing of children from Leningrad by the villagers in Beslenei aul, in today's Republic of Karachay-Cherkessia. Efim Fainer, Mikhail Potapov, Tatyana Yakubovskaya, Yakov Frenkel', Dmitrii Bekker, local Jewish community leaders in the North Caucasus, shared their experiences, their achievements and failures in their attempts to preserve Holocaust memory. Yurii Teitelbaum gave me access to his personal archive, which he had collected as a leader of the Jewish community in Krasnodar and which he took with him when he moved to Israel. Aleksandr Kozhin provided me with unique documents pertaining to the lawsuit over a Holocaust memorial in Rostov-on-Don, which alongside my interview with him helped me to unravel the issues at stake in the case. I spent so many fascinating hours talking with Holocaust survivors and their second-generation relatives. At first, I felt shy when they invited me to share a meal with them after the interview, but then it became a tradition, closing the circle of the traumatic past and bringing us back to the present. It was a unique opportunity to be able to identify so fully with the experiences of survivors, as happened, for example, with Yakov Krut and Evelina Ekonomidi.

All my research trips were made possible by innumerable scholarships and travel grants which I was privileged to receive. A stipend granted by the ZfA at the beginning of my doctoral research enabled me to spend this first year learning German and reading scholarly literature. My background in the Russian academic world had not yet equipped me to conduct research in the European scientific tradition. A lack of English-language literature in Russian provincial libraries, access to international databases, and knowledge of foreign languages, long teaching hours at the university and little time for research are among the main obstacles facing provincial Russian scholars. At the same time, the continued use of Soviet terminology and the predominance of out-of-date methods, especially those applied to the study of World War II, hinder a renewal of the field in Russia. This is why I was so happy in this new setting to read new books, attend seminars, workshops, conferences, and summer schools, and devote time to self-development and professional growth. In 2014 I took up a scholarly residency at the Hadassah-Brandeis Institute and the Brandeis-Genesis Institute (HBI-BGI) at Brandeis University, Massachusetts. During that stay I visited the United States Holocaust Memorial Museum (USHMM) for the first time. The incredible variety of sources available and their ease of access to researchers and the general public impressed me and simplified my archival research. I have subsequently been able to spend several weeks almost every year at the USHMM – and also at Yad Vashem in Israel – participating in thematic summer schools devoted to studying the Holocaust in the Soviet Union. Perhaps most affirming of the value of my research, and certainly a personal high point, was being honoured with a fellowship at the Claims Conference Kagan Fellowship in Advanced Shoah Studies during the 2015–2017 academic years. This prestigious fellowship gave me the opportunity to discuss my thesis with colleagues, prominent scholars, and young researchers. I learnt about the doctorial work of another fellow, Maris Rowe-McCulloch, who also became a friend.

My research trips to the North Caucasus were supported by Sefer, a Russian center for researchers sponsored by the Genesis Philanthropy Group. The opportunity to participate and present at their annual conferences and winter schools on Jewish studies since 2015, as well as to make a research trip to Yad Vashem in 2016 as a fellow of the Eshnav Program, furnished me with new knowledge and sources, and led to important discussions with colleagues, and of course to new friends. Thanks to webinars offered by Sefer, I learnt enough Yiddish to read and understand articles in the Soviet media.

Having used interviews of the University of South California Shoah Foundation (USCSF) and having spent the best part of academic year 2014–2015 listening to and transcribing interviews with Holocaust survivors, I was very happy to benefit from a month-long scholarship in 2017 from the

USCSF Center for Advanced Genocide Research and to have access to videos of training seminars for interviewers in Moscow and Kyiv in the mid-1990s. I managed to find several former interviewers and they kindly agreed to respond to my survey, which I prepared together with Mikhail Tyaglyy, whose manner of conducting interviews inspired me greatly. All these sources helped me better to understand the context, characteristics, and uniqueness of these interviews with survivors of the Holocaust on Soviet soil conducted after the fall of the Soviet Union.

As a fellow of the Zentrum für Holocaust-Studien at the Institut für Zeitgeschichte (IfZ) in 2016 I continued my archival work at the institute and in the state archive and libraries of Munich, benefiting also from the eyes of colleagues who inspired me to look at my thesis from a new perspective. Participation in a number of summer schools brought me many new ideas and friends, with whom I continue to work on new projects and ideas. This is the most wonderful thing that participating in scholarly activities can afford. My last scholarship, a STIBET Degree Completion Grant (funded by the German Academic Exchange Service, DAAD) came from my home university, TU Berlin. It feels very symbolic to close the circle, returning to the starting point but now at a completely new level.

It has taken me almost one-and-a-half years to submit my thesis for publication. I took this time to gain distance from my arguments, and to compare them with the outcomes of my exhibition project dealing with the history and memory of another Nazi victim group – people with disabilities – under the German occupation of the North Caucasus. During this time, Holt Meyer and I discussed many of the particularities of my case study, which resulted in my decision to submit my manuscript to the Spatiotemporality book series. I would like to express my gratitude to the ZfA and the Erfurt University research group Erfurter RaumZeit Forschung (ERZ – Spatio-Temporal Studies Erfurt) for financially supporting for this publication. I was happy to reunite with my old friend Tristam Barrett, who edited the final text and gave me valuable feedback. Many friends, former fellow students, colleagues, and finally my own students in Russia encouraged me to take on this project. My new friends and colleagues in Berlin, with whom I shared an office at the ZfA or attended a variety of academic events, including book presentations, thematic exhibitions (Tatiana Manykina, in particular) expressed their support throughout this journey.

I address my deepest thanks, though, to my relatives and family – my mother, father, brother – and of course Denis: without your crazy idea back in 2012 this project would never have happened. You all know me well and I am so glad of your support over these past years, alleviating my doubts, and giving me the energy to go forward. I love you deeply, although I do not often say it directly to each of you.

I am submitting this manuscript to the publisher at a time of "social distancing," when all of us need to stay at home and isolate ourselves from each other. In these days we understand the value of personal relations, love, support, and community. Many Holocaust survivors were able to survive only because of such individual support. In the 21st century we face another catastrophe, but I hope it will make us more sensitive, attentive to each other, and thankful. And that this book will find its reader in the new world. Be healthy.

Belin, 31 March 2020

Contents

Acknowledgements —— V

1 **Introduction** —— 1
    1.1   Memory Dispositif Theory and its Regional Dimension —— 7
    1.2   Excavating the Holocaust Memory Dispositif in the North Caucasus —— 21

2 **The Holocaust in the North Caucasus** —— 29
    2.1   The North Caucasus: A Socio-Economic, Political, and Ethnic Overview —— 30
    2.2   Jewish History in the North Caucasus before and during World War II —— 39

3 **(Re)writing Holocaust Memory in Official Soviet Wartime Sources** —— 55
    3.1   Collecting Evidence: The Production and General Characteristics of Soviet Official Wartime Documents —— 57
    3.2   From Personal Statement to Final ChGK Report: Changes in the Rhetoric of Testimony —— 64
    3.3   The Place of Official Wartime Sources in the Holocaust Memory Dispositif —— 83

4 **Memorial Sites: Re-framing Holocaust Memory** —— 92
    4.1   "Memory in Stone"? The Commemoration of Holocaust Victims in the North Caucasus in Historical Perspective —— 98
    4.2   Between Official Ideology and Private Memory: Is Zmievskaya Balka the Largest Holocaust Site in Russia? —— 128
    4.3   Monuments as Social Agents in Translating the Memory of Holocaust Victims —— 137

5 **Depicting Jewish Fates: Artistic Responses to the Holocaust in the North Caucasus** —— 145
    5.1   "My Jews:" The Occupation and Holocaust History in the North Caucasus in Soviet/Russian Imagery —— 150
    5.2   "Retribution against Fascists and Collaborators" in Soviet Literature and Cinema —— 159
    5.3   "Accidental Survival:" Images of Holocaust Survivors and their Rescuers —— 168

5.4 A Look from Tomorrow into Yesterday: A Reflection on Artistic Representations of the Holocaust —— 175

6 Reifying Memory? Representations of North Caucasus Holocaust History in Museum Collections —— 181
6.1 Holocaust History without its Story: (Mis-)representation of the North Caucasian Case in International Holocaust and Russian State Museums —— 185
6.2 Regional Practices of Museification of Holocaust History in the North Caucasus —— 201

7 We Don't Need No . . . Holocaust Education? —— 209
7.1 Holocaust History in School Textbooks: From Depersonalisation to Recognition —— 213
7.2 The Spark that Starts the Fire: Regional History Teachers' Initiatives to Teach about the Holocaust —— 228
7.3 Initiatives "from below:" Schoolchildren's Research on Holocaust History in the North Caucasus —— 240

8 Revisiting Experience: Personal Narratives of the Holocaust in the North Caucasus —— 248
8.1 "We used to have a big family; now I am the only survivor:" Historical Overview and Thematic Analysis of the Personal Narratives of Holocaust Survivors —— 252
8.2 "The city is destroyed. All the Jews have been shot:" Eyewitness Narratives of the Holocaust in the North Caucasus —— 267

9 Conclusion. Key Features of the Holocaust Memory Dispositif —— 286

Bibliography —— 303

Tables —— 341

List of Abbreviations —— 351

Index —— 355

1 Introduction

Ust-Labinsk is a sunny provincial town[1] buried in the verdure of Russia's south, 200 km from the foothills of the Western Caucasus, the mountain range that nowadays separates Russia from its southern neighbours. It is situated at the confluence of the Laba and the Kuban rivers, the second of which has given its name to the area, and informally to the larger Krasnodar region itself.[2] The town's history stretches back more than two hundred years to the first construction of military fortifications by General Aleksandr Suvorov during Catherine the Great's Russo-Turkish Wars. It was subsequently occupied by Don Cossacks and became one of their capitals [*stanitsa*] for many long decades. The principal turning points of Russian and later Soviet history all left their mark on the body of this Kuban village, altering its appearance and affecting the lives of its inhabitants. The Second World War left the deepest wound: a wound that is still bleeding.

Jews were uncommon in the Cossack villages of the south. Yet from July 1941, freight cars full of evacuees and refugees started arriving from the western regions of the Soviet Union, many Jews among them. The family of the surgeon Vladimir (Wolf) Pinkenson came from the Bessarabian town of Bălți (Beltsy) and found their new home in Ust-Labinskaya in the autumn of 1941. Pinkenson's only son Musya[3] was fond of music and learned to play the violin. Musya's new schoolmates remembered him as a talented musician who often gave concerts in the Soviet military hospital.[4] He graduated with honours from the fifth grade in the spring of 1942, and his teachers remembered him as "an exemplary schoolboy and pioneer. He had the bright imprint of a happy childhood."[5] However, in the summer of 1942, war broke into the village and Ust-Labinskaya fell under Nazi occupation for more than six months between August 8, 1942 and February 2, 1943. During this period, as the official Soviet account has it, "more than 400 innocent Soviet citizens were tortured and shot, including 116 men, 194 women, [and] children aged

---

[1] The village of Ust-Labinskaya was founded in 1794 and granted town status as Ust-Labinsk in 1958.
[2] The Russian system of federal subdivision includes such federal subjects as *krai* and *oblast'*. Legally identical, these are both understood as relatively large economic and political regions.
[3] Musya's full name is given as Abram or Moses in different sources.
[4] Memories of Musya Pinkenson's former classmate Anna Bakieva, teachers Elena Sakhno and Nadezhda Bannova, neighbour Maria Repeshchuk (likely written in the 1970s), in *personal archive of Irina Svetlichnaya*. Unless otherwise specified, all translations are by the author. The style and language of the original sources have been retained.
[5] Irina Kononenko, "Kak pogib Musya Pinkenson," *Sovetskaya Kuban'*, May 8, 1943.

from 2 months to 15 years. 35 of those murdered were Russians, the rest were Jews."[6] Fleeing from the Nazi genocide in their native towns, these Jews found their deaths in exile in the steppes and mountains of the Caucasus. The Pinkensons – father, grandfather, mother and Musya himself – were all shot "in a ditch prepared in advance in the fortress northeast of the Krasnyi Forshtadt collective farm."[7] After the war, the story of the shooting of this twelve-year-old violinist was lauded as an almost epic example of the courage of young heroes:

> Everyone looked on in terror as machine-gun fire wiped out the famous doctor's family: his parents, his wife, the doctor himself. And here comes the boy's turn already.
> – "Officer, sir, let me play, one last time!" asked Musya.
> – "Play? Here? Why not! It is even funny, only *schnell*, hurry up!"
>
> Putting the violin case down, the boy removed the violin carefully and began tuning it. Hundreds of amused and anxious eyes followed his every movement. Facing the enemy's machine guns aimed at his chest, the boy waved his bow and began to play with inspiration . . . the *Internationale*. People seemed to stand still for a moment, and then it resounded all of a sudden, at first timid, then bolder, then bolder still, and louder: "This is the final . . ."[8] Enraged fascists, coming to themselves, opened fire. Musya fell. The violin fell out of his hands . . .[9]

After the mass killing of Jews in Ust-Labinskaya was registered in a report of the Extraordinary State Commission [*Chrezvychainaya gosudarstvennaya komissiya*], hereafter ChGK, which had been working in the village in the summer of 1943, the story of the brave pioneer[10] made headlines in local and national

---

6 "Extraordinary State Commission to Investigate German-Fascist Crimes Committed on Soviet Territory (hereafter ChGK) report for Ust-Labinskaya village of July 30, 1943," in *State Archive of Krasnodar Krai* [*Gosudarstvennyi arkhiv Krasnodarskogo kraya*], hereafter *GAKK*, f. R-897, op.1, d. 2–1, l. 151; Idem, in *State Archive of the Russian Federation* [*Gosudarstvennyi arkhiv Rossiiskoi Federatsii*], hereafter *GARF*, f. R-7021, op. 16, d. 8, l. 161.
7 Ibid.
8 "This is the final struggle . . ." is first line of the chorus of the Internationale. The Internationale was the official anthem of the Soviet Union between 1922 and 1944, translated from French into Russian by the Soviet poet and translator Arkadii Kots.
9 "Musya Pinkenson," in *Evreiskie deti v bor'be s natsizmom: vsemirnyi dokumental'nyi sbornik*, ed. Maks Privler, Anna Kremyanskaya, and Pavel Kremyanskii, vol. 1 (Tel-Aviv: Biblioteka Matveya Chernogo, 2001), 25. According to another version, a Nazi officer saw a violin in the hands of the boy and forced him to play. In any case, numerous stories attest that it was the Internationale that Musya was playing.
10 It remains unclear whether Musya had time to be enrolled in the pioneer movement. However, the Soviet media treated him as one of the hero-pioneers to whom numerous novels

newspapers.[11] But the boy's ethnicity and cause of death were gradually removed from official heroic Soviet discourse about the "Great Patriotic War."[12] The image of a young hero, the boy-violinist, became the topic of many Soviet literary works: short stories, essays, and lyrics.[13]

A modest wartime obelisk built to commemorate this event was replaced in 1971 by an imposing monument adorned with a bas-relief of a smiling boyish face (Figure 1) on the site of the mass grave of 370[14] civilians. The Ust-Labinsk artist Viktor Smetanin created this monument. It was erected to commemorate the 50th anniversary of the All-Union Pioneer Organisation named after V.I. Lenin. The inscription "Eternal Memory to Ust-Labinsk citizens shot in 1942" was carved at the foot of the monument. Another inscription on the other side of the monument reads: "Sleep well, it will never happen again! Eternal memory to the fallen in the struggle for freedom and independence of our Motherland." The return to the ideals

---

and short stories were dedicated. See: *Pionery-geroi: Al'bom-vystavka* (Moscow: Malysh, 1969), 25–27; Albert Likhanov et al., *Pionery-geroi* (Moscow: Malysh, 1980), 32–38.

11 See, for example, Elena Kononenko, "Slava sovetskim detyam," *Pravda*, May 21, 1945; "Ego oruzhiem byla skripka," *Izvestiya*, February 22, 1968; Aleksei Makarenko, "Pioner-geroi iz Moldavii," *Vechernii Kishinev*, March 2, 1968; Tatyana Stepanova, "Pioner – otvazhnyi borets za kommunizm," *Sel'skaya nov'* [town of Ust-Labinsk], February 7, 1974; Tatyana Gendler, "Zvuchi, prostrelyannaya skripka," *Molodezh' Moldavii*, May 5, 1981. See also materials about Musya in the Yad Vashem Archive, hereafter YVA, RG-O.33, file 5039.

12 The term the "Great Patriotic War" [*Velikaya Otechestvennaya voina*, lit. the "Great Fatherland War"] is widely used in Soviet and Russian historiography. It is the official name of the conflict between Nazi Germany and the Soviet Union from June 22, 1941 to May 9, 1945 along the many theatres of the Eastern Front in World War II. For today's Russian society the war is more connected with the existential struggle for Soviet Union and the fates of Soviet citizens than with the danger of Nazism for Europe and the world. For this reason, the war is perceived not as a world war, but as a patriotic one. Russians celebrate the Great Victory on May 9th, although in the rest of the world it is commemorated on May 8th. See: Boris Dubin, "'Krovavaya' voina i 'velikaya' pobeda," *Otechestvennye zapiski* 5 (2004): 86. I use the term "Great Patriotic War" (with or without quotation marks) to highlight specifically Soviet (and Russian) discourse practices of the war.

13 Saul Itskovich, *Musya Pinkenson* (Moscow: Malysh, 1981); Vasilii Velikanov, "Ranenaya skripka," in *Put' otvazhnykh: rasskazy*, ed. Sergei Baruzdin (Moscow: Detgiz, 1962), 101–110; "Musya Pinkenson," in *Deti Kubani v Velikoi Otechestvennoi voine*, ed. Tatyana Khachaturova (Krasnodar: Traditsiya, 2008), 45–47.

14 The ChGK report reads "more than" and "up to" 400 Soviet civilians became Nazi victims in Ust-Labinskaya. The monument's inscription indicates 370 victims. To establish the exact number of Nazi victims, including Jews as a special victim group, in the studied region is a most complicated task: there are almost no German wartime documents and the estimates of the ChGK were in many cases too high or too low. See: Marina Sorokina, "Lyudi i protsedury. K istorii rassledovaniya natsistskikh prestuplenii v SSSR," in Viktor Suvorov and Dmitri Khmelnitski, ed. *Voenno-istoricheskii al'manakh Viktora Suvorova*, vol. 2 (Moscow: Dobraya Kniga, 2013), 98.

Figure 1: Monument at the "mass grave of 370 civilians and the pioneer hero Musya Pinkenson, killed by fascist invaders," by Viktor Smetanin, Ust-Labinsk, 1971 (Photo by the author, 2015).

of Leninism in the 1960s and 1970s was partially linked to the de-Stalinisation [*destalinizatsiya*] of the Khrushchev era (1953–1964), but also to the desire of public authorities to control the official memory of the Great Victory by preventing the penetration of various "counter-memories," especially of Jews as the main victims of the Nazis, and of Soviet prisoners of war (POWs).[15] The same year (1971) a cartoon – "Pioneer's Violin" [*Skripka pionera*] – was made, based on Musya's story.[16] The young cartoon hero is shown as a brave pioneer challenging an evil-looking Nazi, while playing the first notes of the Internationale instead of the merry song he was asked for. Since that time, teachers and schoolchildren of Ust-Labinsk Secondary School No.1 started to collect evidence and relics of Musya Pinkenson's life. The

---

15 For more on the formation of this war myth in Soviet official discourse, see: Nikolai Koposov, *Pamyat' strogogo rezhima: istoriya i politika Rossii* (Moscow: Novoe Literaturnoe Obozrenie, 2011), 91–110.
16 *Skripka pionera*, directed by Boris Stepantsev, written by Yurii Yakovlev, camera Mikhail Druyan (Soyuzmultfilm, 1971), 8 min.

collected memories, the restored biography of the boy, his portrait and other objects quickly became the property of a school museum. More than 15 of the best Pioneer units in the district of Ust-Labinsk were named in honour of Musya.[17] The internationalisation of Musya's history is shown in the correspondence between Soviet pioneers and East German schoolchildren; several groups of East German teenagers visited Musya's monument in the 1970s.[18]

Thus, in official Soviet memorial culture the mass killing of Jews as the main victims of the Nazis was symbolically minimised and replaced, in this particular case, by the heroic image of a pioneer with a violin in his hands before his execution. The violin symbolised a weapon in the struggle with the enemy. Although in a society used to reading between the lines of official narratives, those so inclined could read Musya's ethnicity from his name and even the violin in his hand (a quintessential marker of Jewishness in Soviet and Russian stereotypes), the reasons for the shooting of Musya and other "Soviet citizens" were of no importance to official Soviet memorial culture, the boy's defiance itself was canonised; the story grew like a snowball.

Musya's story was not forgotten in post-Soviet Russia. On the contrary, it has assumed a new significance in the context of Holocaust studies. Following the collapse of the Soviet Union, attention has shifted in the memories of eyewitnesses, historians, and broadcasters to turn this event into a story about the tragic fate of Soviet Jews during World War II, and to reinterpret Musya's heroic deed as an intrepid act of Jewish defiance. Numerous newspaper articles,[19] scholarly papers,[20] and the writings and personal investigations of bloggers[21] as well as

---

17 "Report on a research project on the Kuban hero pioneer Musya Pinkenson, March 1975, Secondary School No.1, Ust-Labinsk," in *personal archive of Irina Svetlichnaya*.
18 Johann Becker, "Er starb wie ein Held," *Neues Leben*, May 4, 1977.
19 See, for example, Mikhail Tkachenko, "Kholokost – chuzhaya bol'," *Sel'skaya nov'*, January 27–February 2, 2006; Mirra Gunitskaya, "Ya znala Musyu Pinkensona," *Vozrozhdenie: gazeta evreiev Kubani*, December 1, 2006–January 31, 2007; Yuliya Sister, "Podvig yunogo skripacha," *Vremya evreyev*, February 28, 2008; Dariya Yuzhnaya, "Zverstva gitlerovtsev v Ust-Labinskom Raione," *Sel'skaya nov'*, May 8, 2010; Andrei Sidorchik, "Muzyka kak oruzhie. Poslednii kontsert Musi Pinkensona," *Argumenty i fakty*, March 31, 2014.
20 Il'ya Al'tman, *Zhertvy nenavisti. Kholokost v SSSR 1941–1945* (Moscow: Kollektsiya "Soversheno sekretno," 2002), 315; Lev Bondar, *Detyam detei rasskazhite* . . . (Beltsy: Assotsiatsiya evreiskikh organizatsii goroda Beltsy, 2006).
21 Mark Kagantsov, "Rasstrelyannaya skripka," online project *"Sem' na sem' – gorizontalnaya Rossiya,"* http://7x7-journal.ru/post/32683; Lyuba Raskin, "Oni upali ryadom: malchik i skripka," *Livejournal*, October 23, 2010, http://abu-tir.livejournal.com/1590635.html; Yuliya Medvedeva, "Musya Pinkenson," *Forum "Molodaya Gvardiya,"* October 10, 2013, http://molodguard.ru/forum/viewtopic.php?f=28&t=623 (accessed March 20, 2020).

broadcasts on national and local TV channels[22] represent the young violinist Pinkenson's story as an integral part of the history of the Holocaust on Soviet soil.[23] If in the Soviet period schoolchildren laid flowers to Musya's monument mostly on Pioneer's Day (May 19), after the fall of the USSR, the monument is visited on Victory Day (May 9), and recently also on International Holocaust Remembrance Day (January 27). Ust-Labinsk's local history museum hosts a permanent exhibition "Kuban during the Great Patriotic War" with a display dedicated to the young violinist's fate. The museum collection also includes witness testimonies about the shooting of Jews as well as recollections of Musya's personality. Local researchers and professional historians have had access to most ChGK files in the local archives since the late 1980s. These shed light on Holocaust history in the USSR: mostly the victims' names but occasionally also their fates. The schoolchildren of Ust-Labinsk Gymnasium No.5 are actively engaged in research work that started in Soviet times. The only difference is that their research is now devoted to the history of the mass killing of Jews evacuated to Ust-Labinsk, and Musya is perceived as a hero of an individual Jewish resistance. Annual school events often focus on the tragedy of war, and children write letters to Musya's relatives.[24] Mirra Gunitskaya, a living witness to the tragedy and a Holocaust survivor, was a frequent guest at school events and academic seminars organised by a Russian Research and Educational Holocaust Center (hereafter, Holocaust Center) in Krasnodar Krai in the early 2010s.[25] Musya's story and the honour of getting to know him are the main topics of her numerous interviews and publications.[26]

The development of Musya Pinkenson's story is a classic example of how Holocaust remembrance leads its own life in Soviet and post-Soviet Russia. The history of the mass killing of European Jews has been recalled differently in various historical periods. In the Soviet period the stories were mythologised: they were more abstract, Jewish attributes were diminished, and the focus was on the

---

[22] Aleksandra Proskurina, "Istoriki Ust-Labinska rasskazali o podvige evreiskogo mal'chika-skripacha," *Kuban' 24*, aired April 16, 2015.

[23] I will use the term "Holocaust" in its original and narrow sense: The Holocaust was the systematic, bureaucratic, state-sponsored persecution and murder of six million Jews by the Nazi regime and its collaborators. "Introduction to the Holocaust," *Holocaust Encyclopaedia*, https://www.ushmm.org/wlc/en/article.php?ModuleId=10005143 (accessed March 20, 2020).

[24] See correspondence with Musya's cousin, Boris Gendler and his unpublished memories in personal archive of Irina Svetlichnaya.

[25] Interview with Irina Svetlichnaya, in *author's archive*, code 15-MEM-KK02, May 29, 2015, Ust-Labinsk.

[26] Interview with Mirra Gunitskaya, born in 1933, in *author's archive*, code 13-X-KK03, March 6, 2013, Krasnodar; Gunitskaya, "Ya znala."

glorification of heroic Soviet citizens, who, as the saying went, "could be killed but not broken." Musya thus became not a helpless Jewish boy faced with death in a ditch, but instead a fervent believer in the Soviet Communist doctrine. Although Russia after the fall of the USSR faced the task of demythologising Soviet practices of remembrance, the tendency to remember all victims of the war as "Soviet citizens" is still preserved. At the same time the Holocaust as a historical fact has been accepted by the Russian authorities, which has gradually led to the first steps in creating a national culture of Holocaust memory. This book seeks to trace Holocaust remembrance from the bottom-up in the North Caucasus during the existence of the Soviet Union and afterwards. Historically, in terms of its ethnic and cultural diversity, the North Caucasus is unique in Russia, a factor which influenced both the unfolding of the Holocaust in this region and its subsequent remembrance. Yet the North Caucasus is also significant because it became the last resting place for approximately half of the Jews who were killed on the territory of the present-day Russian Federation. While recent scholarship has helped to refigure our understanding of the Holocaust, its unfolding and its subsequent remembrance in the "Bloodlands"[27] of Eastern Europe, the study of this region sheds light on these issues in an important, yet overlooked site of the Holocaust, and, most importantly, the factors affecting its memorialisation in the regional and national memorial culture of Soviet and post-Soviet Russia.

1.1 Memory Dispositif Theory and its Regional Dimension

Holocaust Memory in Postmodern Memory Studies

In spite of its rather short history as an interdisciplinary scholarly field, memory studies still generates lively debate: from the critical understanding of the very concept of memory to theorisation of different levels of memory production and memory function, to meta-critiques of the "memory boom"[28] at the

---

[27] The term "Bloodlands" was coined by Yale historian Timothy D. Snyder to describe the political, cultural and ideological context of the mass killing of Jews in Central and Eastern Europe in the war "between Hitler and Stalin." Timothy Snyder, *Black Earth: The Holocaust as History and Warning* (London: The Bodley Head, 2015).
[28] "Memory boom" refers to a development in which the prominence and significance of memory has risen both within the academy and society, primarily in the Western world since the 1970s. It is associated with the works of the French political scientist Pierre Nora. Jay Winter, "The Generation of Memory: Reflections on the 'Memory Boom' in Contemporary Historical Studies," *Archives and Social Studies: A Journal of Interdisciplinary Research* 1 (2007): 363–397.

turn of the century and the "memory industry"²⁹ of the last two decades. Using the terminology of the French sociologist Bruno Latour, there are several important theoretical "uncertainties"³⁰ in the study of memory.

The main uncertainty consists of the variety of terminology and the ongoing debates between specialists about memory affiliation: in other words, whose memory are we trying to understand and study? The terms "social," "collective," or "public" memory often stand in contrast with "private," "individual," or "personal" memory. In practice, the differences in terminology point less to diverging definitions of memory itself than to different approaches to its study. The French sociologist and philosopher Maurice Halbwachs, father of the "collective memory" concept, chose an approach based on the sociological categories of family, class, and religion.³¹ Scholars of "social memory" tend to focus more on the social environment of memory and ask how individual stories about the past interact with existing narratives and other forms of commemoration.³² The concept of "cultural memory" is shared outside the hallways of formal historical discourse yet is entangled with cultural products and imbued with cultural meaning.³³ German scholars Jan and Aleida Assmann further developed this

---

29 See: Kerwin Lee Klein, "On the Emergence of memory in historical discourse," *Representations* 69 (Winter 2000): 127–159; Friederike B. Emonds, "Revisiting the Memory Industry: Robert Thalheim's 'Am Ende kommen Touristen,'" *Colloquia Germanica* 44, no. 1 (2011): 55–78.

30 Latour uses the term "uncertainty" to launch a series of critiques against conventional sociology. Bruno Latour, *Reassembling the Social: An Introduction to Actor-Network Theory* (Oxford: Oxford University Press, 2005).

31 Maurice Halbwachs, *On Collective Memory*, ed., transl. Lewis A. Coser (Chicago: University of Chicago Press, 1992). For the further evolution of this concept in memory studies and its later criticism, see: Noa Gedi and Yigal Elam, "Collective Memory – What is It?," *History and Memory* 8, no. 2 (1996): 30–50; Jeffrey K. Olick, "'Collective Memory': A Memoir and Prospect," *Memory Studies* 1, no. 1 (2008): 23–29; Idem, "Collective Memory: The Two Cultures," *American Sociological Association* 17, no. 3 (2009): 333–48; Jeffrey K. Olick, Vered Vinitzky-Seroussi, and Daniel Levy, ed., *The Collective Memory Reader* (New York: Oxford University Press, 2011).

32 James Fentress, *Social Memory* (Oxford: Blackwell, 1992); Barbara Miszta, *Theories of Social Remembering* (Maidenhead: Open University press, 1993); Jeffrey K. Olick and Robbins Joyce, "Social Memory Studies: From 'Collective Memory' to the Historical Sociology of Mnemonic Practices," *Annual Review of Sociology* 24 (1998): 105–140; Francis X. Blouin and William G. Rosenberg, ed., *Archives, Documentation, and Institutions of Social Memory: Essays from the Sawyer Seminar* (Ann Arbor: University of Michigan Press, 2007).

33 Jan Assmann and Tonio Hölscher, ed., *Kultur und Gedächtnis* (Frankfurt am Main: Suhrkamp, 1988); Marita Sturken, "The Remembering of Forgetting: Recovered Memory and the Question of Experience," *Social Text* 57 (1999): 103–125; Astrid Erll, "Literatur und kulturelles Gedächtnis: Zur Begriffs- und Forschungsgeschichte, zum Leistungsvermögen und zur literaturwissenschaftlichen Relevanz eines neuen Paradigmas der Kulturwissenschaft,"

concept in their numerous publications referring to objectified and institutionalised memories that can be stored, transferred, and reincorporated across generations.[34]

At the same time, the core of any "collective" level of memory converges on the individual. "Private" or "individual" memory is understood as memory located in an individual mind through which an individual obtains knowledge of things that come across his or her personal experience. Memory of this kind is an integral part of the mental functioning of the person and is closely linked to concepts of personality and selfhood.[35] According to the philosopher Walter Benjamin, memory does not give any advantage in the interpretation of the past but it gives an advantage of a different kind: the interpretation of personal experience and of history.[36] The American feminist philosopher Nancy Fraser demonstrated in her research that the question of responsibility for past events and deeds arises within individuals thinking about their own existence. It also affects ethnic groups, nations, states involved in the process of globalisation, and public spheres in general.[37] Thus, the conditional character of boundaries between "non-individual" and individual memories becomes obvious.[38] In practice, we observe an interconnection between all memory levels using case studies of different groups and individuals as bearers of knowledge and memory of the past, and we posit a link between memory and its "speaking through" to explain why an individual memory can become real only as a part of collective memory. Ludwig Wittgenstein's thesis about language[39] stands equally for memory: as there is no private language for their expression, there can be no absolutely private memories.[40] This demonstrates an interconnection between all memory

---

Literaturwissenschaftliches Jahrbuch 43 (2002): 249–276; Astrid Erll and Ansgar Nünning, ed., *A Companion to Cultural Memory Studies* (Berlin and New York: Walter De Gruyter, 2010).
34 See, for example, Aleida Assmann, *Mnemosyne: Formen und Funktionen der kulturellen Erinnerung* (Frankfurt am Main: Fischer-Taschenbuch-Verlag, 1991); Jan Assmann, *Das Kulturelle Gedächtnis* (München: C.H. Beck, 2007); Aleida Assmann, *Cultural Memory and Western Civilisation: Functions, Media, Archives* (New York: Cambridge University Press, 2011).
35 Geoffrey Cubitt, *History and Memory* (Manchester: Manchester University Press, 2007), 14.
36 Walter Benjamin, *Moskauer Tagebuch* (Frankfurt am Main: Suhrkamp Verlag, 1980), 45.
37 Nancy Fraser, "Transnationalizing the Public Sphere: on the Legitimacy and Efficacy of Public Opinion in a Post-Westphalian World," *Theory, Culture and Society* 24, no. 4 (2007): 7–30.
38 Elena Trubina, "Uchast' vspominat': vektory issledovanii pamyati," in *Vlast' vremeni: sotsial'nye granitsy pamyati: sbornik statei*, ed. Valentina Yarskaya and Elena Yarskaya-Smirnova (Moscow: OOO "Variant," 2011), 26.
39 On the impossibility of a private language, see Ludwig Wittgenstein, *Philosophical Investigations*, 2nd ed. (Oxford: Blackwel, 1997).
40 Halbwachs, *On Collective Memory*, 173–174.

levels. On the one hand, individual memory is limited by collective memory and, on the other, collective memory is embodied in (and expressed by) individuals. The universalistic approach that characterised modern scholarly pursuit perceived history as progress in a unified, homogenous and linear form while being convinced that fragmentation, heterogeneity and different paces of temporality experienced by various individuals, groups, and ethnic communities can be ignored as relics. The postmodern break with this tradition has seen scholars, influenced notably by Michel Foucault, begin to study the "rhetoric" of history and "counter-memory," including voices of the past that are only now entering the public space. The insights of postmodernity have led to an acceptance that "discourses" often "speak" through people in the same way as collective memory "remembers" through them.[41]

Contributing to ongoing discussions that conceptualise various forms of memory, Holocaust memory has become a central topic in memory studies in Western scholarship. According to the historian Omer Bartov, the Holocaust has generated new and particularly intense forms of memory.[42] When someone speaks about a memory boom, they are indeed speaking in part – though far from exclusively – of the vast terrains of Holocaust memory and other terrains of memory modelled on it.[43] Research on trauma and regret, silence and untold life experience, which has appeared at the intersection of memory and Holocaust studies,[44] shows the diversity and complexity of Holocaust memory studies. Despite the European and North American achievements in understanding the mechanisms of Holocaust memory

---

41 The Russian philosopher Elena Trubina critically explored the concepts of cultural and social memory and their links to individual memory. She concluded that "individuals do remember, but do they really control the content and the form of their remembrances? Even if they are sure, it can be only a form of self-defeating, or rather a saving illusion." Trubina, "Uchas' vspominat", 25–42.
42 Omer Bartov, *Murder in Our Midst: The Holocaust, Industrial Killing, and Representation* (New York: Oxford University Press, 1996).
43 See more about the Holocaust as the central theme in memory studies in the introduction to *The Collective Memory Reader* by Jeffrey K. Olick, Vered Vinitzky-Seroussi, and Daniel Levy: idem, *The Collective Memory Reader*, 31–36.
44 See: Saul Friedlander, *Memory, History, and the Extermination of the Jews of Europe* (Bloomington: Indiana University Press, 1993); James E. Young, *The Texture of Memory: Holocaust Memorials and Meaning* (New Haven: Yale University Press, 1993); Manfred Gerstenfeld, "The multiple distortions of Holocaust memory," *Jewish Political Studies Review* 19, no. 3/4 (2007): 35–55; Jeffrey C. Alexander, *Remembering the Holocaust: a debate* (New York: Oxford University Press, 2009).

in general, the remembrance of the mass killings of Jews on Soviet soil is a relatively new research field in memory studies.[45]

Another "uncertainty" can be outlined in Holocaust memory studies. Auschwitz became virtually synonymous with mass genocide and the Holocaust itself throughout the world. The Holocaust is taken as synonymous with the death camps in which the systematic murder of Europe's Jews[46] took place. However, Holocaust history in the USSR does not fit into this generally accepted West European framework. The term "Holocaust by Bullets" [47]– a more appropriate characterisation of Nazi policies towards Jews on occupied Soviet territories – has been recently introduced in historical studies thanks to the work of the French organisation Yahad-In Unum. This organisation aims to identify the forgotten sites of mass graves of Jewish victims killed by Nazi mobile killing units [*Einsatzgruppen*] on Soviet soil.[48] Recent research has made it possible to identify the vast extent of mass killing sites

---

45 See: Karel Berkhoff, *Harvest of Despair: Life and Death in Ukraine under Nazi Rule* (Cambridge-London: The Belknap Press of Harvard University Press, 2004); Yitzhak Arad, *The Holocaust in the Soviet Union, The Comprehensive History of the Holocaust* (Jerusalem: Yad Vashem, 2009); John-Paul Himka and Joanna Beata Michlic, ed., *Bringing the Dark Past to Light: The Reception of the Holocaust in Postcommunist Europe* (Lincoln-London: University of Nebraska Press, 2013); Michael David-Fox, Peter Holquist, and Alexander M. Martin, ed., *Holocaust in the East. Local perpetrators and Soviet responses* (Pittsburgh: University of Pittsburgh Press, 2014); Snyder, *Black Earth*.
46 Giorgio Agamben, *Remnants of Auschwitz: The Witness and the Archive*, transl. Daniel Heller-Roazen (New York: Zone books, 1999).
47 This term was introduced in the mid-2000s and is connected to Yahad-In Unum's activities conducting oral interviews with Holocaust eyewitnesses in distant villages of the former Soviet Union in order to mark the sites of Holocaust mass graves. See: Patrick Desbois, *The Holocaust by Bullets: A Priest's Journey to Uncover the Truth Behind the Murder of 1.5 Million Jews* (New York: Palgrave Macmillan, 2008); Idem, *In Broad Daylight: The Secret Procedures behind the Holocaust by Bullets* (New York: Arcade Publishing, 2018).
48 See the map of Yahad-In Unum fieldtrips: "The map of the Holocaust by bullets," *Yahad-In Unum website*, http://yahadmap.org/#map/ (accessed March 21, 2020). If Holocaust history by means of documenting, mapping, and researching ghettos and concentration camps, including death camps, became known due to the publications of the United States Holocaust Memorial Museum (hereafter, USHMM), then the identification, mapping, and researching of all killing sites of Jews on Soviet soil is still awaiting its researchers. Geoffrey P. Megargee, ed., *Encyclopaedia of Camps and Ghettos*, 2 vol. (Bloomington: Indiana University Press, 2009, 2014).

of Jews in the former USSR,[49] such that Babi Yar[50] may become a recognisable symbol of the Holocaust on Soviet soil.

One of the main characteristics of memory is its ability to change constantly in response to political, social, cultural and other processes in the course of time. Among the obvious attributes of memory as a temporal process are the reconstruction of memory politics, the rewriting of history textbooks, the avoidance of some and appearance of other memory symbols on monuments, shifting narratives and identities, the appearance of technologically new places for preserving people's memory, and conflicts between the polarisation of official state and local memory.[51] Changes in the content of memories over the course of time can be clearly observed in the case of Holocaust memory in Russia. Throughout the post-war history of the Soviet Union, Soviet authorities would consciously conceal the ethnicity of war victims among the civilian population.[52] The "Great Patriotic War" of 1941 to 1945 (i.e., World War II, reckoned

---

[49] See, for example, Christian Gerlach, *Krieg, Ernährung, Völkermord: Forschungen zur deutschen Vernichtungspolitik im Zweiten Weltkrieg* (Hamburg: Hamburger Edition, 1998); Andrej Angrick, *Besatzungspolitik und Massenmord. Die Einsatzgruppe D in der Südlichen Sowjetunion 1941–1943* (Hamburg: Hamburger Edition, 2003); Wendy Lower, *Nazi Empire-Building and the Holocaust in Ukraine* (Chapel Hill: University of North Carolina Press, 2005); Ilya Altman, *Opfer des Hasses. Der Holocaust in der UdSSR 1941–1945* (Zürich: Muster-Schmidt Verlag, 2008); Dieter Pohl, *Die Herrschaft Der Wehrmacht. Deutsche Militärbesatzung und Einheimische Bevölkerung in Der Sowjetunion 1941–1944* (München: Oldenbourg Verlag, 2008); Christoph Dieckmann, *Deutsche Besatzungspolitik in Litauen 1941–1944*, 2 vol. (Göttingen: Wallstein, 2011); Waitman Wade Beorn, *Marching into Darkness: The Wehrmacht and the Holocaust in Belarus* (Cambridge, London: Harvard University Press, 2014).

[50] Babi Yar is a ravine in the Ukrainian capital Kyiv and one of the largest killing sites of Jews in the former Soviet Union. More than 33,000 Jews were killed by German forces with the help of local collaborators during September 29–30, 1941. See: Tatyana Evstaf'eva and Vitalii Nakhmanovich, ed., *Babii Yar: chelovek, vlast', istoriya: dokumenty i materialy*, vol. 1, *Istoricheskaya topografiya i khronologiya sobytii* (Kyiv: Vneshtorgizdat, 2004); Jeff Mankoff, "Babi Yar and the struggle for memory, 1944–2004," *Ab Imperio* 2 (2004): 393–415; Aleksandr Kruglov, *Tragediya Babego Yara v nemetskikh dokumentakh* (Dnepropetrovsk: Tsentr "Tkuma", ChP "Lira LTD", 2011); Vladyslav Hrynevych and Paul Robert Magocsi, ed., *Babyn Yar: History and Memory* (Kyiv: Duch i Litera, 2016); Karel Berkhoff, "The Dispersal and Oblivion of the Ashes and Bones of Babi Yar," in *Lessons and Legacies XII: New Directions in Holocaust Research and Education*, ed. Wendy Lower and Lauren Faulkner Rossi (Evanston: Northwestern University Press, 2017), 256–276.

[51] Elena Rozhdestvenskaya et al., ed., *Collective Memories in War* (London-New York: Routledge, 2016), 2.

[52] Al'tman, *Zhertvy nenavisti*, 373–453; Kiril Feferman, "Pamyat' o voine i Kholokoste v sovetskom i postsovetskom kollektivnom soznanii," in *Istoricheskaya pamyat': protivodeistvie otritsaniyu Kholokosta: materialy 5-i mezhdunarodnoi konferentsii "Uroki Kholokosta i sovremennaya Rossiya,"* ed. Il'ya Al'tman (Moscow: MIK, 2010), 76–87.

from the point at which the Soviet Union entered the war) gradually became one of the foundations upon which the legitimation of power was constructed in Soviet society. The myth of the national unity, "non-ethnicity," and internationality of the Soviet people was maintained throughout the history of the USSR. This official war history was in effect a kind of "barrier myth" from the very beginning. The military defeat of the Red Army and the killing of soldiers and civilians in the USSR at the beginning of the war were incorporated into a heroic story of the great feat of the Soviet people, who consciously sacrificed themselves for their Motherland. This myth works as a unifying national myth because it allows the crimes of the Soviet regime to be hidden behind the heroic fanfares of the Victory Day parade.[53] This memory of "Victory" in present-day Russia still represents the keystone of national identity. In no way does this memory emphasise the victims of the war: it instead underlines the role of the powerful state in achieving this victory. In this form, war memory cannot incorporate Holocaust memory or the memorialisation of Nazi victims on Soviet soil in general. Attempts to talk about the Holocaust within the frames of the Great Patriotic War during the Soviet period – and even subsequently – were treated, and even legislated, as "a distortion of historical truth."[54] Thus it is commonly understood that the main victims of the war were "peaceful Soviet citizens." All alternative memories of the war, particularly those pertaining to the repression of ethnic groups, were ruthlessly suppressed, but not eradicated.[55]

The first official information about mass killings of Jews appeared in the late Soviet Union during Gorbachev's Glasnost.[56] Research centers were established and Jewish communities and organisations were revived in post-Soviet Russia. One of their main aims was to commemorate Holocaust victims all over the country. Their activity has become steadily more visible at both national and regional levels, and has begun progressively to break down the monolithic Soviet Victory narrative and to obtain an appropriate memorialisation of Holocaust victims within public memory of World War II.

---

53 Dina Khapayeva, *Goticheskoe obshchestvo. Mifologiya koshmara* (Moscow: Novoe Literaturnoe Obozrenie, 2008), 12.
54 The "Commission to Prevent Attempts to Falsify History to the Detriment of Russia" was active between 2009 and 2012. It was established by decree of the president of Russian Federation Dmitrii Medvedev on May 15, 2009. Sergei Solov'ev, "Komissiya i istoriya," *Skepsis: nauchno-prosvetitel'nyi zhurnal*, May 25–June 1, 2009, http://scepsis.net/library/id_2476.html (accessed March 20, 2020); Koposov, *Pamyat'*, 228–255.
55 Koposov, *Pamyat'*, 92.
56 Zvi Y. Gitelman, "Politics and Historiography of the Holocaust in the Soviet Union," in *Bitter Legacy. Confronting the Holocaust in the USSR*, ed. Zvi Y. Gitelman (Indiana University Press: Bloomington and Indianapolis, 1997), 14–42.

At the same time, despite the state suppression of Holocaust-related topics within Soviet society, individual and group memories of Holocaust victims continued to exist within individuals and families, through which stories about persecutions and killings of Jews during the war were transmitted from generation to generation. In this context, the study of decentralised, local practices of Holocaust commemoration that were common in the USSR at the initiative of particular individuals or remembering communities[57] can enrich understanding of Holocaust memory in the former Soviet Union. According to Foucault's terminology, such practices belong to the realm of "counter-memory:" they are discursive practices, due to which memory changes constantly.[58] Paying attention to the multiplicity of discourses enriches the detailed study of the memory of the past and permits the detection of its different agents, not only memory creators but also memory keepers, mediums, and translators.

Along with theoretical and historical "uncertainties," a methodological "uncertainty" can also be identified in memory studies. The problem of defining different forms and types of representation of the past shows the complexity of any kind of classification of memory in social science. American historian Yosef Hayim Yerushalmi distinguished between different "vehicles of memory" such as books, films, museums, and commemorations,[59] while Aleida Assmann applied a range of traditions (mnemotechnic and forms of identity), perspectives (individual, collective, and cultural memory), media (texts, images, places), and discourses (literature, history, art, psychology, etc.).[60] The most famous concept, "sites of memory," [*lieux de mémoire*] was introduced by the French historian Pierre Nora.[61] This category is very broad: there are concrete memory sites, such as cemeteries, museums and anniversaries, and more "intellectually elaborated

---

[57] Remembering communities are not controlled by the authorities and yet have the power to create memorials and to organise commemorative meetings. For more on this concept, see: Jay Winter, "Remembrance and Redemption. A Social Interpretation of War Memorials," *Harvard Design Magazine* 9 (1999): 1–6; Idem, *Sites of Memory, Sites of Mourning: The Great War in European Cultural History* (Cambridge: Cambridge University Press, 1995); James V. Wertsch, "Collective Memory," in *Memory in Mind and Culture*, ed. Pascal Boyer and James V. Wertsch (Cambridge: Cambridge University Press, 2009), 119.
[58] Michel Foucault, "Nietzsche, Genealogy, History," in *Language, Counter-Memory, Practice: Selected Essays and Interviews*, ed. Donald F. Bouchard (Ithaca: Cornell University Press, 1980), 152–154.
[59] Yosef Heyim Yerushalmi, *Zakhor: Jewish History and Jewish Memory (The Samuel and Althea Stroum Lectures in Jewish Studies)* (Seattle: University of Washington Press, 1996).
[60] Assmann, *Cultural Memory*, 9.
[61] Pierre Nora, *Realms of Memory: Rethinking the French past*, vol. 1, *Conflicts and Divisions*, ed. Lawrence D. Kritzman (New York: Columbia University Press, 1996).

ones," such as the notions of generation and lineage. They can be portable, topographical, monumental, public, and private sites, sites that are dominant and sites that are dominated, and so forth. The unifying characteristic of all these sites is their purpose to stop time, to block the work of forgetting, to establish a state of things, to immortalise death, and to materialise the immaterial.[62] According to Nora, memory is non-reflective, vital, always actual in the present, symbolic and mythological, connected with the sacral, reliable, multiple and indivisible, collective and individual. History, on the contrary, is always relative, problematic, analytic, critical, and representative. It is strictly entrenched in the past, connected with secular things, it claims to be universal and objective, it is not attached to anything except time, and, of all things, history tends to trust memory the least.[63] Thus, Nora's concept of memory does not tolerate a pluralism of interpretations, making it difficult to apply in the era of dialogical postmodern discourse.

Conceptual critiques of Nora's *lieux de mémoire* have led memory researchers to develop a new approach, encapsulated by the term "*noeuds de mémoire*," or knots of memory. American researcher Michael Rothberg has suggested, "that 'knotted' in all places and acts of memory are rhizomatic networks of temporality and cultural reference that exceed attempts at territorialisation (whether at local or national level) and identitarian reduction. Performances of memory may have territorialising or identity-forming effects, but those effects will always be contingent and open to re-signification."[64] Rothberg's research is an attempt to determine and to study the "multidirectionality" of memory, that is, encounters between a diverse past and a conflictual present, between different agents or catalysts of memory.[65] Rothberg thus proposes to study memory on a global level, ignoring local dimensions of remembering the past.

The above-mentioned "uncertainties" in memory studies in general and Holocaust memory in Soviet and post-Soviet Russia in particular encourage me to apply the concept of "memory dispositif," which was introduced into memory studies several years ago. A memory dispositif is understood as a network of interactions that exist between different forms or elements of memory. Cultural scholar Laura Basu has pioneered the application of the concept in a study that examines

---

62 Pierre Nora, "Between Memory and History: 'Les Lieux de Mémoire,'" *Representations* 26 (1989), 22.
63 Ibid, 8–9.
64 Michael Rothberg, "Introduction: Between Memory and Memory: From Lieux de Mémoire to Noeuds de Mémoire," *Yale French Studies* 118/119 (2010): 7.
65 Michael Rothberg, *Multidirectional Memory: Remembering the Holocaust in the Age of Decolonisation* (Stanford: Stanford University Press, 2009).

cultural memory of the nineteenth-century Australian outlaw Ned Kelly as a memory dispositif.[66] Media, time, power, and the development of Australian identities formed the main elements of the memory dispositif surrounding Ned Kelly. Yet Basu's chronological approach to the study, according to the main historical periods of the development of this memory dispositif, means that it not so theoretically ambitious as it could be as a piece of poststructuralist scholarship. An example of a more successful study is one by the Polish sociologist Magdalena Nowicka, where the category of collective memory dispositif is applied to analyse public and political debates that erupted in Poland following the publication of the book *Neighbours: The Destruction of the Jewish community in Jedwabne, Poland*, by Polish-born American historian and sociologist Jan T. Gross, in which he recounts instances of Poles murdering their Jewish neighbours and discusses antisemitism in Polish society.[67] Nowicka shows that an important way of subduing a "shameful" past is its "self-cancellation" by means of discursive and non-discursive practices of invalidation or media regulation of public space.

The advantage of memory dispositif as an analytical tool for the study of Holocaust memory in a particular region is that it permits one to conduct empirical research and to trace the development of Holocaust memory in all its variety, given a wide range of active and passive factors in its formation and the interaction between discursive and non-discursive memory practices. Memory dispositif also permits the researcher to bring the study of memory formation and its coexistence with the heroic narrative of Victory in the Great Patriotic War at regional level into a single analytical framework. Its main task is the research of multiple practices – both discursive and non-discursive – in their interrelationships. An empirical study of the life of Holocaust memory within a postmodern paradigm also permits us to work through the underlined "uncertainties" in memory studies as a whole, and in Holocaust memory studies in particular. The aim of dispositif analysis is to study the network established between different discursive and non-discursive (material) practices of remembering. By using the concept of memory dispositif, I aim to shed new light on the political, subjective, temporal, ethnic, and cultural dimensions of Holocaust memory, and the mutual influence of different Holocaust memory elements in a particular region, as well as in the USSR and post-socialist Russia more broadly.

---

66 Laura Basu, *Ned Kelly as Memory Dispositif: Media, Time, Power, and the Development of Australian Identities* (Berlin, Boston: Walter de Gruyter, 2012).
67 Magdalena Nowicka, "Zur Diskurs- und Dispositivanalyse des kollektiven Gedächtnisses als Antwort auf einen öffentlichen Krisenzustand. Zwischen Habermas und Foucault," *Forum: Qualitative Sozialforschung* 15, no. 2 (2014), http://nbn-resolving.de/urn:nbn:de:0114-fqs 1401228 (accessed March 17, 2020).

The Theory of Memory Dispositif

Since the poststructuralist concept of dispositif has hitherto rarely been applied in memory studies, it is useful to trace the emergence and philosophical understanding of this concept. The crucial point for poststructuralist thinking is that fixed structures are impossible as they are constantly subject to change. Furthermore, the analysis of such structures occurs inevitably within, and is indeed caused and biased by, a cultural framework that cannot be separated from its context. To understand an object necessitates the study of the object itself along with the systems of knowledge that produce it.[68] The poststructuralist turn concerns not only knowledge itself – for instance, the contents of the cultural, social, or individual memory – but also processes of knowledge production: that is the interaction of its elements, or its dispositif.

The concept of dispositif entered scholarly discourse in the 1960s and 1970s through the works of French theorists Jean-Louis Baudry and Michel Foucault.[69] The French term *dispositif* is derived from the Latin word *dispositio*, meaning "arrangement, distribution," and is commonly translated into English as "apparatus." This translation is problematic due to its connotations of mechanicity and fixity: it implies a mechanical relation between the elements of a dispositif, as if it operates like a machine. In the 1990s German media theorists Knut Hickethier and Nicole Gronemeyer[70] drew attention to the inadequacy of this translation, which to some extent had limited the uptake of the concept in English language scholarship. The last decade has seen increasing attention to this concept and attempts to revive it.[71] The term "dispositif" usually refers to a constellation of

---

68 See: Stuart Sim, "Postmodernism and Philosophy," in *The Routledge Companion to Postmodernism*, ed. Stuart Sim (London, New York: Routledge, 2001), 3–14; Daniel Chaffee and Charles Lemert, "Structuralism and Poststructuralism," in *The New Blackwell Companion to Social Theory*, ed. Turner Bryan (Chichester: Wiley-Blackwell, 2009), 133–139.
69 Jean-Luis Baudry, "The Apparatus: Metapsychological Approaches to the Impression of Reality in the Cinema," in *Narrative, Apparatus, Ideology: a Film Theory Reader*, ed. Philip Rosen (New York: Columbia University Press, 1986), 299–318.
70 Knut Hickethier, "Kommunikationsgeschichte: Geschichte der Mediendispositive Ein Beitrag zur Rundfrage 'Neue Positionen zur Kommunikationsgeschichte,'" *Medien und Zeit* 2 (1992), http://medienundzeit.at/kommunikationsgeschichte-geschichte-der-mediendispositive-ein-bei trag-zur-rundfrage-neue-positionen-zur-kommunikationsgeschichte/ (accessed March 17, 2020); Nicole Gronemeyer, "Dispositiv. Apparat. Zu Theorien visueller Medien," *Medienwissenschaft. Rezensionen/Reviews* 1 (Hamburg: Universität Hamburg, 1998): 9–21. In German texts the germanised 'Dispositiv' is most frequently used; the spelling 'dispositif' is more common in French and English texts.
71 Andrea D. Bührmann and Werner Schneider, "Mehr als nur diskursive Praxis? – Konzeptionelle Grundlagen und methodische Aspekte der Dispositivanalyse," *Forum Qualitative Sozialforschung* 8,

heterogeneous elements within a system and relationships between them, which produces a particular "tendency." Although the analytical concept of dispositif is frequently applied in media studies,[72] psychotherapy, sociology, gender- and cultural studies,[73] its conceptualisation is a relatively new but very promising avenue in memory studies. In this sense it is useful to make a close analysis of Foucault's concept of dispositif, its further development by French, Italian, and German philosophers and sociologists, and its implementation in social studies using a range of ideas drawn from memory studies.

---

no. 2 (2007), http://www.qualitative-research.net/index.php/fqs/article/view/ 237/526 (accessed March 17, 2020); Giorgio Agamben, *What is an Apparatus? And other essays* (Stanford: Stanford University Press, 2009), 1–24.

72 German media theorists developed the concept of media dispositif in media studies during the latest decades. See: Heinrich Brinkmöller-Becker, "Kino und die Wahrnehmung von Filmen: Das Kino-Dispositiv in 100järiger Entwicklung," *Medien und Erziehung: Zeitschrift für Medienpädagogik* 38, no. 6 (1994): 327–332; Jan Hans, "Das Medien-Dispositiv," *Tiefenschärfe* (winter semester 2001/ 2002): 22–28; Arndt Neumann, "Das Internet-Dispositiv," *Tiefenschärfe* 2 (2002): 10–12; Knut Hickethier, *Einführung in die Medienwissenschaft* (Stuttgart: Metzler Verlag, 2003); Idem, "Der politische Blick im Dispositiv Fernsehen. Der Unterhandlungswert der Politik in der medialen Republik," in *Die Politik der Öffentlichkeit – Die Öffentlichkeit der Politik. Politische Medialisierung in der Geschichte der Bundesrepublik*, ed. Bernd Weidbrod (Göttingen: Wallstein, 2003), 79–96; Peter M. Spangenberg, "'Weltempfang' im Mediendispositiv der 60er Jahre," in *Medienkultur der 60er Jahre. Diskursgeschichte der Medien nach 1945*, ed. Irmela Schneider, Thorsten Hahn, and Christina Bartz, vol. 2 (Wiesbaden: Westdeutscher Verlag, 2003), 149–158; Rüdiger Steinmetz, *Das digitale Dispositif Cinéma: Untersuchungen zur Veränderung des Kinos* (Leipzig: Leipziger Universität Verlag, 2011); Philipp Dreesen, Łukasz Kumięga, and Constanze Spie, ed., *Mediendiskursanalyse. Diskurse – Dispositive – Medien – Macht* (Wiesbaden: Springer VS, 2011); Karl Othmer, ed., *Medien – Bildung – Dispositive: Beiträge zu einer interdisziplinären Medienbildungsforschung* (Wiesbaden: Springer VS, 2015). Vytautas Michelkevičius should be mentioned as a non-German author who works with the concept of media dispositif in media studies. Vytautas Michelkevičius, *The Lithuanian SSR Society of Art Photography (1969–1989): An Image Production Network* (Vilnius: Vilnius Academy of Arts Press, 2011).

73 Andrea D. Bühmann, "Die Normalisierung der Geschlechter in Geschlechterdispositiven," in *Das Geschlecht der Moderne*, ed. Hannelore Bublitz (Frankfurt am Main: Campus, 1998), 71–94; Siegfried Jäger and Margret Jäger, "Das Dispositiv des Institutionellen Rassismus. Eine diskurstheoretische Annäherung," in *Konjunkturen des Rassismus*, ed. Alex Demirovic and Manuela Bojadziev (Münster: Westliches Dampfboot, 2002), 212–224; Klaus R. Schroeter, "Pflege als Dispositiv: zur Ambivalenz von Macht, Hilfe und Kontrolle im Pflegediskurs," in *Soziologie der Pflege. Grundlagen, Wissensbestände und Perspektiven*, ed. Klaus R. Schroeter and Thomas Rosenthal (Weinheim: Jeventa, 2005), 385–404; Ulrike Vedder, *Das Testament als literarisches Dispositiv: kulturelle Praktiken des Erbes in der Literatur des 19. Jahrhunderts* (München: Fink, 2011); Joannah Caborn Wengler, Britta Hoffarth, and Lukasz Kumiega, ed., *Verortungen des Dispositiv-Begriffs: analytische Einsätze zu Raum, Bildung, Politik* (Wiesbaden: Springer VS, 2013).

While Jean-Louis Baurdy was historically the first person to introduce the notion of dispositif, Foucault's interpretation had the greatest influence on the further development and use of the concept.[74] Foucault defined a dispositif as a social configuration:

> . . . a thoroughly heterogeneous ensemble consisting of discourses, architectural forms, regulatory decisions, laws, administrative measures, scientific statements, philosophical, moral and philanthropic propositions – in short, the said as much as the unsaid . . . The dispositif [Note: the interpreter used the term 'apparatus'] itself is the system of relations that can be established between these elements . . . and it is precisely the nature of the connection that can exist between these heterogeneous elements.[75]

Foucault therefore considers the dispositif as a mechanism for knowledge production, which in turn influences the dispositif itself. The relations of knowledge and power between the elements of a dispositif constitute subjectivities. As the Italian philosopher Giorgio Agamben has explained, a dispositif "always implies a process of subjectification, that is to say, it must produce its subject."[76] Furthermore, Foucault stated that relations within a dispositif, though strategic, could however have utterly unpredictable outcomes, which must be adapted to or reappropriated within the dispositif.[77] Moreover, a dispositif is a historical formation and relations composed by it change over time. According to Foucault, between the ensemble of elements "whether discursive or non-discursive, there is a sort of interplay of shifts of position and modifications of function which can also vary widely."[78] But if the dynamics within the dispositif in Foucault's perspective remained "relatively stable," the French philosopher Gilles Deleuze introduced the idea of lines to convey its relational nature. To Deleuze, dispositif is understood as:

> a tangle, a multilinear ensemble. It is composed of lines, each having a different nature . . . the lines in dispositif do not outline or surround systems, which are each

---

74 Foucault discussed, *inter alia*, the dispositif of sexuality and the medico-legal dispositif, whereby psychiatry and the penal system became co-dependent. Michel Foucault, *The History of Sexuality*, vol. 1, *An Introduction*, transl. Robert Hurley (New York: Pantheon Books, 1978), Idem, *Discipline and Punish: The Birth of the Prison*, transl. Alan Sheridan (New York: Pantheon Books, 1977).
75 Michel Foucault, *Power/Knowledge: Selected interviews and other writings 1972–1977*, ed. and transl. Colin Gordon (New York: Pantheon Books, 1980), 194–195.
76 Agamben, *What is an Apparatus?*, 11.
77 Foucault described the emergence of the dispositif of mental illness and its treatment as fulfilling the function of assimilating a floating population, which had become "burdensome for an essentially mercantilist economy." Foucault, *Power/Knowledge*, 195, quoted in Basu, *Ned Kelly*, 5.
78 Foucault, *Power/Knowledge*, 195.

homogeneous in their own right . . . but follow directions, trace balances, which are always off balance, now drawing together, and then distancing themselves from one another.[79]

The different lines, which make up a dispositif, are, therefore, always moving, they can change direction, double back on themselves and can branch off, multiply, or break.

Speaking about the influence of sociology of knowledge on the dispositif concept, the German sociologist Reiner Keller argues that "social actors who mobilise a discourse and who are mobilised by discourse establish a corresponding infrastructure of discourse production and problem solving, which can be identified as a dispositif."[80] The notion of "actors" is very useful to characterise memory dispositif: individuals, social groups, and nations are the main agents, becoming simultaneously subjects – who possess an active role and create or establish a culture of memory – and objects, occupying a passive role and involved in the process of memorialisation. It is important to underline that the concept of dispositif is broader and more extensive than the concept of discourse, which was also introduced in Foucault's works.[81]

German sociologists Andrea Bührmann and Werner Schneider, while working on the philosophical bases of dispositif analysis as a research approach or style,[82] proposed four analytical (cross-cutting) relations that order the dispositif within itself and which make it accessible to empirical research. These cross-cutting relations are the following: the relationship between discursive and non-discursive practices; the role of the subject in the dispositive "environment" (subjectivations); the relationship between discursive practices and knowledge (objectifications); and the relationship of the dispositif to social change.[83] The first cross-cutting relation seeks to conceptualise the link between practices that are discursively

---

79 Gilles Deleuze, "What is a Dispositif?" in *Michel Foucault Philosopher: Essays*, ed. and transl. Timothy J. Armstrong (New York: Routledge, 1992), 159.
80 Reiner Keller, "Diskurse und Dispositive analysieren. Die Wissenssoziologische Diskursanalyse als Beitrag zu einer wissensanalytischen Profilierung der Diskursforschung," *Forum Qualitative Sozialforschung* 8, no. 2 (2007), http://www.qualitative-research.net/index.php/fqs/article/view/243/538 (accessed March 17, 2020).
81 Michel Foucault, *The Archaeology of Knowledge and the Discourse on Language* (New York: Pantheon Books, 1972).
82 Andrea D. Bührmann und Werner Schneider, "Vom 'discursive turn' zum 'dispositive turn'? Folgerungen, Herausforderungen und Perspektiven für die Forschungspraxis," in *Verortungen des Dispositiv-Begriffs: analytische Einsätze zu Raum, Bildung, Politik*, ed. Joannah Caborn Wengler, Britta Hoffarth, and Lukasz Kumiega (Wiesbaden: Springer VS, 2013), 24–28.
83 Andrea D. Bührmann und Werner Schneider, *Vom Diskurs zum Dispositiv. Eine Einführung in die Dispositivanalyse* (Bielefeld: Transcript Verlag, 2008).

solidified – and therefore at the disposal of the unsaid, the unsayable, and the cannot-be-said (the forbidden) – to practices, which are discursive, said, and therefore amenable to discussion. The second cross-cutting relation touches on the question of subjectification, which addresses the role of the actor in the dispositif. The third analytical cross-cutting relation concerns the field of tension between discursive knowledge and things, how objects are constituted in interaction with discourses. The fourth cross-cutting relation finally integrates a temporal dimension into the analysis, how dispositifs change and are changed through social process.[84] In this way the dispositif can be operationalised as a describable socio-historical arrangement of discourses, practices, objectifications, and subject constructions.

Applied to memory studies, the concept of dispositif can be defined as a network of interactions that exists between its constitutive elements in response to an historical fact that is remembered. The localisation of the present research in one region (the North Caucasus) and the study of memory dispositif as a combination of discursive and non-discursive practices, material objects and different groups of acting subjects leads this research into qualitatively new terrain, in which the state-sponsored mode of memorialising the war is important and valuable solely if it is confirmed in local practices, which in turn appear to be wider and more multi-layered than the official ideology. This grassroots focus on initiatives from below seeks to reshape understanding of remembrance of World War II in the Soviet Union and modern-day Russia, moving it away from the assumed predominance of an official politics of memory towards an exploration of its personal and unofficial dimensions, using the example of a single Russian region. From this perspective individual meaning becomes a valuable resource in the struggle for social change. Memory dispositif thus helps to reconstruct a network of grassroots initiatives to remember Holocaust victims in a country with (almost) no developed Holocaust memorial culture.

1.2 Excavating the Holocaust Memory Dispositif in the North Caucasus

In this study, I analyse Holocaust memory dispositif in the region of the North Caucasus in Russia's southwest. I trace Holocaust remembrance from its shaping immediately after the German occupation of the region ended in 1943 – when the

---

[84] Ibid; Georg Winkel, "'Dispositif turn' und Foucaultsche Politikanalyse," in *Verortungen des Dispositiv-Begriffs: analytische Einsätze zu Raum, Bildung, Politik*, ed. Joannah Caborn Wengler, Britta Hoffarth, and Lukasz Kumiega (Wiesbaden: Springer VS, 2013), 167–168.

ChGK began collecting evidence of Nazi crimes – through the whole post-war Soviet period and into present-day Russia. A close examination of the network of different elements of commemoration and remembrance processes in a relatively small area allows for the analysis of power-laden interactions and relations between different social actors in the commemoration process, the influence of state ideology on victims' memories about World War II, and the functioning of a Holocaust memory dispositif as a whole. The bottom-up perspective reveals memory as a process. Considering data collected in the region, I analyse the emergence of different elements of this memory dispositif, the ways in which they have developed, and how they have interacted with each other. The different periods of Soviet and post-Soviet Russian political history set the general framework within which the component elements of the Soviet and Russian Holocaust memory dispositif took shape and direction. Thus, my key research questions are the following:

1. What relationship exists between the dominant all-state memory dispositif of the Victory in World War II (or rather, the "Great Patriotic War") with the Holocaust memory dispositif at a regional level? As mentioned above, a heroic-patriotic attitude prevailed in official Soviet memorialisation of the war, and did not pay special attention to the victims outside of their putative heroic sacrifice for the Soviet fatherland. A specific Holocaust memory culture was excluded from official memory policy because of the process of unification and consolidation of different victim groups into a single supranational category of "peaceful Soviet citizens." Nevertheless, some practices of remembering and commemoration of Holocaust victims could be found in remote regions of the USSR. These practices did not significantly change the official war memory culture but they were important elements in establishing a regional dimension to the Holocaust memory dispositif.
2. How has the Holocaust memory dispositif taken shape in a region with a relatively small proportion of Jews in its population? The North Caucasus was historically outside of the Pale of Jewish settlement.[85] The insignificant presence of Jews in the big southern cities from the late 19th century (the region itself remained predominantly rural during the whole of the 20th

---

85 The Pale of Settlement was a western region of the Russian Empire with varying borders that existed from 1791 to 1917, in which permanent residency by Jews was permitted and beyond which Jewish permanent residence – and for a period even temporary stay – was mostly forbidden. Yuri Slezkine, *The Jewish Century* (Princeton: Princeton University Press, 2004), 40–104; Anke Hilbrenner, "Invention of a Vanished World: Photography of Traditional Jewish Life in the Russian Pale of Settlement," *Jahrbücher für Geschichte Osteuropas* 57, no. 2 (2009): 173–188; Nathaniel Deutsch, *The Jewish Dark Continent: Life and Death in the Russian Pale of Settlement* (Cambridge: Harvard University Press, 2011).

century) did not seriously affect the ethnic structure of the region. During the wartime, most of the Jews who came to region – mainly as refugees and evacuees – became Holocaust victims during the first weeks of the Nazi occupation of the North Caucasus. Holocaust survivors returned home to the western regions of the USSR at the end of the war. That is why Holocaust memorial culture in the North Caucasus was established not only by representatives of local Jewish communities, which were very weak and continuously suppressed by the state during the Soviet times, but also with the support of regional authorities, local historians, and activists.

3. How does the Holocaust memory dispositif coexist with the memory of the Soviet repression of indigenous ethnic groups of the North Caucasus? During the war, Stalin conducted a series of deportations of Chechens, Ingush, Balkarians, and Karachays[86] that could be considered acts of genocide.[87] In the aftermath of the Russian Civil War, in which the Bolshevik Red Army was pitched against a monarchist White Army with strong Cossack elements, the Cossacks were liquidated as a social and political elite in the North Caucasus.[88] Public representation of Cossack ethnicity and culture became possible again after the collapse of the USSR. The Cossacks nowadays play a notable role in the geopolitical space of the North Caucasus. At the same time, memory formation about 20th-century Cossack history continues in the region at present. It is too early to talk about fully formed memory dispositifs of the deportation of the Peoples of the Caucasus[89] or the victimisation of the Cossacks because these topics are still considered as taboo in public Russian discourse after the end of the Soviet Union. Moreover, the Russian Federation as the

---

86 Nikolai Bougai, *The Deportation of Peoples in the Soviet Union* (New York: Nova Science Publications, 1996); Nikolai Pobol and Pavel Polyan, ed., *Stalinskie deportatsii, 1928–1953* (Moscow: Materik, 2005); Igor' Pykhalov, *Za chto Stalin vyselyal narody? Stalinskie deportatsii – prestupnyi proizvol ili spravedlivoe vozmezdie?* (Moskow: Yauza-Press, 2008).
87 For example, on February 26, 2004 the European Parliament gave the status of genocide to the deportations of the Chechens during World War II, see: "Chechnya: European Parliament recognises the genocide of the Chechen People in 1944," February 27, 2004, http://www.unpo.org/article/438 (accessed March 17, 2020).
88 Andrei Venkov, *Donskoe kazachestvo v grazhdanskoi voine (1918–1920)* (Rostov-on-Don: Izdatel'stvo Rostovskogo universiteta, 1992); Igor' Kutsenko, *Pobediteli i pobezhdennye. Kubanskoe kazachestvo: istoriya i sud'by* (Krasnodar: Diapazon-V, 2010); Igor' Erokhin, *Kubanskoe kazachestvo v emigratsii XX v.* (Ekaterinburg: Izdatel'stvo knizhnyi perestok, 2013).
89 Under the term "Peoples of the Caucasus" I understand all the ethnic non-Slavic groups who were the indigenous population of this region. Among them are Abkhaz, Cherkess, Kabardians, Ossetians, Ingush, Chechens, peoples of Dagestan, Armenians, Georgians, and others. Those who lived predominantly in the mountain regions of the Caucasus are also known as Mountaineers.

successor state to the USSR has not yet assumed responsibility for these and other blank spots in its history. The coexistence and mutual influence of a memory culture of the repressions of the local population and of the Holocaust has thus become an important feature, inflecting the regional dimension of the Holocaust memory dispositif.

The main elements of the Holocaust memory dispositif in the North Caucasus can be studied according to the theoretical scheme applied by Nowicka on the basis of Bührmann und Schneider's theory.[90] I will refer to different functions of the memory of Musya Pinkenson in Ust-Labinsk, which I mentioned at the very beginning, to illustrate certain memory dispositif elements:

1. The situation of social change or the moment of "ignition," which initiates the mechanisms of the dispositif: in this case, the fact of the Holocaust during World War II.
2. Discourse formations (the variety of statements that constitute the Holocaust discourse in general and at particular places along the timeline in particular). This is the 'heart' of the memory dispositif and includes official and private sources about the Holocaust in the region. It consists of official wartime sources (the wartime Soviet press, Red Army reports from liberated cities and villages, ChGK files, materials of the Commission on the History of the Great Patriotic War established by the Soviet Academy of Sciences) and sources from the post-war period (Soviet and Russian news media, and transcripts of interviews with witnesses in Soviet prosecutions of Nazi criminals). Personal narrations of the wartime period (letters and diaries, primarily by non-Jewish authors, written during the occupation and consisting of stories of mass killing of Jews in a particular locality) and of the post-war period (memoirs and oral history interviews with Holocaust survivors, Holocaust eyewitnesses, local historians, activists, and members of Jewish communities who become the main actors in establishing and developing a Holocaust memorial culture in the region) are also an important part of the discursive formations in the Holocaust memory dispositif. Moreover, there are literary works (novels, stories, poems, etc.), documentaries, movies, and music compositions dealing with an artistic response to the regional history of the Holocaust. Among the authors of these pieces are Holocaust survivors, representatives of the second and third generations, and sensitive regional artists who are interested in local aspects of the history of World War II and the fate of Jews in the Holocaust. In Musya's case, these are published ChGK reports and unpublished testimonies

---

[90] Nowicka, "Zur Diskurs- und Dispositivanalyse."

of Holocaust eyewitnesses, memories of his classmates and teachers, the cartoon *Skripka pionera*, etc. All these sources form a discourse memory field about the young war pioneer-hero, who subsequently became the hero of an individual Jewish resistance.

3. Non-discursive practices or institutional procedures (e.g., in the media or legal system) as well as non-verbal social behaviour. These are papers relating to the installation of monuments to what were termed "peaceful Soviet citizens" in the Soviet period, and memorial plaques and inscriptions mentioning the number of victims, their ethnicity and sometimes even the names of the victims in post-Soviet Russia. This element also includes such items as the resolution adopted by the General Assembly of the United Nations on International Holocaust Remembrance Day in 2005, guidance for teachers on teaching the Holocaust, regulations of the Holocaust Center, and so forth. It also includes the manual on patriotic lessons at school developed by local schoolteacher Irina Svetlichnaya, or the plan approved by the Ust-Labinsk local history museum for an exhibition about Musya Pinkenson, and the monument at the Holocaust site, protected at regional level as an object of historical and cultural importance.

4. Symbolic and material objectivations. These are the mass killing sites of Jews, post-war monuments and memorial complexes, inscriptions on Soviet monuments and post-Soviet plaques, the personal belongings of Holocaust survivors, such as armbands with the Star of David, toys, or family photos that have been a part of a family archive. The monument at the Holocaust site, Musya's portrait, and the collections of the school in Ust-Labinsk and the local history museums can also be referred to as such material objectivations.

5. Subject formations or actors within the Holocaust memory dispositif at regional level. These are local historians who study the history of their hometown and highlight its blank spots, Jewish organisations, the experts at the Holocaust Center, school teachers and museum workers, even certain schoolchildren can become actors within the memory dispositif. Schoolchildren attend the memorial with Musya's bas-relief at the killing site, teachers initiated the collecting of artefacts concerning the boy's life and death for the school museum, local historians and journalists publish articles, and authors write novels and verses about the Soviet pioneer and the hero of Jewish resistance.

6. Medialisation, insofar as strategic relations between discourse and non-discursive practices increasingly occur in the media sphere. The elements of medialisation include printed media materials, TV broadcasts of events commemorating certain dates (International Holocaust Remembrance Day, Day of Regional Liberation from the Nazi Occupation, the anniversaries of particular mass killings of Jews in each locality). Internet discussions, bloggers' reflections, and

social media accounts dedicated to important historical Holocaust-related events have recently become visible and are important elements in the medialisation of Holocaust memory. Ust-Labinsk schoolchildren corresponded with Musya's relatives during the Soviet period by means of an international network both during the Soviet period and now. Special broadcasts about commemorative events, the history of the killing of Jews in Ust-Labinsk, and Musya as a hero of individual Jewish Holocaust resistance are now more often seen on national and local TV channels.

7. Expansion and clarification of the omission of certain questions and topics of discussion, which appears to have been the most important element in the Holocaust memory dispositif of the Soviet Union. Established by the Soviet authorities the impersonal concept of "peaceful Soviet citizens" could in fact refer to a specific category of Nazi victims, such as Jews or underground activists and partisans, or other groups of victims. Musya's story is a good example to illustrate the omissions of the state memory culture of World War II. A modest obelisk was installed on the killing site of 370 Jews in Ust-Labinsk at the end of the war. In 1971 it was replaced by a bas-relief of the pioneer hero Musya Pinkenson. Commemoration events were held on Pioneer's Day but the locals visited the monument on both Victory Day and Ust-Labinsk Liberation Day. Not all the sayable may be said, and not all the sayable needs to be said because it appears to be self-evident or because as something said it would lose its effect as opposed to 'being done.'[91] Sometimes something implied but not said has a more important semantic meaning than something said, especially in the Soviet Union with its state oversight of all spheres of life and the permanent fear of stepping out of line.

The elements of a memory dispositif rarely exist in a pure independent form, being often a part of other political and social systems or communicative networks. For instance, a museum is an institution in which historical, artistic or scientific objects are exhibited, preserved, or studied. It is a site where the past is represented according to an official memory politics or from an individual perspective. It is also a space where different historical objects and installations put into relation with each other can produce new meaning of the past in the present. Conversely, the individual elements of a memory dispositif can be analysed to reveal the system's functioning.

---

91 Werner Schneider, "Dispositive . . . – überall (und nirgendwo)? Anmerkungen zur Theorie und methodischen Praxis der Dispositivforschung," in *Medien – Bildung – Dispositive: Beiträge zu einer interdisziplinären Medienbildungsforschung*, ed. Karl Othmer (Wiesbaden: Springer VS, 2015), 37.

1.2 Excavating the Holocaust Memory Dispositif in the North Caucasus — 27

Holocaust memory can be treated as a memory dispositif that reveals the variety of practices of remembering the mass killing of Jews in one particular region and their incorporation into official (national state) and international memorial culture.

Bearing in mind the above-elaborated theoretical outline of how a memory dispositif functions, the following book chapters each take up one of the elements that are relevant to the regional dimension of Holocaust memorial culture.

The following chapter (Chapter 2) gives an overview of the geography, history and ethnic composition of the studied region and how this inflects the memory dispositif. Special attention is paid to the historical presence of Jews in the North Caucasus and their fates during the Holocaust. Each of the other chapters is devoted to an element of the Holocaust memory dispositif, which is examined in the variety of interactions of discursive and non-discursive practices, commemorational acts, acting objects, and inclusion/exclusion of remembering communities and ordinary locals in remembrance of Holocaust victims over time. Sometimes it was problematic to identify commemorative practices relating solely to Holocaust victims. The Soviet memorial war culture remembers "Soviet citizens" impersonally, distinguishing primarily between fallen soldiers and civilians. At the same time, a close look at the deeds of local actors, detailed study of the ceremonies attending commemorative events, history schoolbooks, museum artefacts, and so forth, enable us to trace a specific memory culture surrounding Holocaust victims in this particular region of Russia. Thus, Chapter 3 focuses on the analysis of official Soviet wartime sources as the main element shaping the Holocaust memory dispositif. The key source for this chapter is the regional collection of the ChGK records, the linguistic analysis of which evokes the topic of mass killing of Jews, amongst other Nazi crimes in the occupied areas of the North Caucasus. In order to analyse the role of official Soviet wartime sources in shaping the regional dimension of the Holocaust memory dispositif, it also describes the primary uses to which these sources were put in post-war and post-Soviet Russian society. Symbolic and material objectification of memory in the form of monuments and memorials, memorial plaques, and avenues of the Righteous Among the Nations is studied in the Chapter 4. It chronologically distinguishes individual, collective, and state practices of remembering victims of the Nazis and examines the semantics of Soviet, post-Soviet, and Jewish monuments to Holocaust victims in Russia. This chapter also studies discursive practices of commemorative acts at Holocaust memorial sites: who visits them, when, and for what purpose? How are memorial meetings organised and performed? Artistic responses to Holocaust history in the North Caucasus are discussed in Chapter 5 with regard to the thematic analysis of different types and genres of imagery, mainly novels and documentaries. It also underlines the diverse perspectives of the authors, readers, viewers, and critics of such pieces, that is, the actors within the Holocaust memory dispositif. Where possible, it traces the fate of a particular work of art after its publication or release, because not only the fact of its availability, but also the public reaction permits a better understanding of the functioning of the regional dimension of the Holocaust memory dispositif. A separate chapter, Chapter 6, is devoted to analysing the activities of museums representing regional Holocaust history. It ascertains what kind of Holocaust history in the North Caucasus is represented in museum exhibitions – central and regional, public and private – around Russia and in the permanent exhibitions of international Holocaust museums. It seeks to understand how and through what means Holocaust history in the North Caucasus is (or is not) being written into the pan-European Holocaust narrative, as well as into the Russian grand narrative of the Great Patriotic War. Chapter 7 examines teaching about the Holocaust in Russian history courses, and regional and local history in secondary schools in the North Caucasus from a historical perspective. It traces the place given to the study of mass killings of Jews on Soviet soil in lessons on the history of World War II and the Great Patriotic War. In what context is this topic revealed, and what aims are pursued through teaching about the Holocaust in secondary school? Content analysis of regional history textbooks allows me to determine the place of the Holocaust on Soviet soil in history courses, and to assess whether it echoes in the dramatic wartime history of various Caucasian peoples, such as the deportation of the Chechens, Ingush, Balkars, Karachays and other groups. Surveys and thematic interviews with regional school teachers reveal the investment (or lack thereof) of teachers in teaching about the Holocaust in the classroom. The last chapter (Chapter 8) deals with ego documents of Holocaust survivors and eyewitnesses. It analyses personal narratives of the Holocaust in the North Caucasus, as well as the circumstances of their emergence. It also highlights how personal narratives of the Holocaust in the region are or are not being used to commemorate Holocaust victims and teach about the Holocaust in the region. Key features of the regional dimensions of the Holocaust memory dispositif are distinguished in the Conclusion (Chapter 9).

2 The Holocaust in the North Caucasus

Appeal
to the Jewish population of the city of Voroshilovsk[1]

In the city of Voroshilovsk a Committee of Jewish Elders has been organised for the regulation and management of all matters concerning the Jewish population, for which the Committee assumes responsibility.

The orders of the Committee of Elders are binding on the entire Jewish population and must be unconditionally implemented by the latter.

As a first measure, the following is to be implemented: All Jews without exception (and those who do not belong to any religious society, as well as those belonging to a religion other than the Jewish religion, as well as their wives and children) who arrived in the city of Voroshilovsk after June 22, 1941 are obliged to gather on August 12, 1942 at 7 a.m. on Yarmorochnaya Square (Ordzhonikidze Sq., near the railway station).

It has become necessary for all the above-mentioned persons to relocate to places that have become free of population due to military actions.

Everyone is obliged by that day, i.e., August 12, 1942, to have the necessary things for personal use, such as bedding, kitchen and tableware, clothing and food, for at least 2–3 days. For your own peace of mind, we recommend taking money and valuables with you. The weight of brought items should not exceed 30 kg per person. All measures have been taken to ensure that the property left behind is safe. Especially responsible for the safety of the remaining property are homeowners, house managers, apartment owners and neighbours. It is advisable to leave a list of things with house managers, etc. In this case, the house management, etc., is obliged to issue receipts.

The population of Voroshilovsk is informed that those responsible for the theft of Jewish property will be immediately sentenced to death without any investigation.

All heads of family are obliged to give a note stating: surname, name and patronymic, as well as the exact address where the property is kept, to the task force of the Council of Elders. The note should also include the name of the person with whom the property inventory has been left.

Failure to do all of the above will be subject to high fines, imprisonment and, in particularly severe cases, the death penalty.

German Administration
Einsatzkommando 12 Voroshilovskii Prospekt 11.[2]

Appeals [sing. *vozzvanie*] like the one reproduced above were the beginning of the end for most of the Jewish population in the occupied regions of the North Caucasus. The Nazi administration set about implementing a well-prepared model for the "final solution to the Jewish question," as it was euphemistically

---

[1] The name of the city of Stavropol between 1935 and 1943.
[2] "Appeal to the Jewish population of the city of Voroshilovsk," in *GARF*, f. R-7021, op. 17, d. 1, l. 95.

known among its architects. Almost all Jews followed the instructions of the occupying authorities, they gathered at the assembly points believing they were to be resettled. Once there, reality would dawn on them quickly but by then they had almost no chance of escaping the killing. Those few who, by accident or design, managed to not be present on the appointed days lived in constant fear of being captured by Nazi forces or Soviet collaborators. Only a few Jews would survive the Holocaust in the region.

This chapter discusses the history of the mass killing of Jews in the North Caucasus, setting the spatial and historical frame for the research. But before reconstructing a general picture of the Holocaust in the region and identifying its peculiarities, I would like to present the region itself. The North Caucasus is unique within the Russian Federation. An area of spectacular natural beauty, it boasts a favourable climate, high mountains, fast rivers, verdant foothills and lowlands, and popular seaside resorts. Its population is also highly diverse ethnically; linguistic and ethnological maps of the region reveal an intricate patchwork of languages and customs overlaying its mountains and valleys. The beauty and renowned hospitality of the region played a bad trick on most local and evacuated Jews during its occupation by the Nazi regime.

2.1 The North Caucasus: A Socio-Economic, Political, and Ethnic Overview

Geographically and economically the Caucasus is traditionally divided into the North Caucasus (or Ciscaucasia) and the South Caucasus (Transcaucasia),[3] with the borderline running along the major Greater Caucasus mountain range. The North Caucasus is the southernmost region of the Russian Federation and possesses natural boundaries on three sides: the Black and Azov Seas to the northwest, the Caspian Sea to the southeast, and the Caucasus Mountains to the south. It includes Krasnodar Krai, Stavropol Krai, Rostov Oblast, the Republics of Adygea, Dagestan, Ingushetia, Chechnya, Kabardino-Balkaria, Karachay-Cherkessia, and North Ossetia-Alania[4] (Figure 2). This territory was an object

---

[3] The territory of Transcaucasia nowadays comprises the independent states of Azerbaijan, Armenia, and Georgia, partially recognised Abkhazia and South Ossetia, and the unrecognised Nagorno-Karabakh Republic.
[4] The Western outskirts of the Republic of Kalmykia also fall within the geographical boundaries of the North Caucasus, however economically, politically, and ethno-culturally, Kalmykia has never formed part of the North Caucasus and belongs to the Povolzhsky [Volga] economic region. Valentin Zheltikov, *Ekonomicheskaya geografiya* (Rostov-on-Don: Phoenix, 2001), 327–344.

Figure 2: North Caucasus physical map (Edward W. Walker, Eurasian Geopolitics website, https://eurasiangeopolitics.com/north-caucasus-maps/).

of the colonial policy of the Russian Empire since the first half of the 18th century, when the lands of the Don Cossacks were annexed to the Empire. The atamans, leaders of the Cossack hosts, were deprived of their authority.[5] As a result of the Russo-Turkish wars in the second half of the 18th century, the Empire's borders were considerably expanded, as lands on the right-

---

Some historians have unreasonably considered Republic of Kalmykia to be part of the North Caucasus while excluding, for example, Rostov Oblast. See: Kiril Feferman, *The Holocaust in the Crimea and the North Caucasus* (Jerusalem: Yad Vashem, 2016). When Kalmykia is included in the studied region, it is appropriate to refer to it not as the North Caucasus, but as the South of Russia. This term has been used since 1991 for the designation of the Southern Federal District and the North Caucasian Federal District, see: "Regiony yuga Rossii," *Portal SouthRu.info*, http://www.southru.info/ (accessed March 23, 2020). For example, Elena Voitenko's PhD thesis is devoted to the study of the Holocaust in the South of Russia, that is, the geographic scope of the study is all the occupied areas of southern Russia in World War II, including Kalmykia. Elena Voitenko, "Kholokost na yuge Rossii v period Velikoi Otechestvennoi voiny (1941–1943 gg.)" (cand. of science diss. in history, Stavropol State University, 2005).

5 Aleksandr Pronshtein, "Rol' kazachestva i krest'yanstva v zaselenii i khozyaistvennom osvoenii Dona i stepnogo Predkavkaz'ya v XVIII – pervoi polovine XIX veka," *Izvestiya Severo-Kavkazskogo nauchnogo tsentra vysshei shkoly: Obshchestvennye nauki* 1 (1982): 55–59; Vasilii Sukhorukov, *Istoricheskoe opisanie zemli voiska donskogo* (Rostov-on-Don: Novyi Biznes, 2001).

bank of the Kuban river and the northern Black Sea coast were annexed. The lands of Chechnya, mountainous Dagestan, and the Northwestern Caucasus finally became part of Russia in 1864.[6] With the expansion of its possessions, the Russian Empire colonised the annexed North Caucasian territories by granting lands to the Cossacks – the Terek and Black Sea (later Kuban) Cossack hosts were founded here – and by resettling peasants from the central and western governorates [guberniya].[7] The North Caucasus was, in the Russian imagination, a frontier: a site of perpetual conflict with its indigenous inhabitants, an as yet "uncivilised" population, and a place of exile for the intelligentsia who opposed the government (Decembrists, Petrashevists) during the 19th century.[8] It later became a destination for the forceful relocation of certain ethnic groups, for example, Poles.[9]

---

6 For more details on the wars of the Russian Empire in the Caucasus, see: Moshe Gammer, *Muslim Resistance to the Tsar: Shamil and the Conquest of Chechnia and Daghestan* (London: Frank Cass, 1994); Marie Broxup, ed., *The North Caucasus Barrier: The Russian Advance Towards the Muslim World* (London: Hurst, 1992); Mark Bliyev, *Kavkazskaya voina (1817–1864)* (Moscow: TOO "Roset," 1994); Vladimir Bobrovnikov and Irina Babich, ed., *Severnyi Kavkaz v sostave Rossiiskoi Imperii* (Moscow: Novoe Literaturnoe Obozrenie, 2007), 112–135; Shakhrudin Gapurov, Abdula Bugaev, and Viktor Chernous, "K 150-letiyu okonchaniya Kavkazskoi voiny: o khronologii, prichinakh i soderzhanii," *Nauchnaya mysl' Kavkaza* 4 (2014): 90–100.

7 The strengthening of feudal exploitation in Russia had intensified the escape of serf to the south – to the rivers Terek, Kuma, and Kuban and Stavropol Krai – long before the mass government-sponsored colonisation. Galina Malakhova, "Stanovlenie i razvitie rossiiskogo gosudarstvennogo upravleniya na Severnom Kavkaze v XVIII – XIX vv." (doctor of science diss. in history, Russian Academy of Public Administration under the President of the Russian Federation, 2001); Andrei Kaznacheev, *Puti kazach'e-krest'yanskogo pereselencheskogo dvizheniya na Severnyi Kavkaz* (Moscow: Molodaya Gvardiya, 2005); Natalia Varivoda, "Zaselenie Severnogo Kavkaza slavyanskim naseleniem v XVIII veke," *Istoricheskii vestnik* 3 (2006): 176–197.

8 Andrei Popov, *Dekabristy-literatury na Kavkaze* (Stavropol: Knizhnoe izdatel'stvo, 1963); Melitsa Nechkina, *Dekabristy* (Moscow: Nauka, 1982); Elena Mirzoyan, "Sibirskaya i kavkazskaya ssylka dekabristov, 1826–1856 gg.: opyt sravnitel'nogo issledovaniya" (cand. of science diss. in history, Siberian branch of the Russian Academy of Sciences, 2002); Laurence Kelly, *Lermontov: Tragedy in the Caucasus* (London: Tauris Park Paperbacks, 2003). For discussion of Russian representations of the Caucasus, see also Bruce Grant, *The Captive and the Gift: Cultural Histories of Sovereignty in Russia and the Caucasus* (Ithaca, NY: Cornell University Press, 2009); David Schimmelpenninck van der Oye, *Russian Orientalism: Asia in the Russian Mind from Peter the Great to the Emigration* (New Haven, CT: Yale University Press, 2010).

9 Irina Tsifanova, "Pol'skie pereselentsy na Severnom Kavkaze v XIX veke: osobennosti protsessa adaptatsii" (cand. of science diss. in history, Stavropol State University, 2002); Aleksandr Bogolyubov, *Polyaki na Severnom Kavkaze v XIX-XX vv.* (Krasnodar: Kuban State University, 2008).

During the Soviet period, the Peoples of the Caucasus gained statehood within the Russian Soviet Federative Socialist Republic (RSFSR) as Autonomous Soviet Socialist Republics (ASSRs) and Autonomous Oblasts (AOs). In 1924, the North Caucasus Krai was established according to a plan for the regionalisation of the USSR. It included the present-day territories of the European South of Russia, with the exception of Dagestan.[10] Historians often regard the idea of the "North Caucasus" as a derivative of the North Caucasus Krai. For the purposes of this study, I also adopt such an historical interpretation of the North Caucasus, because this was the territory that was occupied by German armed forces during World War II and on which the Holocaust occurred.[11]

The North Caucasian region has a number of unique features, which played an important role (not always positive) during World War II. The North Caucasus boasts good agroclimatic resources: plenty of sunlight, warmth, enough moisture, fertile chernozem and kastanozem soils (the famous black earth of the Caucasus), and natural pastures, which make it a predominantly agricultural region. On the eve of the war, the area under cultivation amounted to 12.6 million hectares in the North Caucasus,[12] and the number of livestock of all types totalled some 13.1 million (including 3.8 million cattle).[13] The region is also rich in natural resources.[14] The following mineral resources are of value: oil from Chechnya and

---

[10] In 1934, the Azov-Black Sea Krai was formed out of the North Caucasus Krai. In 1937, it was divided into Krasnodar Krai and Rostov Oblast. The North Caucasus Krai was renamed to Ordzhonikidze Krai in 1937 and to Stavropol Krai in 1943. For the purposes of this study, I refer only to Stavropol Krai to facilitate geographical orientation.

[11] The exception is the Chechen-Ingush ASSR, which was part of the North Caucasus Krai but did not fall under German occupation; therefore, the mass killing of Jews was not recorded in this territory. Detailed information on the administrative territorial structure and demographic development of the North Caucasus at the outbreak of the war is presented in Table 1.

[12] Cereal crops constituted 8.6 million hectares, and industrial crops nearly 1.5 million hectares. Elena Malysheva, *V bor'be za pobedu: sotsial'nye otnosheniya i ekonomicheskoe sotrudnichestvo rabochikh i krest'yan Severnogo Kavkaza v gody voiny 1941–1945* (Maykop: Adygeiskoe knizhnoe izdatel'stvo, 1992), 40–41.

[13] Stavropol Krai was one of the top regions in the country for the production of Merino sheep, growing to nearly 4 million head by the beginning of the 1940s. Sergei Boiko, ed., *Stavropol'e v Velikoi Otechestvennoi voine 1941–1945 gg: sbornik dokumentov i materialov* (Stavropol: Knizhnoe izdatel'stvo, 1962), 6.

[14] Stanislav Kolesnik, ed., *Sovetskii Soyuz*, vol. 3, Petr Abramov et al., *Rossiiskaya Federatsiya, evropeyskii yugo-vostok, Povolzh'e, Severnyi Kavkaz* (Moscow: Mysl', 1968); Vladimir Gnilovskoi, ed., *Landshafty i ekonomicheskaya geografiya Severnogo Kavkaza: sbornik statei* (Stavropol: Stavropol'skii gosudarstvennyi universitet, 1977); Mikhail Belikov, *Severnyi Kavkaz: realii sotsial'no-ekonomicheskoi sfery na poroge tysyacheletiya: geograficheskii aspekt* (Krasnodar: Kubanskii gosudarstvennyi universitet, 2002).

Adygea, gas from Krasnodar and Stavropol Krais, non-ferrous and polymetallic ores from mountainous areas, and bituminous coal from Rostov Oblast, Karachay and Circassia.[15] By the summer of 1942, the North Caucasus and Transcaucasia accounted for 86.5 per cent of all-Union oil production, 65 per cent of natural gas, and 56.5 per cent of manganese ore.[16] Deposits of lead and zinc ores are found in North Ossetia, copper and molybdenum in Kabardino-Balkaria.[17] Most major rivers of the North Caucasus (the Kuma, the Kuban, and the Terek) flow from the mountains through valleys and ravines in the upper reaches; in the lower reaches they flow more calmly through broad valleys.[18] Water from many rivers is used to irrigate the arid areas of Ciscaucasia. A distinctive feature of the North Caucasus at the beginning of the 20th century was the development of rail,[19] sea,[20] and river[21] transport, which accounted for the bulk of the country's freight and passenger transportation. After the outbreak of World War II on Soviet soil, enterprises and people from the front lines were evacuated in large numbers to the North Caucasus by river and rail.[22] However, the railway in the vicinity of the resorts of Mineralnye Vody District, Stavropol Krai is a dead end; that is, it comes up against a massif, which, along with the swift advance of

---

[15] The miners of Karachay AO and Cherkess AO extracted 310,000 tons of coal for the country in 1940. Vladimir Seliunin, *Promyshlennost' i transport yuga Rossii v voine 1941–1945 gg.* (Rostov-on-Don: Knizhnoe izdatel'stvo, 1997), 28.

[16] Andrei Grechko et al., ed., *Istoriya Vtoroi mirovoi voiny 1939–1945 gg.*, vol. 5 (Moscow: Voenizdat, 1975), 199.

[17] The first part of the tungsten-molybdenum plant was launched on October 1, 1939, and after three months daily ore extraction constituted 1,200 tons. The yield of the finished concentrate amounted to 2,800 kg for tungsten and 2,600 kg for molybdenum. Selyunin, *Promyshlennost'*, 29.

[18] Only the Don river is flat, and it is the fourth longest river in Europe.

[19] The North Caucasian and the Ordzhonikidzevskaya railways had a total length of 4,917 km. Sergey Linets, "Severnyi Kavkaz nakanune i v period nemetsko-fashistkoi okkupatsii: sostoyanie i osobennosti razvitiya, iyul' 1942 – oktyabr' 1943 g." (doctor of science diss. in history, Pyatigorsk State Technological University, 2003), 41.

[20] At the beginning of the war the Black, Azov and Caspian Seas had nine ports. About 800 vessels of various classes sailed the three seas. Selyunin, *Promyshlennost'*, 36.

[21] In Rostov Oblast and Krasnodar Krai alone, the Don, Kuban, Protoka, and other rivers had over 600 units of steamships, barges, boats, etc. Selyunin, *Promyshlennost'*, 36.

[22] 37,165 evacuees (mainly from the Ukrainian SSR) arrived in Krasnodar Krai from July 19 to July 25, 1941. The first wave of refugees arrived by water through Novorossiysk and by rail through Rostov-on-Don. "Information and memoranda of the authorised regional committee members, secretaries of district party committees on the evacuation of citizens," in *Center for the Documentation of Contemporary History of Krasnodar Krai* [Tsentr dokumentatsii noveishei istorii Krasnodarskogo kraya], hereafter TsDNIKK, f. 1774-A, op. 2, d. 166, l. 157–158; d. 271, l. 2–3.

German armed forces through the Caucasus, hampered the further evacuation of people and enterprises from this region in the summer of 1942.[23]

The North Caucasus is also famous for its recreational resources: sandy beaches, mineral water springs, and downhill skiing resorts. By the end of the 1930s, the region had become an all-Union sanatorium possessing world-famous unique natural balneological and climate recourses, which were used to treat tens of millions of Soviet citizens.[24] An important hospital base for the treatment of wounded and sick Red Army soldiers was established on the premises of health resorts in the town of Mineralnye Vody and the city of Sochi in the summer of 1941.[25]

Rich in natural resources, the North Caucasus was a strategically important region in Nazi German planning during World War II. Access to the Black Sea might have solved the political task of pushing Turkey to fight on Germany's side, and the seizure of fertile lands and oil fields might have helped meet the economic demands of the war.[26] However, the Caucasus Mountains were a

---

23 For the miscalculations of the Soviet command and the unsuccessful re-evacuation of the population from the North Caucasus, see: Georgii Kumanev, "Evakuatsiya naseleniya SSSR: Dostignutye rezul'taty i poteri," in *Lyudskie poteri SSSR v period Vtoroi mirovoi voiny: materialy konferentsii, 14–15 marta 1995 goda* (St. Petersburg: Russko-Baltiiskii informatsionnyi tsentr "Blitz," 1995), 137–146; Linets, "Severnyi Kavkaz," 81–111; Idem, "Evakuatsiya s territorii Severnogo Kavkaza naseleniya letom 1942 goda: Otsenka rezul'tatov," *Nauchnaya mysl' Kavkaza* 6 (2003): 89–96. Many Holocaust survivors mention the impossibility of getting out of the area of Mineralnye Vody by rail in August 1942 in their oral interviews. See: Interview with Larisa Apininskaya, born in 1932, in *University of South California Shoah Foundation Visual History Archive*, hereafter *VHA USCSF*, code 45214–55, May 22, 1998, Kiev, Ukraine.

24 About 160,000 people rested and improved their health in Stavropol Krai in the resorts of Mineralnye Vody in 98 sanatoriums and guesthouses all year round, while the resort city of Sochi in Krasnodar Krai had a capacity of 103,000 holidaymakers a year. Boiko, *Stavropol'ye*, 7; Aleksandr Belyaev et al., ed., *Krasnodarskii Krai v 1937–1941 gg.: dokumenty i materialy* (Krasnodar: "Edvi," 1997), 533.

25 See: Sergey Linets, *Gorod vo mgle . . . (Pyatigorsk v period nemetsko-fashistskoi okkupatsii. Avgust 1942 g. – yanvar' 1943 g.)* (Pyatigorsk: Pyatigorskii gosudarstvennyi lingvisticheskii universitet, 2010); Aleksandr Cherkasov, "K nekotorym aspektam raboty sochinskoi gospital'noi bazy (1941–1945 gg.): periodizatsiya i effektivnost'," *Bylye gody* 2 (2008): 19–28; Olga Chekeres, "Sanatorium Ordzhonikidze during the Great Patriotic War (1941–1945 years)," *Gardarika* 4, no. 3 (2015): 91–100.

26 According to the order of the High Command of the Armed Forces [*Oberkommando der Wehrmacht*], hereafter OKW, no. 45 of July 23, 1942, Army Group A (under the command of Field Marshal Wilhelm List from July 7 to September 10, 1942; at which point Hitler himself assumed command until November 22, 1942, when he passed the command to Field Marshal Ewald von Kleist) had to solve with Operation Braunschweig a supposedly crucial strategic task in the campaign for further warfare, namely to conquer the North Caucasus oil sources

natural obstacle to conducting military operations in the region. Despite the successful offensive of Army Group A into the territory of Rostov Oblast and further into the Caucasus in the summer and the autumn of 1942, by December of that year the Transcaucasian Front of the Red Army under the command of General Ivan Tyulenev had fought the Wehrmacht to a standstill east of Mozdok, on the outskirts of Ordzhonikidze, on the passes of the Major Caucasus Range, and in the southeastern part of Novorossiysk.[27]

Another important feature of the North Caucasian region is its ethnically diverse society. By the beginning of World War II, the Caucasus had the most complex ethnic, linguistic and confessional composition of the entire Soviet Union. Here dwelled more than 50 ethnic groups who spoke the languages of three linguistic families: Iberian-Caucasian, Indo-European, and Altaic. Slavs and some of the Peoples of the Caucasus profess Orthodox Christianity (Russians, Ukrainians, Greek, Armenians, most Ossetians, Mozdok Kabardians, Abkhazians), but the majority of Peoples of the Caucasus practice Sunni Islam (e.g., Circassians, Kabardians, Karachais, Balkars, Abazins).[28] According to the data of the All-Union Population Census of 1939, the population of the North Caucasus was 10,332,000,[29] more than 70 per cent of whom lived in rural areas.[30]

---

and the crossings over the Caucasus. See the documents and the chronology focusing on military operations and battles in the Caucasus region during World War II: Andrei Grechko, *Bitva za Kavkaz* (Moscow: Ministerstvo oborony SSSR, 1971); Wolfgang Schumann, ed., *Deutschland im Zweiten Weltkrieg*, vol. 2, *Vom Überfall auf die Sowjetunion bis zur sowjetischen Gegenoffensive bei Stalingrad (Juni 1941 bis November 1942)* (Berlin: Akademie-Verlag, 1975), 338–343; Pohl, *Die Herrschaft der Wehrmacht*, 299–300; David M. Glantz, "The Struggle for the Caucasus," *The Journal of Slavic Military Studies* 22, no. 4 (2008): 588–711.

27 Joachim Hoffmann, *Kaukasien, 1942–1943: Das deutsche Heer und die Orientvölkern der Sowjetunion* (Freiburg: Rombach Verlag, 1991), 63–66, 80–81; Vladimir Gurkin and Aleksandr Kruglov, "Oborona Kavkaza: 1942 god," *Voenno-istoricheskii zhurnal* 10 (1992): 11–18; Joel Hayward, "Hitler's Quest for Oil: The Impact of Economic Considerations on Military Strategy, 1941–1942," *The Journal of Strategic Studies* 18, no. 4, (1995): 64–135; Joel Hayward, "Too Little, Too Late: An Analysis of Hitler's Failure in August 1942 to Damage Soviet Oil Production," *The Journal of Military History* 64 (2000): 769–794; Daniel Müller, "Besetzte Kaukasus-Gebiete," in *Handbuch zum Widerstand gegen Nationalsozialismus und Faschismus in Europa 1933/39 bis 1945*, ed. Gerd R. Ueberschär (Berlin: Walter de Gruyter, 2011), 235–242.

28 Mark Kosven et al., ed., *Narody Kavkaza*, vol. 1 (Moscow: Akademiya nauk SSSR, 1960); Timur Dzeranov, "Etnokonfessional'nye razlichiya naseleniya Severnogo Kavkaza," *Fundamental'nye issledovaniya* 3–4, (2014): 861–865.

29 Including the Dagestan ASSR and the Chechen-Ingush ASSR.

30 Rostov Oblast and the North Ossetian ASSR reached the highest level of urbanisation: the urban population was 44 per cent and 43 per cent respectively of the total population. Meanwhile, Stavropol Krai had the lowest level of urbanisation – 20 per cent. See: *Naselenie*

In its calculations for the rapid seizure of the Caucasus, German army command relied on inciting ethnic strife among the peoples, the collapse of the republics, and the weakening of the Soviet rear in the south of the USSR. In German designs, *völkisch* ideology was intertwined with economic policies and military objectives. Some Peoples of the Caucasus had conflicts with the Soviet authorities and the Germans wanted these groups to collaborate.[31] The soldiers and officers of Army Group A were preliminarily instructed to show respect for the traditions and customs of the Peoples of the Caucasus.[32] The plans of the Nazi administration included extensive use of Islam in the war against the USSR. The occupation authorities restored the positions of mullahs in many Muslim settlements of the North Caucasus, which had been previously restricted by the Soviet administration. The Nazi administration used mullahs to justify the mass shootings of civilians, especially Jews. This occurred especially in the winter of 1942–1943 in the territory of the Kabardino-Balkarian ASSR.[33] At the same time, the complex multi-ethnic character of the region rendered invaluable service to

---

SSSR: 1973: statisticheskii sbornik (Moscow: Statistika, 1975), 16; Elena Malysheva, *Ispytanie: sotsium i vlast': problemy vzaimodeistviya v gody Velikoi Otechestvennoi voiny 1941–1945* (Maykop: Adygeya, 2000), 45; Pavel Polyan, *Mezhdu Aushvitsem i Bab'im Yarom: razmyshleniya i issledovaniya* (Moscow: ROSSPEN, 2010), 132.

31 For example, there were no German garrisons in the mountain villages [*auly*] of Adygea and local policemen had authority over the settlements. The police presence in the aul of Hachemzii, for instance, consisted of eight Adygeans. Aleksandr Belyaev and Irina Bondar', ed., *Kuban' v gody Velikoi Otechestvennoi voiny: 1941–1945: rassekrechennye dokumenty: khronika sobytii*, vol. 1, *Khronika sobytii 1941–1942 gg.* (Krasnodar: Sovetskaya Kuban', 2000), 606–607. Historian Sergei Linets explains this policy as simple pragmatism: the German command lacked enough troops to man military garrisons, posts and outposts in numerous settlements of the North Caucasus. Linets, "Severnyi Kavkaz," 429–430.

32 For example, Field Marshal Ewald von Kleist mentioned this in his proclamation to the personnel of the 1st Panzer Army, which was under his command. The former German military attaché in Moscow, General Ernst Köstring who specialised in Caucasian questions was attached to Army Group A, which advanced on the territory of the North Caucasus in the summer and autumn of 1942. See: Peter Kleist, *Zwischen Hitler und Stalin: 1939–1945* (Bonn: Athenäum-Verlag, 1950); Hermann Teske, *General Ernst Köstring: Der militärische Mittler zwischen dem Deutschen Reich und der Sowjetunion 1921–1941* (Frankfurt am Main: Mittler und Sohn, 1965). Another example: The Commander of the 44th Army Corps of the 17th Army, General Maximilian de Angelis, emphasised in an order of August 8, 1942: "To act differently here than on the Don. An uprising of Caucasian mountain peoples [. . .] can result in serious consequences for us." Mikhail Semiryaga, *Kollaboratsionizm: priroda, tipologiya i proyavleniya v gody Vtoroi mirovoi voiny* (Moscow: ROSSPEN, 2000), 239.

33 Islam Aliskerov, "Vliyanie religioznykh i etnicheskikh faktorov na voenno-politicheskuyu obstanovku na Severnom Kavkaze v gody Velikoi Otechestvennoi voiny 1941–1945 gg.," in *Religioznye organizatsii Sovetskogo Soyuza v gody Velikoi Otechestvennoi voiny 1941–1945 gg.*:

Jews, especially evacuees and refugees. Jews could impersonate other Peoples of the Caucasus or Slavs in order to survive during the occupation. Changing appearance and disguise (for example, wearing headscarves or clothes with elements of Slavic embroidery) or copying the behaviour of local people (usually in the context of religion: celebrating Orthodox or Muslim holidays, saying prayers, etc.) were the most common survival strategies of Jews in the North Caucasus, according to oral interviews with Holocaust survivors.[34]

Besides the Peoples of the Caucasus, the Cossacks were traditionally one of the most important ethnic groups in the North Caucasus with whom the Nazis also had high hopes of cooperation.[35] The Germans regarded the Cossacks as irreconcilable fighters against Bolshevism. They had gained this reputation during the Russian Civil War of 1917–1922. The most common way of revitalising the old Cossack traditions was the reinstatement of the ataman's authority in German-occupied former Cossack villages and settlements [*khutor*], permission to wear the Cossack uniform and old tsarist awards, and the entrenchment of elements of Cossack lifestyle and culture in everyday life. There were cases when the Cossacks switched to serving the occupation authorities as policemen and participated in identifying and killing Jews.[36] This is reflected in the memories of Holocaust survivors about the Don and Kuban Cossacks, in which they are often negatively coloured and accompanied by stories of the constant fear of being exposed or found by the Cossacks.[37]

---

materialy 'kruglogo stola', posvyashch. 50-letiyu pobedy, 13 aprelya 1995 g., ed. Nikolai Trofimchuk, (Moscow: Rossiyskaya akademiya gosudarstvennoi sluzhby, 1995), 91.

34 Irina Rebrova "Traumatische Kindheit: Holocaust und Überlebenspraktiken jüdischer Kinder in den besetzten Gebieten des Nordkaukasus im Zweiten Weltkrieg," in *Kindheiten im Zweiten Weltkrieg*, ed. Francesca Weil, Andre Postert, und Alfons Kenkmann (Halle: Mitteldeutscher Verlag, 2018), 393–410.

35 The Cossacks (just like Peoples of the Caucasus) did not receive the official status of "honorary Aryans" in the racial concept of the Third Reich. Nevertheless, Berlin's trust reached its peak when they were given the status of "equal allies" in the struggle against the partisans and the Red Army in April 1942. However, the initiative for this decision came not from Hitler but from Major Claus Graf Schenk von Stauffenberg, and this "privilege" did not concern the peoples, but their representatives in the Wehrmacht. Rolf-Dieter Müller, "Gebirgsjäger am Elbrus: Der Kaukasus als Ziel nationalsozialistischer Eroberungspolitik," in *Wegweiser zur Geschichte Kaukasus*, ed. Bernhard Chiari (Paderborn: Ferdinand Schöningh, 2008), 61.

36 Linets, "Severnyi Kavkaz," 410–415; Petr Krikunov, *Kazaki: mezhdu Stalinym i Gitlerom: krestovyi pokhod protiv bol'shevizma* (Moscow: Yauza: Eksmo, 2006), 207–235; Feferman, *The Holocaust*, 433–448.

37 See: Interview with Yakov Krut, born in 1928, in *author's archive*, code 16-H-RO01, June 4, 2016, Ashkelon, Israel; Interview with Mikhail Achkinazi, born in 1930, in *VHA USCSF*, code

The above-detailed political, economic, and ethno-confessional features of the region testify to its uniqueness within Russia, as well as to the role of the North Caucasus during World War II. Due to its geographical location in the southernmost part of Russia and its relative proximity to the western republics of the Soviet Union, the region had attracted a large number of Jewish refugees and evacuees by the summer of 1942. Most of them found themselves in the occupied territory and were murdered in these beautiful mountains and fertile lands.

2.2 Jewish History in the North Caucasus before and during World War II

Despite the ethnic heterogeneity of the North Caucasus, Jewish residents accounted for less than one per cent of the total population in the pre-war period (excluding the Dagestan and Chechen-Ingush ASSRs). Historically this territory was beyond the Pale of Jewish settlement, the territory of the former Russian Empire upon which Jews were permitted to settle.[38] Therefore, apart from small privileged groups, Ashkenazi Jews were forbidden to live in the region before 1917.[39] Those permitted residence faced a Tsarist anti-Jewish policy conducted with particular vigour in the Cossack areas.[40] After the Bolshevik Revolution,

---

37 429–40, December 15, 1997, Yalta, Ukraine; Interview with Faina Babitskaya, born in 1933, in *VHA USCSF*, code 49602, October 8, 1998, Simferopol, Ukraine.

38 The only exceptions were a very few cities in today's Rostov Oblast. Thus, since 1802, Rostov-on-Don had been an *uyezd* [county] capital in the Ekaterinoslav governorate, which was included in the Pale of Settlement. In 1888, it became part of the Don Host Oblast, the territory of the Don Cossacks. According to the law of 1880, certain categories of Jews, namely graduates of universities or those in government service, were permitted to inhabit the Don Host Oblast. After Rostov-on-Don was incorporated into the Don Host Oblast, it was subjected to residence restrictions for Jews. However, Jews residing in the territory given over to the Don Host (Rostov Uyezd and the city of Taganrog) retained right of residence (about 21,000 Jews, including over 10,000 Jews in Rostov), but lost the right of free movement and residence in the rest of the Don Host Oblast. See: "Rostov-na-Donu," in *Kratkaya evreiskaya entsiklopediya*, ed. Oren (Nadel') Itskhak and Mikhael' Zand (Jerusalem: Keter, 1994), 7:403–404; Evgenii Movshovich, *Ocherki istorii evreiev na Donu* (Rostov-on-Don: ZAO "Kniga," 2011), 61.

39 Sergei Kuleshov et al., ed., *Natsional'naya politika Rossii: istoriya i sovremennost'* (Moscow: "Russkii Mir," 1997), 151.

40 Nikolai Kirei, "Evrei Krasnodarskogo kraya," *Byulleten': antropologiya, men'shinstva, mul'tikul'turalizm* 2 (2000): 134–144; Ekaterina Norkina, "The Origins of Anti-Jewish Policy in the Cossack Regions of the Russian Empire, Late Nineteenth and Early Twentieth Century," *East European Jewish Affairs* 43, no. 1 (2003): 62–76.

some Jews – mainly representatives of the intelligentsia: doctors, teachers and scientists – migrated to the North Caucasus mostly to industrial and resort centers.[41] During the early years of Soviet power, Jewish newcomers usually chose assimilation,[42] which later helped many of them survive the occupation and the Holocaust.[43]

Taking the ethnic diversity of the North Caucasus as one of its defining features, it should be noted that Jews in the region did not form a homogenous group here either. Representatives of several Jewish ethnicities resided in the region in the pre-war period. Mountain Jews[44] and Karaites[45] lived here along with

---

41 For example, in the pre-war period about 31 per cent and 34 per cent of all the Jews of Krasnodar Krai lived in Krasnodar and Novorossiysk, respectively. Vadim Rakachev, "Natsional'nyi sostav naseleniya Kubani v XX veke: istoriko-demograficheskii aspekt" (cand. of science diss. in history, Kuban State University, 2003), 85. Out of a total population of 510,212 people, 27,053 Jews (or 5.3 per cent) lived in Rostov-on-Don, according to the census of 1939. Evgenii Movshovich, "Kholokost v Rostove," *Yakhad* 3 (1999): 3.

42 Kirei, "Evrei," 140. Jewish assimilation in Krasnodar Krai was due to the absence of a strong Jewish community here. Rostov Oblast had large Jewish communities (more than 500 people) in Rostov, Taganrog, Nakhichevan-on-Don, Azov and in some big settlements, so Jews resided there more or less compactly, leading to a weaker assimilation in the demography of the region. See: Movshovich, *Ocherki istorii*, 60–105.

43 For example, in Stavropol Krai, citizens of mixed origin (with one Jewish parent and the other Russian, Ukrainian, or of another ethnicity) may not have been registered as Jewish. ChGK report for Stavropol Krai, in *State Archive of Stavropol Krai* [*Gosudarstvennyi arkhiv Stavropol'skogo kraya*], hereafter *GASK*, f. R-1368, op. 1, d. 84, l. 16.

44 The term "Mountain Jews" appeared in the first half of the 19th century during the conquest of the territory of Dagestan and Azerbaijan by the Russian Empire. The endonym of Mountain Jews or Caucasus Jews is Juhuro. According to the linguistic and indirect historical data, it can be assumed that the community of Mountain Jews was formed due to the constant immigration of Jews from Northern Iran (mid 7th–11th centuries), and also, possibly, from the nearby Byzantine empire to Transcaucasian Azerbaijan, where they settled (in its eastern and north-eastern regions) among the Tat-speaking population and switched to this language. "Gorskie Evrei," in *Kratkaya evreiskaya entsiklopediya*, ed. Oren (Nadel') Itskhak and Mikhael' Zand (Jerusalem: Keter, 1982), 2:182. For more detailed information about Mountain Jews, see: Svetlana Danilova, ed., *Gorskie evrei v Kabardino-Balkarii* (Nalchik: El'brus, 1997); Valerii Dymshits, ed., *Gorskie evrei: istoriya, etnografiya, kul'tura* (Jerusalem, Moscow: DAAT/Znanie, 1999).

45 The term Karaite derives from the Hebrew *bnei haMiqra* (בני ארקמה), meaning "Son of Mikra." Mikra is an acronym for the Hebrew Bible, also called Tanakh. According to the dominant account, they appear to have stemmed from various Jewish groups in Mesopotamia that rejected the Talmudic tradition. The Karaites could also be the remnant of the Sadducees, the Second Temple priestly class who denied the Jewish "Oral Law." In the 13th century, a considerable number of Karaites, primarily from the Byzantine Empire, settled in the Crimean Peninsula. In the 14th century, a portion of them moved further north and settled in Lithuania.

Ashkenazi Jews. Mountain Jews are a population indigenous to the Eastern Caucasus. The area of their traditional residence spans northern Azerbaijan and southern Dagestan, from where they have settled throughout the North Caucasus, primarily in the territory of modern Kabardino-Balkaria and Stavropol Krai.[46] Mountain Jews lived in separate settlements within the mountainous region; their main occupation was rural economy. Several collective farms [*kolkhozy*] of Mountain Jews were founded in the second half of the 1920s within the framework of a Soviet programme for settling Jews on the land (OZET).[47] Thus, in the North Caucasus (now Stavropol) Krai, two Jewish collective farms were established in Bogdanovka village and in Ganshtakovka farm [*khutor*] (in the late 1930s it was renamed Menzhinskii farm). The population was 105 families or 518 people in Ganshtakovka and 125 families or 682 people in Bogdanovka in 1929. By the end of the 1930s Mountain Jews had left the collective farms, but many Jewish collective farms remained even after World War II; in the early 1970s about 10 per cent of the community were collective farmers.[48] Only a few Karaite communities were known in the North Caucasus in the pre-revolutionary period.[49] The Soviet regime recognised the Karaites as a separate nationality.

The Soviet Karaites defined their identity in two ways: ethnically they belonged to Turkic peoples, but religiously they considered themselves followers of the Old Testament. So, certain researchers distinguish between the original Karaites and the European Karaites. See: "Karaimy," in *Kratkaya evreiskaya entsiklopediya*, 4:90; Dan Shapira, "Beginnings of the Karaite Communities of the Crimea Prior to the Sixteenth Century," in *Karaite Judaism: A Guide to its History and Literary Sources*, ed. Meira Polliack (Leiden: Brill, 2003), 709–728; Tapani Harviainen, "The Karaites in Eastern Europe and the Crimea: An Overview," in Polliack, *Karaite Judaism*, 633–656.

46 There is no precise information on the number of Mountain Jews in the North Caucasus before World War II. The 1939 census provides data only about Jews without division into Ashkenazi and Mountain Jews. Referring to the study of the Israeli historian Mordechai Altshuler, Feferman points to 34,000–35,800 Mountain Jews who lived in the territory of the entire Caucasus before World War II. See: Feferman, *The Holocaust*, 293.

47 OZET [*Obshchestvo zemleustroistva evreiskih trudyashchikhsya*] was a public Society for Settling Toiling Jews on the Land in the USSR during 1925–1938. See: Jonathan L. Dekel-Chen, *Farming the Red Land: Jewish Agricultural Colonisation and Local Soviet Power, 1924–1941* (New Haven: Yale University Press, 2005).

48 Dymshits, *Gorskie evrei*, 69–89; Natalia Bulgakova, "Sel'skoe naselenie Stavropol'ya vo vtoroi polovine 20-kh – nachale 30-kh gg. XX veka: izmeneniya v demograficheskom, khozyaistvennom i kul'turnom oblike" (cand. of science diss. in history: Stavropol State University, 2003), 76–82; Polyan, *Mezhdu Aushvitsem*, 137–140.

49 A large part of the Karaites were found in Russia since 1795, when Catherine II exempted them from double taxation, which had previously concerned only the Jewish population. This started the long tradition in the Russian Empire of treating the Karaites like Christians. The official Russian terminology then changed from "Karaite Jews" to "Karaites." During the Russian Civil

According to the 1926 census, 8,324 Karaites resided in the Soviet Union, including 337 in the North Caucasus Krai (mainly in Krasnodar and Novorossiysk).[50] Further on, I treat Mountain Jews and Karaites as separate groups and when I refer to Jews, I refer solely to the Ashkenazi sub-ethnic group.

According to the All-Union Population Census of 1939, the Jewish population in the regions of the North Caucasus was small: from 0.2 per cent in Krasnodar Krai up to 1.1 per cent in Rostov Oblast, where the strongest and oldest Jewish community in the whole region existed.[51] The rights of Jews as well as all other ethnicities in the early Soviet Union were proclaimed to be equal.[52] In reality the doctrine of Russians as the 'elder brother' was implemented throughout the country, and state antisemitism had become visible by the end of 1930s.[53]

When Nazi Germany attacked the USSR on June 22, 1941, the North Caucasus became one of the important evacuation zones for the state programme to evacuate industrial enterprises, cultural and scientific institutions, food supplies, raw materials, as well as the population, including skilled workers.[54] For these

---

War, a number of Karaites fought on the side of the White Army, which supported the monarchy. After the White Army was defeated, hundreds of Karaites left Russia for France, Germany, Poland, Turkey, Romania and other countries. See: Mark Kupovetskii, "Dinamika chislennosti i rasselenie karaimov i krymchakov za poslednie dvesti let," in *Geografiya i kul'tura etnograficheskikh grupp tatar v SSSR*, ed. Igor' Krupnik (Moscow: Geograficheskoe obshchestvo, 1983), 78–81.

50 "Natsional'nyi sostav naseleniya Severo-Kavkazskogo kraya po perepisi 1926 g.," in *Demoskop Weekly*, http://demoscope.ru/weekly/ssp/rus_nac_26.php?reg=807 (accessed March 23, 2020); Kupovetskii, "Dinamika Chislennosti," 80.

51 Table 1 shows the total population of the regions of the North Caucasus according to the results of the 1939 census and the share of the Jewish population in each region. For the history of the Jewish community in Rostov-on-Don, see: Movshovich, *Ocherki istorii*; Mikhail Gontmakher, *Evrei na donskoi zemle: istoriya, fakty, biografiya* (Rostov-on-Don: Rostizdat, 2007).

52 See Stalin's speech at the Extraordinary 7th Congress of Soviets of the Soviet Union, where the new addition to the Soviet Constitution was proclaimed: "Doklad tovarishcha Stalina I.V. o proekte konstitutsii soyuza SSR," *Pravda*, November 26, 1936.

53 Matthias Vetter, *Antisemiten und Bolschewiki: zum Verhältnis von Sowjetsystem und Judenfeindschaft 1917–1939* (Berlin: Metropol-Verlag, 1995), 299–395; Gennadii Kostyrchenko, "Doktrina 'starshego brata' i formirovanie gosudarstvennogo antisemitizma v SSSR v svete ideologicheskoi i etnopoliticheskoi transformatsii stalinskogo rezhima v 1930-e gg.," in *Sovetskiye natsii i natsional'naya politika v 1920–1950-e gg.: materialy VI mezhdunarodnoi nauchnoi konferentsii: Kiev, 10–12 oktjabrja 2013 g.* (Moscow: Politicheskaya entsiklopediya, 2014), 44–48.

54 On June 27, 1941, the Central Committee of the All-Union Communist Party Bolsheviks [*Tsentralnyi komitet vsesoyuznoi kommunisticheskoi partii bol'shevikov*], hereafter TsK VKP(b), and the Council of People's Commissars [*Soviet narodnykh kommissarov*], hereafter SNK, of the USSR determined the order of evacuation from the front-line areas in a joint decree. Accordingly, important industrial assets including the most important machinery and vehicles, valuable raw materials (non-ferrous metals and fuel), food (first of all grain), and other

purposes the State Defence Committee [*Gosudarstvennyi komitet oborony*], hereafter GKO, was established on June 24, 1941.[55] Alongside the more or less organised evacuation process controlled by the GKO, streams of refugees rushed from the front line to the South of Russia.[56] Soviet authorities considered the North Caucasus to be a relatively safe shelter for evacuees from other regions.[57] Therefore, many evacuees and refugees from the territory of the Ukrainian, Belorussian, and Moldavian SSRs, Crimean ASSR, and from the northwest of the RSFSR continued to arrive in the North Caucasus until the summer of 1942. In particular, thousands of residents of the besieged Leningrad, including a number of children's homes were evacuated along the "Road of Life" [*doroga zhizni*][58]

---

assets of national importance had to be evacuated first. Various categories of Soviet citizens ("skilled workers, engineers, and employees along with the enterprises evacuated from the front, youth eligible for military service, and responsible Soviet and party workers") could be evacuated afterwards. "Correspondence with party committees, institutions and organisations on military-mobilisation issues," in *TsDNIKK*, f. 1774-A, op.1 dop., d. 15, l. 33.
55 Georgii Kumanev, "Voina i evakuatsiya v SSSR: 1941–1942 gody," *Novaya i noveishaya istoriya* 6 (2006): 38.
56 For example, over 218,000 refugees arrived in Krasnodar Krai from the front line as of August 15, 1941, fleeing from the Wehrmacht offensive. There were 160,000 Jews among them, or 73 per cent of the total number of evacuees. "Information and memoranda of the authorised regional committee members, secretaries of district party committees on the evacuation of citizens," in *TsDNIKK*, f. 1774-A, op. 2, d. 271, l. 7–18.
57 For more information about the evacuation history of Jews in general, see: Ivan Belonosov, "Evakuatsiya naseleniya iz prifrontovoy polosy v 1941–1942 gg.," in *Eshelony idut na vostok: iz istorii perebazirovaniya proizvoditel'nykh sil SSSR v 1941–1942 gg.: sbornik statei i vospominanii*, ed. Yurii Polyakov (Moscow: "Nauka," 1966), 15–30; Mordechai Altshuler, "Escape and Evacuation of the Soviet Jews at the time of the Nazi Invasion," in *Holocaust in the Soviet Union: Studies and Sources on the Destruction of the Jews in the Nazi-Occupied USSR, 1941–1945*, ed. Lucjan Dobroszyski (Armonk: Sharpe, 1993), 77–105; Idem, "Evacuation and Escape During the Course of the Soviet-German War," *Dapim: Studies on the Holocaust* 28, no. 2 (2014): 57–73. For more information about the Jewish refugees and evacuees in the North Caucasus see: Kiril Feferman, "A Soviet Humanitarian Action?: Center, Periphery and the Evacuation of Refugees to the North Caucasus, 1941–1942," *Europe-Asia Studies* 61, no. 5 (2009): 813–831; Idem, *The Holocaust*, 80–109; Irina Rebrova, "Evakuatsiya evreiev na Severnyi Kavkaz: motivatsiya i puti sledovaniya (po dannym ustnykh interv'yu s perezhivshimi Kholokost)," in *Trudy po evreyskoi istorii i kul'tury: materialy XXII mezhdunarodnoi ezhegodnoi konferentsii po iudaike*, ed. Viktoriya Mochalova, vol. 52 (Moscow: Sefer, 2016), 108–122.
58 The Road of Life was the ice road winter transport route across the frozen Lake Ladoga, which provided the only access to the besieged city of Leningrad while the perimeter of the siege was maintained by German Army Group North and the Finnish Defence Forces. Each winter during the siege of Leningrad (September 8, 1941 to January 27, 1944) the Lake Ladoga ice route was reconstructed by hand, and built according to precise arithmetic calculations depending on traffic volume. See: John Barber and Andrei Dzeniskevich, ed., *Life and Death in*

and domiciled in Krasnodar and Stavropol Krai in the spring of 1942.[59] Incoming evacuees, including Jews, were dispatched all over the North Caucasus, especially into the resort towns of Stavropol Krai. They were also dispatched on a smaller scale to Russian villages, including Cossack settlements. Most of the evacuees, especially when accommodated in the countryside, got to work on collective farms and were involved in harvesting, for which they received a salary or food rations and sometimes the opportunity to hold a small piece of land for private gardening.[60]

It is well known that the Soviet leadership did not privilege any ethnic group during the evacuation of the population from the front line in 1941–1942. There was no plan for evacuation in case of German invasion, and the whole process of evacuation was implemented under wartime conditions. In the context of ensuing panic and refugee streams, the Soviet authorities did not focus particularly on Jews, but it was necessary to inform people about the Nazi atrocities against Jews. There is a discussion in the post-war literature on the

---

Besieged Leningrad: 1941–44 (Houndmills: Palgrave Macmillan, 2005); Lisa A. Kirschenbaum, *The Legacy of the Siege of Leningrad, 1941–1995: Myth, Memories, and Monuments* (Cambridge: Cambridge University Press, 2006).

59 23,000 evacuees from the besieged Leningrad were sent to Stavropol Krai in February 1942, according to the instructions of the SNK of the USSR. "Correspondence with party, military, administrative bodies on material support for families of servicemen, evacuees, Spanish political emigrants and their children, support to Leningrad of 1942," in *State Archive for Contemporary History of Stavropol Krai* [*Gosudarstvennyi arkhiv noveishei istorii Stavropol'skogo kraya*], hereafter *GANISK*, f. 1, op. 2, d. 251, l. 22; d. 210, l. 25–28. Altogether, over 36,000 evacuees from Leningrad, including more than 10,000 children, were accommodated in Krasnodar Krai in April 1942. "Correspondence of the secretaries of the VKP(b), the chairmen of the executive committees of the district councils in the regional [*krai*] party on supplying, restoring enterprises, etc.," in *TsDNIKK*, f. 1774-A, op. 2, d. 495, l. 7; Idem, d. 626, l. 11–18.

60 Thus, for example, according to a Soviet wartime report, 39,100 out of 51,353 evacuees in Krasnodar Krai were sent to villages as of January 1942. 18,000 adults worked, 11,000 unemployed people were members of military families who received allotments, pensioners, and handicapped persons. To provide the evacuees with material assistance the Krasnodar Krai Executive Committee [*Kraiispolkom*] gave 600,000 roubles twice, in August and in December 1941. "Materials and protocols No. 99–101 of the meetings of the Bureau of the Regional Committee of VKP(b)," in *TsDNIKK*, f. 1774-A, op. 2, d. 373, l. 69–76. Holocaust survivors mentioned that they worked in collective farms and received rations and/or a piece of land. Interview with Roza Lantsman, born in 1929, in *VHA USCSF*, code 36391, September 9, 1997, Haifa, Israel; Interview with Andrei Manto, born in 1927, in *VHA USCSF*, code 42599–55, March 30, 1998, Sevastopol, Ukraine; Interview with Rosa Khailovskaya, born in 1934, in *VHA USCSF*, code 49206–27, November 23, 1998, Volgograd, Russia.

possible prioritising of Jews during the evacuation,[61] but no "convincing documented evidence of these statements"[62] has been found. There are no relevant statistics on the ethnic composition of the evacuees, because the registration of evacuees and refugees was an extremely difficult task in wartime.[63] A census of the evacuated population that arrived from the threatened areas of the USSR was held in all rear zones on February 1, 1942 under the direction of the GKO,[64] but there is no accurate count of the evacuees due to possible double counting.[65] Calculating the actual number of individual families and persons in hiding during the subsequent occupation of the North Caucasus is also extremely difficult. Jews who were evacuated to southern Russia in 1941 and early-1942 were mostly women and children, as well as elderly scientists and artists. The vast majority of these people had neither enough energy, funds nor contacts to seek refuge elsewhere or go into hiding. Only a small number of Jewish

---

[61] See the works of Arkadii Vaksberg and Moshe Kahanovich who argue that "the Presidium of the Supreme Council of the USSR issued a special decree on the priority evacuation of the Jewish population": Moshe Kahanovich, *Yidisher onteyl in der partizaner-bavegung fun Sovet-Rusland* [in Yiddish] (Rome: Oysg. fun der Tsentraler historisher komishe baym Partizaner farband P.H.H. in Italye, 1948), 188; Arkadii Vaksberg, *Iz ada v rai i obratno: evreiskii vopros po Leninu, Stalinu i Solzhenitsynu* (Moscow: "Olimp," 2003), 189–190.

[62] Gennadii Kostyrchenko, *Tainaya politika Stalina: vlast' i antisemitizm* (Moscow: Mezhdunarodnye otnosheniya, 2003), 222.

[63] For example, the Russian historian Marina Potemkina studied the data on evacuees to the Ural region and argued that Jews made up an average of 22 per cent of the total number of evacuees. Marina Potemkina, "Evakuatsiya i natsional'nye otnosheniya v sovetskom tylu v gody Velikoi Otechestvennoi voiny," *Otechestvennaya istoriya* 3 (2002): 148–156.

[64] Its data has been stored in classified archival collections for years. Kumanev, "Voina i evakuatsiya," 41. The statistical materials of the GKO on the placement of evacuees in the North Caucasus with regard to their ethnicity have not yet been released to the scientific community despite the fact that these materials have been already declassified. See: GARF, f. A-327, op. 2 "The main resettlement administration under the Council of Ministers of the RSFSR and its predecessors (1942–1956)."

[65] For example, approximately 220,000 evacuees (including 160,000 Jews) arrived in Krasnodar Krai before November 1941. Subsequently, further evacuation to Krasnodar was terminated. In November and December 1941, most of the evacuees left for other regions. By the beginning of 1942 the remaining number was 51,353. See: Belyaev and Bondar', *Kuban' v gody*, 181–182. However, reports to Moscow kept the number nearly unchanged, at 226,700. Thus, 170,000 people were almost counted twice. Vadim Dubson, "Toward a Central Database of Evacuated Soviet Jews' Names, for the Study of the Holocaust in the Occupied Soviet Territories," *Holocaust and Genocide Studies* 26, no. 1 (2012): 100. Moreover, the evacuation process to the North Caucasus continued until the summer of 1942, and since July 1942 evacuation from Krasnodar and Stavropol Krais and Rostov Oblast became more spontaneous and individual-oriented because of the sudden advance of Wehrmacht troops in the south.

families were able to evacuate themselves to the mountainous regions of the North Caucasus. Israeli historian Kiril Feferman estimates the number of evacuated Jews in the entire region at about 50,000 as of July–August 1942.[66]

In the summer of 1942, Army Group A successfully launched the Case Blue [*Fall Blau*] offensive on the territory of the North Caucasus. By the middle of autumn 1942, most of the North Caucasus[67] had fallen under a military administration [*Militärverwaltung*], in which officers of the German Army headed regional and municipal commands (Figure 3).[68] The relative briefness of the

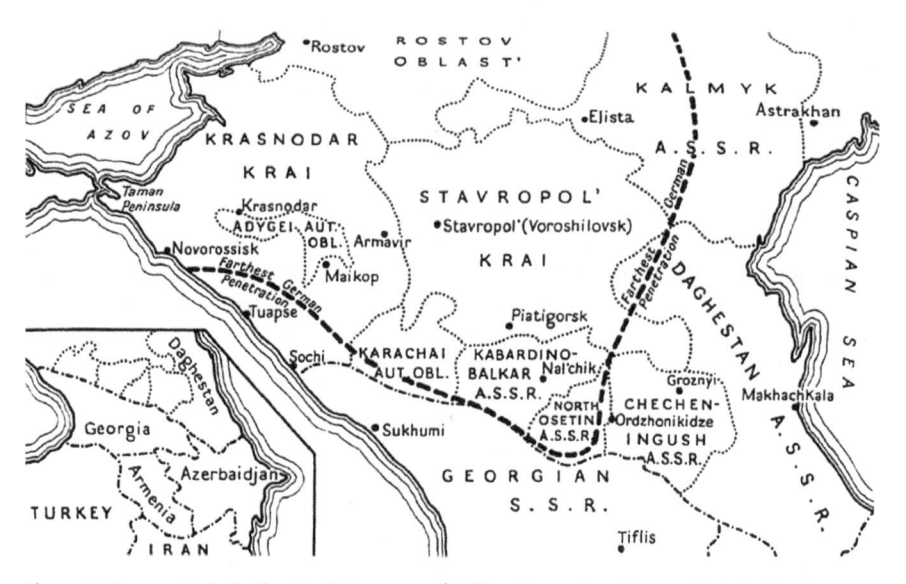

Figure 3: German Rule in the North Caucasus (Dallin, Alexander. *German Rule in Russia 1941–1945. A Study of Occupation Policies*. New York: Palgrave, 1981. S. 245).

---

[66] However, Feferman's calculations are based mainly on the Soviet data primarily in Krasnodar and partly in Stavropol Krais. See: Feferman, *The Holocaust*, 90.

[67] It should be noted that Rostov-on-Don was occupied twice: for one week in the autumn of 1941 and then for a little more than six months from July 1942; Taganrog, Anastasievskii and Fedorovskii districts were under occupation for almost two years: from October 1941 to August 1943. ChGK reports for Rostov Oblast, in *Center for Documentation of Contemporary History of Rostov Oblast* [*Tsentr dokumentatsii noveishei istorii Rostovskoi oblasti*], hereafter TsDNIRO, f. R-1886, op. 1, d. 22, l. 7–9. For more specificity about the period of the occupation in different regions of the North Caucasus, see Table 2.

[68] See more about the occupation of the North Caucasus: Zoya Bochkareva, "Okkupatsionnaya politika fashistskoi Germanii na Severnom Kavkaze" (cand. of science diss. in history: Kuban State University, 1992); German Belikov, *Okkupatsiya, Stavropol, avgust 1942 – yanvar' 1943* (Stavropol:

2.2 Jewish History in the North Caucasus before and during World War II — 47

German occupation (from several weeks to 6–7 months in most cases),[69] as well as the fact that German occupation policy depended on local collaborators and did not install civil administration, affected Holocaust history in this region.

The mass killing of Jews in the region was conducted mostly under the direction of SS Einsatzgruppe D[70] with the help of regular army troops. Recruited Soviet

---

Fond dukhovnogo prosveshcheniya, 1998); Evgenii Krinko, *Zhizn' za liniei fronta: Kuban' v okkupatsii (1942–1943 gg.)* (Maykop: Adygeya, 2000); Evgenii Zhuravlev, "Okkupatsionnaya politika fashistskoi Germanii na yuge Rossii (1941–1943 gg.): tseli, soderzhanie, prichiny krakha," *Nauchnaya mysl' Kavkaza* 1 (2001): 36–43; Angrick, *Besatzungspolitik*, 545–669. For administration in the military occupation zone, see: Pohl, *Die Herrschaft*, 87–116.

69 The withdrawal order was announced on December 28, 1942, while the actual retreat of the German troops began on December 31. See: "Operationsbefehl no. 2 vom 28 Dezember 1942 betr. weitere Kampfführung auf dem Südflügel der Ostfront," in *Kriegstagebuch des Oberkommandos der Wehrmacht (Wehrmachtführungsstab), 1 Januar 1942–31 Dezember 1942*, ed. Percy E. Schramm, 2nd half-band (Frankfurt am Main: Bernard und Graefe, 1982), 1318–19. The Germans had abandoned most of the North Caucasus by February 1943, but retained some areas, most specifically Novorossiysk, until September 1943. See Table 2.

70 The genesis of Einsatzgruppe D is unusual compared to that of Einsatzgruppen A to C. While these were established in the early spring of 1941, initially as police units to help the Wehrmacht fight the main enemy – Judeo-Bolshevism – Einsatzgruppe D war formed only in June 1941 when the involvement of the Romanian kingdom under Antonescu became clearer. Einsatzgruppe D was headed by SS-Standartenführer Otto Ohlendorf (June 1941–July 1942), then SS-Oberführer Walther Bierkamp (July 1942–July 1943). In the North Caucasus, the following Sonderkommandos (SKs) and Einsatzkommando (EK) participated in the mass killing of Jews. SK 10a under the command of SS-Sturmbannführer Kurt Christmann (August 1942–July 1943) was active in Krasnodar Krai and Rostov Oblast. SK 10b under the command of SS-Sturmbannführer Alois Persterer (June 1941–December 1942) and SS-Sturmbannführer Eduard Jedamzik (December 1942–February 1943) was active on the territory of the North Ossetian and the Kabardino-Balkarian ASSR. SK 11b under the command of SS-Sturmbannführer Werner Braune (October 1941–September 1942) and then SS-Obersturmbannführer Paul Schulz (September 1942–February 1943) was active on the territory of Krasnodar Krai and the North Caucasian Autonomous Oblasts: Adyghe, Karachay, and Cherkess. EK 12, under the command of SS-Sturmbannführer Dr. Erich Müller (February–October 1942) and SS-Obersturmbannführer Günther Herrmann (October 1942–March 1943) was active in Stavropol Krai. See: Helmut Krausnick, "Die Einsatzgruppen vom Anschluß Österreichs bis zum Feldzug gegen die Sowjetunion: Entwicklung und Verhältnis zur Wehrmacht," in *Die Truppe des Weltanschauungskrieges: Die Einsatzgruppen der Sicherheitspolizei und des SD 1938–1942*, ed. Helmut Krausnick and Hans-Heinrich Wilhelm (Stuttgart: Deutsche Verlags-Anstalt, 1981), 195–204; Ernst Klee, *Das Personenlexikon zum Dritten Reich: wer war was vor und nach 1945?* (Frankfurt am Main: Fischer, 2003), 49, 72–73, 93, 248, 419; Angrick, *Besatzungspolitik*; Idem, "Die Einsatzgruppe D," in *Die Einsatzgruppen in der besetzten Sowjetunion 1941/42: die Tätigkeits- und Lageberichte des Chefs der Sicherheitspolizei und des SD*, ed. Peter Klein and Andrej Angrick (Berlin: Ed. Hentrich, 1997), 88–110. EK 6 of Einsatzgruppe C, headed by Robert Mohr (November 1941–September 1942) and SS-Sturmbannführer Ernst Biberstein (September 1942–May 1943) also participated in the Holocaust in some localities of

POWs along with local collaborators[71] also participated in these actions in many places in the North Caucasus: local policemen were often involved not only in the apprehension of Jews in towns and villages, but they also escorted detainees, cordoned off execution sites, and participated in the shooting actions.[72]

A standard procedure can be reconstructed for the mass killing of Jews in the occupied regions of the North Caucasus.[73] A few days after occupying a

---

Rostov Oblast (Shakhty, Rostov-on-Don). See: ChGK report for Shakhty, Rostov Oblast of October 25, 1943, in *GARF*, f. R-7021, op. 40, d. 777, l. 231–237; Evgenii Movshovich, "Shakhty," in *Kholokost na territorii SSSR: Entsiklopediya*, ed. Il'ya Al'tman (Moscow: ROSSPEN, 2009), 1087– 88; "Charge of the Wuppertal Prosecutor's Office against Robert Mohr, 23 July 1962," in Bundesarchiv Ludwigsburg, hereafter BAL, B 162/4690, Bl. 59.

71 The question of collaboration in the USSR requires further examination. See the following general studies: Semiryaga, *Kollaboratsionizm*; Boris Kovalev, *Kollaboratsionizm v Rossii v 1941– 1945 gg.: tipy i formy* (Velikii Novgorod: Novgorodskii gosudarstvennyi universitet, 2009); Evgenii Zhuravlev, "Voenno-politicheskii kollaboratsionizm na yuge Rossii v gody nemetsko-fashistkoi okkupatsii (1941–1943 gg.)," *Vestnik RUDN, seriya istoriya Rossii* 4 (2009): 20–34.

72 For example, local collaborators took part in killing actions in Krasnodar Krai; 11 people were brought to justice immediately after the region was liberated. See: *Sudebnyi protsess po delu o zverstvakh nemetsko-fashistskikh zakhvatchikov i ikh posobnikov na territorii g. Krasnodara i Krasnodarskogo kraya v period ikh vremennoi okkupatsii* (Moscow: Gospolitizdat, 1943); Roman Rudenko, ed., *Nyurnbergskii protsess nad glavnymi nemetskimi voennymi prestupnikami: sbornik materialov v 7 tomakh* (Moscow: Gosyurizdat, 1959), 4:604. For the records of first Soviet trials against Soviet collaborators and Nazi perpetrators, see: Central Archive of the Federal Security Service [*Federal'naya sluzhba bezopasnosti*], hereafter FSB, of the Russian Federation, Records Relating to War Crime Trials in the Soviet Union, 1939–1992 (bulk dates 1945– 1947), H-16708, in United States Holocaust Memorial Museum Archive, hereafter USHMMA, RG-06.025, Reel 15–17.

73 The history of the Holocaust in the North Caucasus is the subject of three dissertations: one in Russian (Voitenko, "Kholokost na yuge"), one in Hebrew, which was published by the author in English as a monograph (Feferman, *The Holocaust*), and one in French (Andrej Umansky, "L'extermination des Juifs dans le Caucase du Nord pendant la Seconde Guerre mondiale (1942–1943)" [PhD diss., Doctoral School on Human and Social Sciences of the University of Picardie Jules Verne, 2016]). The German historian Christina Winkler studies the history of the Holocaust in Rostov-on-Don: Christina Winkler, "The Holocaust in Rostov-on-Don: Official Russian Holocaust remembrance versus a local case study" (PhD thesis, School of History at the University of Leicester, 2015); Idem, "Rostov-on-Don 1942: A Little-Known Chapter of the Holocaust," *Holocaust and Genocide Studies* 30, no. 1 (2016): 105–130. The Canadian historian Maris Rowe-McCulloch defended a PhD thesis on mass violence in one southern city: Maris Rowe-McCulloch, "The Holocaust and Mass Violence in the German-Occupied City of Rostov-on-Don, 1941–1943" (PhD thesis, Department of History, University of Toronto, 2020). For separate chapters on Holocaust history in the region, see: Alt'man, *Zhertvy nenavisti*, 272–286; Angrick, *Besatzungspolitik*, 545–669; Yitzhak Arad, *The Holocaust in the Soviet Union* (Jerusalem: Yad Vashem, 2009), 286–297; Polyan, *Mezhdu Aushvitsem*, 137–145; Aleksandr Rafailov, *Zazhivo pogrebennye* (Kislovodsk: MIL, 2004).

locality, the military commandant or other authorised person of the German military administration issued an order concerning the mandatory registration of the entire Jewish population.[74] For this purpose, the Germans established Jewish committees in many cities (for instance, in the cities of Krasnodar, Armavir, and Novorossiysk in Krasnodar Krai, Essentuki and Kislovodsk in Stavropol Krai,[75] and in Rostov-on-Don[76]) under the chairmanship of Jews respected in the city. As part of their policy, Germans installed institutions between them and their victims that allowed them to stay in the background. These Jews were deceived into believing they could help their fellows by

---

[74] It should be noted that the German authorities carried out the registration of all local civilians in the occupied areas as a matter of occupation policy. Registration helped to identify competent professionals who were sent to institutions, organisations and enterprises operating under the occupation regime. Furthermore, the registered population served as a database for selecting citizens to be sent for forced labour in Germany. Thus, for example, the German authorities divided all the residents of Pashkovskaya village, Krasnodar Krai into four categories: "The first and the second categories were workers, peasants, and employees without reasonable grounds for suspicion, admitted to the administration; the third group included 136 Soviet workers who have not participated in bombings and repressions [i.e., who had not committed acts of sabotage, but were considered dubious], could allegedly work everywhere, but could not be admitted to the administration; the fourth group was Jews and active workers who supposedly could only be allotted rough, hard labour." In Krasnodar, the city and district authorities made up three lists of the city's residents on the instructions of the occupation authorities in December 1942. The first list consisted of "all registered citizens who had lived in the city since the pre-war period excluding the politically unreliable elements. The second list included all the persons who settled in Krasnodar after June 22, 1941. The third recorded those, who were subject to arrest and destruction by punitive occupational authorities." The third list was not publicly disclosed anywhere. Belyaev and Bondar', *Kuban' v gody*, 513–514; Anatolii Avramenko et al., *Ekaterinodar-Krasnodar: dva veka goroda v datakh, sobytiyakh, vospominaniyakh [1793–1993]: materialy k letopisi* (Krasnodar: Knizhnoe izdatel'stvo, 1993), 596. In Rostov-on-Don, only Jews were registered in three lists: Jews who were able to work were recorded in one with a special note for doctors and scientists, handicapped persons in the other, and people who had non-Jews or baptised Jews as family members in the third. "ChGK report for Rostov Oblast," in *State Archive of Rostov Oblast [Gosudarstvennyi arkhiv Rostovskoi oblasti]*, hereafter *GARO*, f. R-3613, op. 1, d. 2, l. 2.

[75] The Jewish Committee in Kislovodsk under the chairmanship of the dentist Moisey Samoilovich Beninson, born in 1878, was told to seize valuables from the local Jews and hand them over to the German command. "ChGK report for Stavropol Krai," in *GASK*, f. R-1368, op. 1, d. 84, l. 13ob.

[76] In Rostov-on-Don, this organ was called the Jewish Council, which was headed by the former director of the Health Education House Dr. Lur'e. Yitzhak Arad, Shmuel Krakowski, and Shmuel Spector, *The Einsatzgruppen Reports: Selections from the Dispatches of the Nazi Death Squads' Campaign Against the Jews in Occupied Territories of the Soviet Union July 1941 – January 1943* (New York: Holocaust Library, 1989), 358; "ChGK report for Rostov-on-Don of August 5, 1943," in *GARF*, f. R-7021, op. 40, d. 10, l. 3.

cooperating with the German administration, thus turning them into agents of their own and their community's destruction. Through the mediation of these Jewish committees, Jews were sent to conduct heavy unpaid labour (building airfields, roads or highways; doctors and professors were forced to sweep streets) in some localities and for a short time.[77] In some cities and districts of the North Caucasus, Jews were required to wear distinguishing emblems: a six-pointed star (Star of David) as armbands in Stavropol and in the city of Tikhoretsk, Krasnodar Krai or as sewn to the chest in Kislovodsk. The stars were yellow in Armavir and in Taganrog white in Essentuki and Labinskaya village blue in Meshkovskaya village, Rostov Oblast, and white armbands with red stars were worn in Novo-derevyankovskaya village, Krasnodar Krai.[78]

An "Appeal to the Jewish population" from the German administration, consisting of the order to be registered as a Jew was issued on behalf of the Jewish committees as well.[79] One such appeal was cited at the beginning of this chapter. After registration, all Jews were ordered to come to a designated location carrying luggage from 20 to 32 kg at an appointed day and hour under the pretext of "relocating to a new safer place of residence."[80] At collecting stations, the possessions of the Jews were taken away.[81] They were loaded onto train wagons (Stavropol Krai) or carts in rural areas of the region, crammed in

---

77 "ChGK report for Stavropol Krai," in *GASK*, f. R-1368, op. 1, d. 142, l. 170.
78 "ChGK reports for Krasnodar and Stavropol Krais, and Rostov Oblast," in *GARF*, f. R-7021, op. 16, d. 2, l. 1-2ob.; d. 7, l. 184–189; d. 8, l. 1–3; d. 92, l. 2,3; in *GAKK*, f. R-897, op.1, d.2, l. 123, 124, 179, 480-481ob; in *GASK*, f. R-1368, op. 1, d. 142, l. 169; in *TsDNIRO*, f. R-1886, op. 1, d. 8a, l. 7-7ob.; "Red Army Report for Essentuki of January 13, 1943," in *Central Archive of the Russian Ministry of Defence* [*Tsentral'nyi arkhiv ministerstva oborony Rossiiskoi Federatsii*], hereafter *TsAMO*, f. 51, op. 958, d. 52. l. 17 (access: in *USHMMA*, RG-22.016); "Red Army Report for Armavir of January 28, 1943," in *TsAMO*, f. 51, op. 958, 52, l. 91 (access: in *USHMMA*, RG-22.016); "Order of Bürgermeister of Tikhoretsk no.1 of August 20, 1942," in *GAKK*, f. 487, op. 1, d. 1, l. 1.
79 Ivan Yudin, *Sledy fashistskogo zverya na Kubani* (Moscow: Gospolitizdat, 1943), 10–11.
80 See the texts of the "Appeal to the Jewish population" for Stavropol Krai, in *GASK*, f. R-1368, op. 1, d. 84, l. 16; in *GARF*, f. R-7021, op. 17, d. 1, l. 95; in Pyatigorsk Local History Museum [*Pyatigorskii kraevedcheskii muzei*], Nauchno-vspomogatel'nyi fond, no. 2354; in Grigorii Prozritelev and Georgii Prave Stavropol State Historical-Cultural and Natural-Landscape Museum-Reserve [*Stavropol'skii gosudarstvennii istoriko-kul'turnyi i prirodno-landshaftnyi muzei-zapovednik imeni Grigoriya Prozriteleva i Georgiya Prave*], hereafter SGMZ, of. 23307; for Rostov Oblast, in *GARO*, f. R-3613, op. 1, d. 19, l. 1; in *Staatsarchiv München*, hereafter *StA München*, Staatsanwaltschaft, 35308/32.
81 As a rule, the property of Jewish citizens was plundered and appropriated by the German authorities. See: "Red Army Report for Pyatigorsk of January 14, 1943," in *TsAMO*, f. 51, op. 958, d. 52, l. 76 (access: in *USHMMA*, RG-22.016); "Report of the chief of the Agitation and Propaganda Department of the 37th Army Major Smirnov of March 14, 1943," in *TsAMO*, f. 51, op. 958, d. 52, l. 149 (access: in *USHMMA*, RG-22.016).

small groups into special mobile gas chambers or gas vans [*Gaswagen*], known among the Russian population as "*dushegubka*" [soul-killer] (Krasnodar[82] and Stavropol[83] Krai), and sometimes forced to walk (Rostov Oblast) to the killing site. They were taken to the outskirts of a city or a village where they were shot or poisoned.[84] The corpses were dumped into mass graves, which were often

---

[82] "SK 10a used special gas vans in Krasnodar several times a week from the end of August 1942. These were covered 6–7-ton dark grey vans with a diesel engine. The vehicle was lined inside with galvanised iron and fitted with a hermetically lockable double door in the rear. On the floor of the rear compartment there was a grate, under which ran a pipe with holes, connected to the exhaust pipe of the diesel engine, the exhaust gases – containing high concentrations of carbon monoxide – entered the body of the machine and caused rapid poisoning of the people locked in there. Putting victims in the vans, the Germans undressed them, telling them that they were taking them to a bath, and their clothes were left for disinfection. The vans had a capacity of between 60 and 80 people; men, women, children. After standing for a few minutes in the courtyard of the Gestapo, the van headed for an anti-tank ditch, where the corpses were unloaded and buried." "Reporting notes of the heads of the military departments of the city and district party committees on the work of the military departments," in *TsDNIKK*, f. R-1774, op. 2, d. 1396, l. 26. Such gas vans were known as "Black Ravens" [*chernye vorony*] in other places in Krasnodar Krai. "ChGK report for Ladozhskaya village of September 4, 1943," in *GARF*, f. R-7021, op. 16, d. 8, l. 5–6. See also the materials of the 1946 prosecution of the driver of a gas van in Krasnodar, Otto Enoch: "Records Relating to War Crime Trials in the Soviet Union, 1939–1992 (bulk dates 1945–1947)," in Central Archives of the FSB of the Russian Federation, K-110815 (access: in *USHMMA*, RG-06.025, Reel 42); Mathias Beer, "Die Entwicklung der Gaswagen beim Mord an den Juden," *Vierteljahrshefte für Zeitgeschichte* 3 (1987): 403–413.

[83] EK 12 of Einsatzgruppe D used two such vans, which were also called "Bakeries" [*pekarni*] or "Fly Swatters" [*mukhoboiki*] in Pyatigorsk, Stavropol Krai. "ChGK reports for Stavropol Krai," in *GASK*, f. P-1368, op. 1, d. 142, l. 171; d. 153, l. 7.

[84] Soviet records for Rostov Oblast and Krasnodar Krai contain testimonies about the poisoning of children before execution by lubricating their lips with a poison. There were recorded cases of Jews' bodies dropped into wells in Stavropol and Krasnodar Krais. See: "Untitled Report," in *Informatsionnyi sbornik 22* (Rostov-on-Don: Shtab Severo-kavkazskogo voennogo okruga, 1963), 63; "ChGK reports for Krasnodar and Stavropol Krais and Rostov Oblast," in *GARO*, f. 3613, op. 1, d. 30, l. 1; *GARF*, f. R-7021, op. 16, d. 4, l. 10; d. 7, l. 37; *TsDNISK*, f. 1, op. 2, d. 359. l. 13–15, 20; Valeriya Vodolazhskaya et al., ed., *Stavropol'e v period nemetsko-fashistskoi okkupatsii (avgust 1942 – yanvar' 1943 gg.): dokumenty i materialy* (Stavropol: Knizhnoe izdatel'stvo, 2000), 81–82. See also the following research on this topic: Anna Bogachkova, *Istoriya Izobil'nenskogo raiona* (Stavropol: Knizhnoe izdatel'stvo, 1994), 111; Il'ya Karpenko, "Pyati smertyam nazlo," *Lekhaim* 9 (2006), http://www.lechaim.ru/ARHIV/173/VZR/v02.htm (accessed March 22, 2020); Winkler, "Rostov-on-Don," 106–115; Maris Rowe-McCulloch, "Poison on the Lips of Children: Rumors and Reality in Discussions of the Holocaust in Rostov-on-Don (USSR) and Beyond," in *The Journal of Holocaust Research*, 33 no. 2 (2019): 157–174. There are also testimonies with Holocaust eyewitnesses that mention these killing methods in the North Caucasus: Interview with Valentina G., born in 1930, in *Yahad-In Unum Archive*, code 263R, March 31, 2012, Ladozhskaya village, Krasnodar Krai; Interview with Nikolai T., born in 1930, in *Yahad-In Unum*

dug by the local non-Jewish population, Soviet POWs, or sometimes by the victims themselves. After the mass killing of Jews no later than two to three weeks after the occupation began (in most cases), the Nazi authorities with the help of local collaborators continued to identify and shoot Jews individually.[85] The tactic of the fastest possible "final solution of the Jewish Question" was adopted in the North Caucasus. Therefore, formal ghettos with the structure of an internal Jewish self-administration, which were common in occupied Poland and today's Baltic countries and Belarus, were not implemented here. Nonetheless, there were instances of non-institutionalised and short-term concentration of Jews in temporary and improvised prisons; for instance, the concentration of Jews in one building from a few weeks to one-and-a-half months (Essentuki and Zheleznovodsk in Stavropol Krai; Mikoyan-Shakhar in Karachay AO). Imprisoned Jews were sometimes used as forced unskilled labour in Kislovodsk and Armavir.[86]

The Mountain Jews and Karaites of the North Caucasus were more fortunate, although not everywhere. In those places with mixed Ashkenazi and Mountain Jew populations, for example in Kislovodsk and Pyatigorsk, Stavropol Krai, all the Jews were killed. The same fate befell the population of some Mountain Jew collective farms in Stavropol Krai.[87] Mountain Jews in Nalchik, Kabardino-

---

Archive, code 299R, April 12, 2012, Gulkevichi, Krasnodar Krai; Interview with Dmitrii G., born in 1928, in *Yahad-in Unum Archive*, code 265R, April 1, 2012, Petropavlovskaya village, Krasnodar Krai; Interview with Valentin P., born in 1931, in *Yahad-In Unum Archive*, code 565R, October 21, 2015, Soldato-Aleksandrovskoe village, Stavropol Krai.

85 As a rule, the capture of Jews was accompanied by torture and humiliation. See: "ChGK reports for Krasnodar and Stavropol Krais," in *GASK*, f. R-1368, op. 1, d. 152, l. 7; in *GARF*, f. R-7021, op. 16, d. 2, l. 4; d. 8, l. 8 ob., 82; op. 17, d. 1, l. 1–6.

86 "ChGK reports for Stavropol Krai," in *GARF*, f. R-7021, op. 17, d. 10, l. 195–196; d. 6, l. 1–2; "The letter of the artist L. Tarabukin and his wife, professor of the music conservatory D. Goldstein, to the writer Yurii Kalugin 'On the massacre in Essentuki,'" in *GARF*, f. 8114, op. 1, d. 963, l. 116–118; Idem, in *Unknown Black Book: Evidence of eyewitness accounts of the catastrophe of Soviet Jews (1941–1944)*, ed. Yitzhak Arad (Jerusalem – Moscow: Yad Vashem, 1993), 389–90; "Red Army report for Kislovodsk of January 14, 1943," in *TsAMO*, f. 51, op. 958, d. 52, l. 68 (access: in *USHMMA*, RG-22.016); "Red Army Report for Armavir of January 28, 1943," in *TsAMO*, f. 51, op. 958, d. 52, L. 91 (access: in *USHMMA*, RG-22.016). Some Holocaust survivors remembered that they were kept in prison or in a separate building, for example, in a school building from several weeks to several months (Novoleushkovskaya village, Krasnodar Krai; Milyutinski District, Rostov Oblast). Interview with Tankha Otershtein, born in 1932, in author's *archive*, code 13-X-R011, March 1, 2013, Taganrog, Russia; Interview with Fira Ezerskaya, born in 1925, in *VHA USCSF*, code 15232, May 16, 1996, West Hollywood, California, USA.

87 In Kurskii District, Stavropol Krai, 452 Mountain Jews of Bogdanovka village were killed in September 1942 and 40 families of Mountain Jews from the collective farm Menzhinskii were killed in October 1942. "ChGK report No. 92 for Kurskii village of June 27, 1943," in *GARF*, f. R-7021, op. 17, d. 10, l. 155–162; Ibid, in *GASK*, f. R-1368, op. 1, d. 89, l. 1–10.

Balkarian ASSR miraculously survived: they were left alive by the decision of a specially created German commission. It is possible that this was also thanks to Field Marshal Ewald von Kleist, as well as the chairman of the puppet National Council and the Kabardian national leader Selim Shadov, who asked him not to kill the Mountain Jews or Tats on the grounds that they were not Jews but a mountain tribe.[88] According to another version, the Russian orientalist of German origin Nicholas Poppe – caught by the German occupation in Mikoyan-Shahar and serving as an interpreter for the Germans – was involved as an expert ethnographer on the issue of the origin of the Mountain Jews in November 1942. In his conclusion, Poppe confirmed that Mountain Jews were not viewed as Jews, but as other mountaineers in the times of the Russian Empire, that they are more correctly to be regarded as a variety of Tat, a people of Iranian origin.[89] Thus, according to the estimates of the Russian researcher Pavel Polyan, about 3,200 Mountain Jews were saved during the occupation of the North Caucasus, while about 700 people became Holocaust victims.[90] The Germans were uncertain whether the Karaites should be treated as Jews.[91] Most Karaites, with a few exceptions, managed to survive the German occupation. At the same time, several dozen Karaites were killed alongside local Jews in Krasnodar in mid-August 1942.

---

[88] "Baksheev Iosif Agaevich," in *GARF*, f. 8131, op. 31, d. 32217, l. 90–94.
[89] Svetlana Danilova, ed., *Iskhod gorskikh evreiev: razrushenie garmonii mirov* (Nalchik: Poligrafservis i T, 2000), 65; Polyan, *Mezhdu Aushvitsem*, 137–145; Kiril Feferman, "Nazi Germany and the Mountain Jews: Was There a Policy?," *Holocaust and Genocide Studies* 21, no. 1 (2007): 96–114; Richard Löwenthal, "The Judeo-Tats in the Caucasus," *Historica Judaica* XIV (1952): 61–82; Nicholas Poppe, *Reminiscences*, ed. Henry G. Schwarz (Bellingham: Western Washington University, 1983), 166.
[90] Polyan, *Mezhdu Aushvitsem*, 142.
[91] For Nazi policies towards the Karaites and their fate during World War II, see: Philip Friedman, "The Karaites under Nazi Rule," in *On the Track of Tyranny. Essays Published by the Wiener Library to Leonard G. Montefiore, on the Occasion of his Seventieth Birthday*, ed. Max Beloff (London: The Wiener Library, 1960), 97–123; Warren P. Green, "The Nazi Racial Policy towards the Karaites," *Soviet Jewish Affairs* 8 (1978): 36–44; Warren P. Green, "The Fate of the Crimean Jewish Communities: Ashkenazim, Krimchaks and Karaites," *Jewish Social Studies* 46 (1984): 169–76; Hannelore Müller, *Religionswissenschaftliche Minoritätenforschung: zur religionshistorischen Dynamik der Karäer im Osten Europas* (Wiesbaden: Harrassowitz, 2010), 120–162; Kiril Feferman, "The Fate of the Karaites in the Crimea during the Holocaust," in *Eastern European Karaites in the Last Generations*, ed. Dan Shapira and Daniel Lasker (Jerusalem: Benzvi Institute, 2011), 171–191; Idem, "Nazi Germany and the Karaites in 1938–1944: Between Racial Theory and Realpolitik," *Nationalities Papers* 39, no. 2 (2011): 77–94; Mikhail Kizilov, *The Sons of Scripture: The Karaites in Poland and Lithuania in the Twentieth Century* (Berlin-Warsaw: Walter de Gruyter, 2015), 294–369.

According to the data of the famous Karaite expert Boris Kokenai, who saved the Karaite community of Rostov-on-Don, 96 Karaites perished in Krasnodar.[92]

The total number of Holocaust victims in the North Caucasus is subject to debate among European – primarily German – and Russian researchers, local historians, and activists of Jewish communities. The main source for the counting of Holocaust victims is Soviet documentation,[93] primarily ChGK records. Overestimation and distortion of quantitative data, together with the fact that the final reports refer to "peaceful Soviet citizens" as the main Nazi victim group, makes it difficult for historians to obtain an objective figure of Holocaust victims. The lowest and highest possible number of victims for each region of the North Caucasus is given in Table 3, considering the analysis of available sources and literature. If we consider the data on the number of Holocaust victims in Russia (i.e., within the boundaries of the RSFSR), more than half of them are assignable to the North Caucasus.[94]

---

92 Boris Kokenai, "Dnevnik," in *personal archive of Sergei Shaitanov*, 69–70. For the list of killed and tortured Karaites during the war years, see: *Official site of the Moscow Karaite Society*, http://www.karaims.ru/page.php?cod=ru&page=225&node=215&p=288 (accessed March 23, 2020). Also, the fact that Karaites were killed in Krasnodar on August 21, 1942 was recorded in the ChGK report: "ChGK report for Krasnodar," in *GARF*, f. R-7021, op. 16, d. 435, l. 73. Feferman believes that the destruction of nearly the whole Karaite community of Krasnodar may have to do with the personal proclivities of the new Einsatzgruppe D chief Walter Bierkamp, or it can be argued that the small number of Krasnodar's Karaites did not succeed in presenting their case collectively to the German authorities. Feferman, "Nazi Germany," 288.
93 "Ereignismeldungen USSR" [USSR event reports] and "Meldungen aus den besetzten Ostgebieten" [Reports from the occupied eastern territories] were the main reporting forms of the chief of the SS Sicherheitspolizei and Sicherheitsdienst (the Security Police and the Security Service of the SS), but information about the killing of Jews had practically disappeared from them by the summer of 1942. For example, the report from the occupied eastern territories no. 16 of August 14, 1942, said that 2,000 Jews were located in Rostov-on-Don. There is no information on their further fate. "Meldungen aus den besetzten Ostgebieten, no. 16 (August 14, 1942)", in *Online Database Nationalsozialismus, Holocaust, Widerstand und Exil 1933–1945*, http://db.saur.de/DGO/basicFullCitationView.jsf?documentId=rk1355 (accessed March 23, 2020). See also published documents of the Einsatzgruppen in the USSR with extensive commentaries: Andrej Angrick et al., ed., *Deutsche Besatzungsherrschaft in der UdSSR 1941–1945, Dokumente der Einsatzgruppen in der Sowjetunion*, vol. 2 (Darmstadt: Wissenschaftliche Buchgesellschaft, 2013); Klaus-Michael Mallmann et al., ed., *Deutsche Berichte aus dem Osten. Dokumente der Einsatzgruppen in der Sowjetunion*, vol. 3 (Darmstadt: Wissenschaftliche Buchgesellschaft, 2014).
94 According to Al'tman's data, from 119,210 to 140,350 people were murdered on the territory of Russia in its modern borders without regard to the Republic of Crimea. See: Al'tman, *Zhertvy nenavisti*, 286.

3 (Re)writing Holocaust Memory in Official Soviet Wartime Sources

April 5, 1943
to the Rostov Regional Committee of the VKP(b) from Genya Moiseevna Golos, a resident of the village of Balabanovka, Malchevskii district, Rostov Oblast.

Statement No. 92.

I am Jewish. Before the war, I lived in Zaporozhye [Ukrainian SSR] with my relatives. When the war began, we were supposed to be evacuated to Saratov Oblast, but we only managed to reach Balabanovka in Malchevskii District [Rostov Oblast] and stayed here to work on the collective farm. That was in late September 1941. When the German advance reached our village in July 1942, we, all the Jewish families, decided to leave, because we knew that we were all facing death. With our carts, we managed to reach only the banks of the Don River near Konstantinovka [a village in Donetsk Oblast, Ukrainian SSR] and we ended up surrounded by German troops 10 km from the river crossing. We were not able to proceed further and had to go back. We tried to move along remote roads, through Gypsy tabors,[1] and often ran into the Germans. The Germans found us in one of the Gypsy tabors in Tatsinskii District [Rostov Oblast], seized eight men, and led them away. I don't know what happened to them, but the Germans suggested that we should leave the place immediately [i.e., return for registration], or else they would shoot at us.

Wandering through Gypsy camps was becoming increasingly difficult, we headed for Kryvyi Rih District [in Dnipropetrovsk Oblast, Ukrainian SSR] to stay and live there. The commandant of Kryvyi Rih suggested that we should immediately return to our former place of residence and we went back to Balabanovka. As soon as we arrived, a German officer and village headman [*starosta*] named Slavik approached us. They announced that there would be a meeting in an hour for all those who had returned and everybody had to be there. A vehicle with German soldiers arrived about an hour and a half later and the officer told all the families to get up one after another. I got scared by that and told my mother I wanted to leave. Meanwhile, my two-year-old son started to cry and I stepped aside as if I was going to give him water. I did not return to my people, and then I learnt from the locals that everybody had been put in the vehicle and taken to be shot.

Thereafter, I had to hide from the German authorities and local policemen. They looked for me in nearly every house, that is why I left Balabanovka. I could not [stay] in one place for a long time, because as soon as people discovered I was Jewish, they started to pursue me right away, and I had to flee. I passed through many districts, hiding from the German barbarians. I faced death five times because I fell into the hands of the local police, but Soviet people who fought against the German fascists helped me to escape the

---

[1] The term Gypsy tabor [*tsyganskii tabor*] refers to a group of wandering Roma or their temporary settlements. It is unclear if the author of this statement really lived in such settlements together with Roma people. The term Gyspy [*tsygan*] is still used in contemporary Russia without pejorative connotation.

clutches of the Gestapo² every time. I worked on the Kosharskii State Grain Farm [Rostov Oblast] for several weeks and enjoyed a most sensitive attitude from the director and the administrator. The locals often helped me by warning me of a threat. It was not rare that my son and I had to endure hunger and freezing cold, but I lived on in the hope that our [Soviet] people would come soon and my tortures would end.

I will never forget what my child and I had to go through, I will always remember my comrades who died at the filthy hands of the fascist barbarians and will devote my whole life to the fight against the German invaders, the fight to strengthen the power of the Soviet country.

Genya M. Golos[3]

101 people, members of about 20 Jewish families evacuated from Zaporozhye in September 1941, were shot in Balabanovka, Rostov Oblast in July 1942.[4] The statement above, by a Soviet Jewish woman who managed to survive the occupation of Rostov Oblast, was sent to the ChGK immediately after the region was liberated. There are only a few such statements by Holocaust survivors since only a few Jews managed to avoid the implementation of the Final Solution in the occupied Soviet territories. Such statements are valuable because they are the first official evidence of the Holocaust in the USSR. When the final ChGK reports on every liberated area were drawn up, the generalised categories of "peaceful citizens" and "Soviet citizens" appeared. This depersonalised and generalised definition of all Nazi victims in the occupied Soviet territories blurred the categorical differences between victim groups. The term "peaceful Soviet citizens" is a Soviet propaganda construct, which the Soviet authorities artificially used to avoid talking about Jews and the specifics of what was later called the Holocaust. This was not accidental, but rather a clear decision: Soviet authorities knew precisely that crimes had been committed against the Jewish population specifically.[5]

---

2 Gestapo [*Geheime Staatspolizei*, Secret State Police] was the official secret police of Nazi Germany. Soviet sources, especially personal narratives, usually use this term to describe the whole system of German administration under the occupation.
3 "Statement by Genya Golos from Balabanovka, Rostov Oblast, 5 April 1943," in *GARF*, f. R-7021, op. 40, d. 5, l. 15–16.
4 For the list of full names and ages of the killed Jews, see: "ChGK report No. 110 for Olkhovo-Rogskii village of April 5, 1943," in *GARF*, f. R-7021, op. 40, d. 5, l. 18–19. According to German data, a small Wehrmacht unit was based in Balabanovka since late July 1942. The leader of the unit, Alwin Weisheit, decided on his own initiative to shoot all the Jews of the village; at least 75 women, men and children. When information about that crime reached his superiors, he was brought to trial in Pyatigorsk on September 29, 1942 and sentenced to two years in prison. Else Staff, *Justiz im Dritten Reich: eine Dokumentation* (Frankfurt am Main: Fischer, 1964), 255–256.
5 See more about the creation of the term "peaceful Soviet citizens" in the wartime Soviet media: Karel Berkhoff, *Motherland in Danger: Soviet Propaganda during World War II* (Cambridge, MA: Harvard University Press, 2012), 149–153, 160–164; Idem, "Total Annihilation

The Soviet authorities published final ChGK reports in multiple copies and used them in their accusations against the Nazi regime, including at the Nuremberg trials. This way the Soviets tried to avoid "the inconvenient aspect" of the war's history, i.e., the mass killings of Jews. In its surprising tenacity, this war myth is indebted to the anonymity of the elements that give it structure and significance: the Unknown Soldier [*neizvestnyi soldat*], the Living and the Dead [*zhivye i mertvye*], the Eternal Flame [*vechnyi ogon'*], the Victorious People [*narod pobeditel'*], and so forth. The vast distance separating the myth's dramatis personae from the lives of real people and concrete events guaranteed that a "national amnesia" [*obshchenatsional'noe zabvenie*] would serve as an important element in the political stability of the Soviet regime for decades.[6] Nevertheless, original ChGK personal statements and reports attested to mass killings of Jews and often recorded Holocaust victims by name. Thus, official wartime sources have a double function: they are the main sources recording the history of the occupation in the western and southern regions of the USSR, including that of the Holocaust, and at the same time they are an important element of the Holocaust memory dispositif. In this chapter, I trace the development of official Holocaust memory policy in the Soviet Union on the basis of a discourse analysis of official Soviet wartime documents. Following a general characterisation of the main wartime sources on Holocaust history, I will cover the regional collection of ChGK records in more detail, paying close attention to analysing the language of Soviet documents in so far as it refers to the mass killings of Jews (among other Nazi crimes) in the occupied areas of the North Caucasus. In order to analyse the role of official Soviet wartime sources in the construction of a regional dimension of the Holocaust memory dispositif, I will also analyse how these sources have been taken up in post-war Soviet and post-Soviet Russian society.

3.1 Collecting Evidence: The Production and General Characteristics of Soviet Official Wartime Documents

Red Army forces became the first chroniclers of Nazi crimes against the Jewish population in the occupied territories of the USSR. They were first of all soldiers

---

of the Jewish Population: The Holocaust in the Soviet Media, 1941–45," *Kritika: Explorations in Russian and Eurasian History* 10, no. 1 (2009): 86–87; Arkadi Zeltser, "Differing Views among Red Army Personnel about the Nazi Mass Murder of Jews," *Kritika: Explorations in Russian and Eurasian History* 15, no. 3 (2014): 564.

6 Marina Sorokina, "People and Procedures: Toward a History of the Investigation of Nazi Crimes in the USSR," *Kritika: Explorations in Russian and Eurasian History* 6, no. 4 (2005): 5.

and commissars in fighting units, as well as military police officers and medical personnel. As Soviet territories were liberated, Red Army combatants recorded "hot on the trail" the economic destruction and other damage caused by the Nazi administration. There was a political department within every military unit of the Red Army, which served propaganda and agitation functions and was subordinate to the Main Political Administration of the Workers' and Peasants' Red Army [*Glavnoe politicheskoe upravlenie raboche-krest'yanskoi Krasnoi armii*], hereafter GlavPU KA.[7] Commissars and officers in army political departments drew up detailed reports about the economic and political situation in the liberated territories and sent them to higher authorities. The source materials for these documents were reports that Red Army soldiers drew up with civilians immediately after a village, town, or city, or a concentration or death camp had been liberated. The earliest reports of Red Army soldiers date from November and December 1941 when a number of Soviet regions including the city of Rostov-on-Don were liberated during the counter-offensive of the Red Army. The main body of Red Army evidence, however, dates from the first half of 1943 when units on the North Caucasian Front liberated the region. These sources record Nazi crimes in southern Russia, including evidence of mass killings of Jews, based on the reports of civilians who survived the occupation, and liberated Soviet POWs.[8] The Holocaust was a significant theme in these records, but by no means the only one. Written in challenging wartime circumstances, these documents were highly detailed: they frequently indicated the exact killing site,

---

[7] GlavPU KA was a department of the TsK VKP(b), later the Communist Party of the Soviet Union, hereafter CPSU, between 1924 and 1991. The TsK developed the ideological and propaganda doctrine of the entire Soviet state, which was then imposed on Soviet society. Between 1941 and 1946, GlavPU was officially named the Main Political Administration of the Workers' and Peasants' Red Army. It was headed by Lev Mekhlis, a member of the Orgburo [Organisational Bureau] of the TsK VKP(b), Doctor of Economics, and Deputy of the Peoples Commissariat for Defence (July 16, 1941–June 12, 1942) and then by Aleksandr Shcherbakov, Colonel General and Director of the Soviet Informational Bureau [*Sovetskoe informatsionnoe byuro*], hereafter Sovinformburo (June 12, 1942–May 10, 1945). See: Gennadii Sredin, ed., *Politorgany sovetskikh vooruzhennykh sil: istoriko-teoreticheskii ocherk* (Moscow: Vrenizdat, 1984); Nikolay Pupyshev, "Iz vospominanii o A.S. Shcherbakove," *Druzhba narodov* 2 (1985): 162–174; Yurii Rubtsov, *Mekhlis: ten' vozhdya* (Moscow: Veche, 2011).

[8] The GlavPU KA collection in TsAMO also contains personal records (letters and diaries of Jews from occupied territories), testimonies of Soviet soldiers and officers who escaped from POW camps about the killing of their Jewish comrades, and captured documents and evidence from POWs (servicemen of the Wehrmacht and the Romanian army alike). See: *TsAMO*, f. 32, op. 11302 (access: in *USHMMA*, RG-22.016, box 5).

the names of eyewitnesses, and sometimes recorded the names of victims.[9] In the absolute majority of reports, Jews were represented as a separate victim group alongside Soviet POWs and civilians of other ethnicities, e.g., Russians, Kabardians, Circassians.

At the same time, Red Army reports recorded the facts of the Holocaust itself without detailing the reasons motivating the near total annihilation of Jews by the Nazis, whereas crimes against non-Jewish Soviet citizens were listed along with an explanation, or even shown to result from a cause-and-effect relation. For example, a report of January 7, 1943 recorded that 100 Russian inhabitants of the small village of Bygulov, located 15 km from the town of Mozdok, were shot "because someone killed a German soldier in the village." The report continued, attesting to the total eradication of the Jews of several small villages in Mozdokskii District, Stavropol Krai without stating any reason for these acts.[10] Whereas reports by Red Army soldiers only recorded Nazi crimes,[11] material from Red Army commissars was prepared for printed agitation bulletins. They presented numerous details of "monstrous crimes," "from which the blood in the veins runs cold when one learns about thousands of unlucky Jews." For example, in the village of Ladozhskaya in Krasnodar Krai, "over 3,000 Jews were atrociously shot. There were many cases of young children being tied to their mothers and thrown into the river Kuban. About 60 POWs were shot in the village; their corpses were not allowed to be buried." The head of the Agitation and Propaganda Department of the 37th Army, Major Smirnov, summarised his report: "The compiled material gives a complete picture of the fact that where Hitler's soldiers set foot, a

---

9 For example, a Red Army report of January 7, 1943 consists of a record of the death of 12 Jews in the settlement of Sheinfeld, Stavropol Krai. Another report from September 23, 1943 for the city of Novorossiysk, recorded the Holocaust site: "Jews were shot in the anti-tank ditches near [Sudshukskaya] Kosa and Raievskogo [Street]." See: "Report of the Red Army for the farm Russkii Pervyi, Stavropol Krai of January 7, 1943," in *TsAMO*, f. 51, op. 958, d. 52, l. 51–52 (access: in *USHMMA*, RG-22.016); "Report of the Red Army for Novorossiysk of September 23, 1943," in *TsAMO*, f. 51, op. 958, d. 52, l. 241–248 (access: in *USHMMA*, RG-22.016).
10 "Report of the Red Army for the farm Russkii Pervyi."
11 For example, a report of January 6, 1943 recorded "the systematic killings of the Jewish population" in the village of Stepnoe of Stavropol Krai. Another report for Pyatigorsk of January 4, 1943 stated along with other Nazi crimes and "according to preliminary data, the German command robbed and then shot 2,800 men, women and children, who were all Jewish." See: "Report of the Red Army for the village of Stepnoe of January 6, 1943," in *TsAMO*, f. 51, op. 958, d. 52, l. 43 (access: in *USHMMA*, RG-22.016); "Report of the Red Army for Pyatigorsk of January 4, 1943," in *TsAMO*, f. 51, op. 958, d. 52, l. 76 (access: in *USHMMA*, RG-22.016).

great cultural life froze, dehumanisation and desolation ensued."[12] The agitation texts had plenty of epithets as they aimed to increase the Red Army combatants' hatred of the enemy, which was a fundamental requirement for a successful struggle.[13] The authors of these texts believed that by raising awareness of Nazi crimes against specific ethnic or social groups and/or members of the Communist party, they would give the soldiers stories to which they could relate, because many soldiers were party members or belonged to the ethnic and social groups that were killed by the Germans in places where some of the combatants came from. This agitprop aimed to communicate that the Germans threatened everything in which the soldiers believed, and which they held dear, including their families.

Red Army reports were meant primarily for official use: the information they contained was used as the main evidence for agitation and propaganda in the Army and the Navy and was not widely reproduced. The ChGK records, on the contrary, are the best known and the largest wartime source for Nazi crimes, including those against Jews in the occupied regions. Already in the first months of the war, the Soviet authorities had begun to plan the creation of a specialised institution which would keep "a circumstantial account of all the crimes of the Nazi troops for which the indignant Soviet people justly demand and will achieve retribution."[14] That intention was implemented on November 2, 1942 with the signing of a Decree of the Presidium of the Supreme Soviet of the USSR "On the Establishment of an Extraordinary State Commission on Reporting and Investigating the Atrocities of the German Fascist Occupiers and their Accomplices and the Damages they caused to Citizens, Kolkhozes, Public Organisations, State Enterprises of the USSR," the ChGK.[15] It was led by Nikolai

---

12 "Report to the head of the propaganda and agitation department of the North-Caucasian front of March 14, 1943," in *TsAMO*, f. 51, op. 958, d. 52, l. 147–153 (access: in *USHMMA*, RG-22.016).

13 One of the main tasks of an agitator "is to strengthen in each Soviet citizen the feeling of absolute hatred towards German butchers, to untiringly raise the desire for victory over the mortal enemy, and the craving for revenge on German bastards for all their atrocities." *Pamyatka agitatora* (Moscow: Gospolitizdat, 1943), 29. See also: Sergei Zverev and Nikolai Storozhev, "Vospitanie nenavisti k vragu v sovetskoi voinskoi kul'ture 30–40-kh gg. XX veka," *Vestnik Sankt-Peterburgskogo gosudarstvennogo instituta kul'tury* 3 (2014): 21–28.

14 Sergei Maiorov, ed., *Vneshnyaya politika Sovetskogo Soyuza v period Otechestvennoi voiny*. vol. 1, *22 iyunya 1941 g. – 31 dekabrya 1943 g.* (Moscow: Gospolitizdat, 1946), 228.

15 "Decree of the Presidium of the Supreme Soviet of the USSR 'On the Formation of the ChGK on Reporting and Investigating the Atrocities of the German Fascist Occupiers and their Accomplices and the Damages they caused to Citizens, Kolkhozes, Public Organisations, State Enterprises of the USSR,' November 2, 1942," in *Russian State Archive of Socio-Political History* [*Rossiiskii gosudarstvennyi arkhiv sotsial'no-politicheskoi istorii*], hereafter *RGASPI*, f. 17, op. 125,

Shvernik, head of the Soviet Trade Unions, Andrei Zhdanov, First Secretary of the Leningrad City and Regional Party Committees and a member of the Politburo, Metropolitan Nikolai (born as Boris Yarushevich) of Kiev and Galicia, and Valentina Grizodubova, a female pilot and Hero of the Soviet Union. Six full members of the Soviet Academy of Sciences became leaders of the ChGK as well: historian Evgenii Tarle, engineer Boris Vedeneev, physician Nikolai Burdenko, agrobiologist Trofim Lysenko, writer Aleksei Tolstoi, and the legal scholar Il'ya Trainin.[16] In order to validate Soviet claims, the Soviet government seemingly wished to present the Commission as a public institution that was formally independent of the party and the government. The inclusion of public figures of not only state but also international significance in the Commission was intended to attach global importance and objectivity both to the ChGK itself and to its findings. The Commission needed to increase its "community" representativeness, and so it turned to the Russian Orthodox Church. By that time, the latter had proven itself an undoubted spiritual force, uniting believers and opposing Nazism. This also included a material dimension, namely, raising funds for the front. Yet, when the Jewish Anti-Fascist Committee [*Evreiskii antifashistskii komitet*] (JAC) sought representation in the Commission's department investigating the killing of Jews, its initiative seems to have been rejected.[17] The Commission had extensive powers: it had the right to conduct investigations of Nazi crimes and to determine the material damage suffered by the USSR, to coordinate the activities of all Soviet organisations in this field, to reveal the names of war criminals, and to publish official

---

d. 79, l. 15–20. The decree was published: *Pravda*, November 4, 1942; *Vedomosti Verkhovnogo Soveta SSSR*, November 7, 1942. The Soviet government took numerous preparatory measures before the Commission was formed. For more details about the projects on the formation of the ChGK, see: "Pis'mo akademika Petra Kapitsy Vyacheslavu Molotovu ob 'Organizatsii sledstvennoy komissii' ot 15 sentyabrya 1941 g.," *Rodina* 4 (2005): 43; Sorokina, "People and Procedures," 10–21; Andrej Umansky, "Geschichtsschreiber wider Willen? Einblick in den Quellen der 'Außerordentlichen Staatlichen Kommission' und der 'Zentralen Stelle'," in *Bewusstes Erinnern und Bewusstes Vergessen: Der Juristische Umgang mit der Vergangenheit in den Ländern Mittel- und Osteuropas*, ed. Angelika Nußberger and Caroline von Gall (Tübingen: Mohr Siebeck, 2011), 355–357.

16 Sorokina points out that "despite their differences in age (the age range of the Commission members spanned 35 years: the oldest, Tarle, was born in 1875 and the youngest, Grizodubova, in 1910), social origin, and education, almost all of the ChGK members were in their own way upwardly mobile 'careerists' who owed their rise up the professional ladder to the changes that had taken place in their respective institutions after the October Revolution of 1917, and in this sense they personified the concrete opportunities Soviet power had created for specific people." Sorokina, "People and Procedures," 22.

17 Gennadii Kostyrchenko, ed., *Evreiskii antifashistskii komitet v SSSR, 1941–1948: dokumentirovannaya istoriya* (Moscow: Mezhdunarodnye otnosheniya, 1996), 95.

reports on their findings. The wide scope of activity given to the ChGK testifies to the importance of its work for the Communist Party and state authorities.

In addition to the 10 "active members" of the ChGK plus the staff of the latter, more than 100 auxiliary Commissions operated during the war years in the Union Republics, ASSRs, and regions of the USSR.[18] Under Decree No. 299 "On the Work of the ChGK on Reporting and Investigating the Atrocities of the German Fascist Occupants" of March 16, 1943, republican and regional commissions were established with a view to assisting the work of the ChGK in the liberated areas. These commissions were formed as follows: the First Secretary of the Central Committee of the Communist Party of the Union Republic (Regional Committee) became its chairman; the other committee members (vice-chairmen) were the head of the Regional Executive Committee of the Council of People's Commissars [*Sovet narodnykh kommissarov*, SNK] and the head of the People's Commissariat for Internal Affairs [*Narodnyi komissariat vnutrennikh del*, NKVD] of the USSR. Each commission was ordered to include one or two representatives of the local community.[19] The regional commissions in the North Caucasus were formed under special decrees of the SNK in the liberated cities and villages and operated between 1943 and 1944.[20] The ChGK also sent its commissioners to the regional commissions. Their functions consisted in organising the investigation of Nazi crimes, determining the material damage to a given locality, and preparing associated reports and

---

18 Sorokina, "People and Procedures," 5. According to the calculations of the ChGK, around 32,000 public representatives took part in determining the facts of Nazi crimes, and more than 7 million Soviet citizens directly collected and prepared data for the Commission, which in turn read through more than 54,000 statements and more than 250,000 protocols of witness interrogations and declarations of Nazi crimes, as well as approximately 4 million documents on the damage caused by the Nazis. See: Sergei Kuz'min, "Ikh zakapyvali zhivymi," in *Natsistskikh prestupnikov – k otvetu!*, ed. Dmitrii Pogorzhel'skii (Moscow: Politizdat, 1983), 44; Natalia Lebedeva, *Podgotovka Nyurnbergskogo protsessa* (Moscow: Nauka, 1975), 26–31.
19 By the beginning of 1944, 17 regional and republican Commissions operated in the USSR. "List of members of republican and regional commissions of the ChGK," in *GARF*, f. R-7021, op. 116, d. 326, l. 85–89.
20 Thus, for example, according to Decree No. 299 of the SNK, the Krasnodar Regional Commission was established on April 29, 1943. It was headed by the first secretary of the regional committee of the VKP(b) Petr Seleznev, the Chairman of the Regional Executive Committee Pavel Tyulyayev, the archpriest Nikolai Bessonov, Doctor of Agricultural Sciences Professor Anton Nosatovski, and the Director of the all-Union Research Institute of Tobacco Industry of the USSR Petr Pantikov. At the first meeting of the regional Commission Aleksandr Medvedev, the head of the regional administration of the NKVD became also a member of the Krasnodar Regional Commission. See: "Appendix to the Decree No. 299 of the SNK of March 16, 1943," in *GAKK*, f. R-897, op. 1, d. 10, l. 23 ob., 255.

paperwork. They were entitled to task the appropriate authorities with investigating cases, to interview victims or witnesses, and to collect any other evidence relating to Nazi crimes. Staff of Soviet economic and cooperative institutions, trade unions and other public organisations, workers and employees of enterprises and institutions, members of collective farms, urban and rural intelligentsia, as well as combatants were involved in drawing up the reports.[21]

The evidence of Nazi atrocities was collected according to the instructions of the ChGK.[22] An "Instruction on the Order for Reporting and Investigating the Atrocities of the German Fascist Occupants and their Accomplices" dealt with recording cases of mass killing of civilians and Soviet POWs.[23] According to this, "proof of killings, rapes, abuses, and tortures of defenceless people, including women, children and old people; evidence that Soviet citizens were taken into German slavery [forced labour]; torture and killing of sick and wounded Soviet POWs" were to be recorded.[24] Later in this chapter, I will turn to the discourse analysis of different types of ChGK records that attest to the mass killings of Jews in the North Caucasus.

---

[21] "Appendix to the Decree No. 299," l. 258.

[22] The main instruction on reckoning economic losses was the "Instruction on Accounting the Damage Caused by the German Fascist Invaders and their Accomplices to State, Cooperative and Public Enterprises, Institutions and Organisations" approved by the SNK on May 7, 1943. A report scheme and the following tables were attached to the Instruction: a) "the cost of annihilated, destroyed, ransacked and damaged property in monetary terms," b) "costs connected with evacuation and re-evacuation," c) "a list of individual types of annihilated, destroyed, ransacked and damaged property." See: "'Instruction of the ChGK on Accounting the Damage caused by the German Fascist Invaders and their Accomplices to State, Cooperative and Public Enterprises, Institutions and Organisations' of May 7, 1943," in *TsDNIKK*, f. 1072, op. 2, d. 630, l. 1–8; Sergei Stepanenko, "Deyatel'nost' Chrezvychainoi gosudarstvennoi komissii SSSR po vyyavleniyu voennykh prestuplenii fashistkoi Germanii na territorii Krasnodarskogo kraya" (cand. of science diss. in history, Maykop State University, 2010), 82–112.

[23] On March 15, 1943, at the first meeting of the ChGK, its members Il'ya Trainin, Valentina Grizodubova, and Aleksei Tolstoi were tasked with preparing instructions "on accounting the facts of murders of peaceful civilians, violence by the occupants against defenceless people, women, children, old people, as well as the facts of taking Soviet people into German slavery, and on accounting the damage caused to citizens of the USSR." "Minute No. 1 of the ChGK meeting of March 15, 1943," in *GARF*, f. R-7021, op. 116, d. 2, l. 4–5.

[24] "ChGK Instruction on the Procedure for Establishing and Investigating the Atrocities of the German Fascist Invaders and their Accomplices of May 31, 1943," in *GARF*, f. R-7021, op. 116, d. 7, l. 18–19.

3.2 From Personal Statement to Final ChGK Report: Changes in the Rhetoric of Testimony

As stated in the "ChGK Instruction on the Calculation of Damage caused by German Fascist Invaders and their Accomplices to Collective Farms," in the areas liberated before May 1, 1943 "reports on the damage were to be drawn up and sent to superior organisations no later than July 1, 1943." Henceforth, as occupied areas were liberated, reports were to be compiled within a month.[25] ChGK members collected all relevant data in each liberated locality. They gathered statements from Soviet citizens who survived the occupation with a request to compensate them for economic losses including the loss of the head of the family,[26] took photos of destroyed buildings and killing sites, and recorded interviews with victims, witnesses, and perpetrators of German crimes. These detailed primary sources formed the basis for summarised final reports for the respective city, district, krai or oblast, which were sent to the Central Commission afterwards. They were then partially used in mass media, at the Nuremberg trials, and in other judicial proceedings. With each new level of generalisation, primary evidence acquired ideological features and was translated into the language of official Soviet propaganda. I have studied the ChGK records stored in the central archive (GARF) and regional (GAKK, GASK, GARO) archives, which fully or partially deal with the mass killing of Jews specifically or of "peaceful Soviet citizens" in general during the occupation of the North Caucasian territories. In the following, I examine each type of the ChGK record separately in order to analyse when, how, and why Jews, as the main Nazi victim group were replaced by the depersonalised universal definition of "peaceful Soviet citizens" in official Soviet paperwork. I will thus be able to trace the process by which Holocaust memory was (re)created in the region and in the whole country.

Statements

According to Ushakov's dictionary, a statement [*zayavlenie*] is a public official message in oral or written form.[27] A modern law dictionary adds that a statement is understood as "an official address of a citizen or several persons to a

---

25 "The ChGK Instruction on Accounting for Damage Caused to Collective Farms by the German Fascist invaders and their Accomplices of April 14, 1943," in *GARF*, f. R-7021, op. 116, d. 4, l. 15.
26 The head of the family was always the breadwinner in the patriarchal household.
27 Ushakov's dictionary is one of the main dictionaries of the Russian language edited by the philologist and lexicographer Dmitri Ushakov. The dictionary was published in four volumes over the period 1935–1940. "Zayavlenie," in Dmitri Ushakov, *Bol'shoi tolkovyi slovar' russkogo yazyka:*

state or local authority, an administration of an institution or an organisation, which, as distinct from a complaint, is not connected with a violation of his or her rights and legitimate interests, but is aimed at the realisation of the rights and interests of the person or at the removal of any shortcomings in the activity of enterprises, institutions, and organisations."[28] In fact, the ChGK statements by citizens who survived the occupation were descriptive and declarative. They served as a basis for ChGK members to conduct additional investigations and compile more detailed reports later on. The text of a statement often ends with calls for revenge on the enemy such as in the statement by Golos quoted at the beginning of the chapter. Such calls can be interpreted as a feature of Soviet wartime writing, and not as a form of mournful address to a state body with the demand to solve the problem.

Structurally, the text of a ChGK statement can be divided into three parts. The heading contained the addressee of the statement, which went "to the regional/district/urban Commission for investigating the atrocities of the German fascist invaders." Most statements were addressed personally to the chairman of a local Commission or the chief of the district police, to the local Executive Committees of Working People's Deputies, or to the city/town Soviet, and included the applicant's personal data; their full name and place of residence. Then followed the text of the statement itself, usually written in free form, where the authors described particular cases of the crimes they had witnessed. The repetition of such expressions as "I remember," "I know/I do not know," "I saw/witnessed," etc., demonstrates a direct link between the authors of the texts and the facts they described. The applicants, especially villagers were not always literate, so the local ChGK members sometimes wrote statements "from what they were told." In writing down the verbal statement of an illiterate witness the language was processed, as it is entirely obvious that the verbal expression of a possibly uneducated person would not perfectly conform to the rules of either formal written language or official Soviet terminology. This resulted in a unified document format in some areas. Special typewritten forms were developed, which only had to be filled in with an applicant's personal data and the essence of the statement. Not only did the document format become standardised, but also the statement's text. This standardisation of statements is characterised by the usage of standard phrases and epithets to describe perpetrators and victims. For instance, there are calls for revenge in

---

sovremennaya redaktsiya (Moscow: Dom slavyanskoi knigi, 2008), http://ushakovdictionary.ru/word.php?wordid=19157 (accessed March 23, 2020).
28 "Zayavlenie," in *Slovar' yuridicheskikh terminov*, http://yurist-online.com/uslugi/yuristam/slovar/z/2100.php (accessed March 23, 2020).

nearly every statement.[29] A typed personal statement points, on the one hand, to the availability of technical resources in a particular area, and on the other hand, to the dependence of an applicant on other people to compile his or her message. The place where the statement has been written, the date, and the applicant's signature concluded the statement. Some signatures were notarised. For instance, a notary attested to practically all the signatures in the records of the Stavropol Commission. Another method of confirmation was that several witnesses, whose personal data and signatures can also be found at the end of the document, confirmed the information presented in the statement.

All the statements, which arrived at the regional Commissions, can be divided into the following groups based on genre:

– Statements given by Holocaust survivors

This is the least numerous, but the most interesting group of statements. These statements were the first written accounts of Holocaust survivors, which gave more or less detailed descriptions of the lives of Jews (the authors themselves, their relatives, and acquaintances) in the North Caucasus. Among the main themes of these testimonies are descriptions of the evacuation of Jews from the western regions of the country to the North Caucasus in 1941 and early 1942 and their life in a new place among the local population; the impossibility of further evacuation and life under the occupation; the announcement of orders for the registration of the Jewish population and the subsequent murder of Jews; detailed description of the reasons for the author's survival (avoiding the registration or escaping during the rounding-up of Jews); the story of their further wanderings and descriptions of the places where they found refuge, the help of local citizens and joy of liberation by the Red Army. A good example of such a testimony is the one by Genya Golos, cited at the beginning of this chapter. In spite of a certain sovietisation of the language (the calls for her killed relatives to be avenged, avows her faith in the power of communism and the Red Army), she obviously recognised the Jews as the main victims of the Nazis.

Jewish authors were very emotional in the way they wrote about their fate as Jews surviving the occupation. For instance, Faina Gulyanskaya decided not to attend the registration of the Jewish population in Essentuki. She stated, "not many Jews of the city succeeded in escaping, but my child and I did, wandering like hunted beasts, like lepers for as long as the Germans remained in

---

[29] See, for example, unified texts of statements written according to a developed sample: "Personal statements by the citizens who survived the occupation to the Timashevsk Commission (Krasnodar Krai)," in *GARF*, f. R-7021, op. 16, d. 8, l. 67–74.

the city."[30] Proper attention in all testimonies is paid to the help of local acquaintances, without whom a Jew would not have been able to survive the occupation. This way, Mikhail Fingerut, a Jewish evacuee from Novorossiysk (Krasnodar Krai) to Kislovodsk succeeded in escaping. When the Jews were transported in railway wagons to a glass factory in Mineralnye Vody[31] he managed to hide under a carriage in the panic and chaos of arrival and flee. "Later on, with the help of my friend Romanovski, whom I met in Mineralnye Vody, I got a residence permit in Kislovodsk, which was in his name, and I arrived in Kislovodsk on September 15, 1942 at the house of my old acquaintances the Airapetovs and hid there from the fascist monsters."[32]

The survivor S. Belenkov (his full name is unknown) described in detail not only his own rescue but also the living conditions of Jews in the first weeks of the occupation of Stavropol Krai. Non-Jewish acquaintances helped the Belenkov family to avoid the registration and provided them with false documents. "We found shelter in the farm of Privolny, Nevinnomysskii District of Stavropol Krai with comrade Melnikov, foreman of a tractor brigade, who helped us earn bread and other products. We lived at his place for three and a half months till the Red Army arrived."[33] False documents played a crucial role in the rescue of the Belenkovs, however, they were obviously Jewish in appearance, and the family often had to change their official place of residence to avoid possible suspicion on the part of locals. As Russian historian Tatyana Pavlova notes, "statements by Holocaust eyewitnesses make it possible to reveal both cases in which the local population collaborated with the enemy (whistleblowing, participation in the killing actions) alongside cases in which people selflessly helped to save Jews

---

30 "Statement by Faina Gulyanskaya from Essentuki of August 2, 1943," in *GARF*, f. R-7021, op. 17, d. 4, l. 18.
31 Mineralnye Vody was the site of the mass killing of Jews from the local resort cities of Essentuki, Pyatigorsk, Kislovodsk, and Zheleznovodsk. In early September 1942, over 6,300 Jews were killed in anti-tank ditches around 1 km from the glass factory of the village of Ardzhievskii in the north-western outskirts of Mineralnye Vody under the direction of EK 12 with the participation of the local police. "ChGK report for Mineralnye Vody of July 15, 1943," in *GASK*, f. R-1368, op. 1, d. 97a, l. 57–58; Idem, in *GARF*, f. R-7021, op. 17, d. 2, l. 2–8; "Statistical data about the number of killed, tortured, and poisoned peaceful Soviet citizens and Soviet soldiers by the German fascist invaders in Stavropol Krai," in *GARF*, f. R-7021, op. 17, d. 311, l. 44.
32 "Statement by Mikhail Fingerut from Kislovodsk of February 5, 1943," in *GARF*, f. R-7021, op. 17, d. 5, l. 42–44.
33 "Statement by S. Belenkov from Essentuki of June 29, 1943," in *GARF*, f. R-7021, op. 17, d. 4, l. 58.

from certain death."[34] Only the statements of Holocaust survivors contain the information about potential Righteous Among the Nations[35] in the regional ChGK records. Statements by local citizens who sheltered Jews in the region are not to be found.

Besides their rescuers, the authors of survivor testimonies described the collusion of Soviet collaborators and antisemites, who knew local life better than the Germans and could identify Jews living nearby. Anna Shlaen, an evacuee from the Ukrainian SSR, stated: "poor mother would go and ask for bread and they [the locals] would shout at her: go away, Jew, or else a German will kill you! She would answer calmly: I am Polish. How terrible it is to be a person among people and not to consider yourself a person."[36] Nevertheless, there is little such evidence in the statements by Holocaust survivors. This can be fully explained by the very survival of the authors of these memoirs: they survived because they had the luck to be protected by kind, sympathetic neighbours. Unfortunately, as evidenced by the thousands killed and the handful of Jewish survivors, not all neighbours were so kind during the Holocaust.[37]

The survivor testimonies pay a lot of attention to the description of killed relatives, their authors often felt guilty in relation to them. A Holocaust survivor Raisa Kogan wrote seven months after the mass killing of the Jews of Essentuki: "The image of my mother with strands of grey hair showing out of a shawl still haunts me, running across the field and calling for us children to help save her."[38] Friends rescued Raisa by taking her away from the site where Jews were to be registered, but her mother and other relatives were taken to the glass factory of Mineralnye Vody.

---

34 Tatyana Pavlova, "Dokumenty Chrezvychainoi gosudarstvennoi komissii po ustanovleniyu i rassledovaniyu zlodeyanii nemetsko-fashistskikh zakhvatchikov o Kholokoste," in *Nyurenbergskii protsess: uroki istorii*, ed. Natalia Lebedeva (Moscow: Institut vseobshchei istorii, RAN, 2006), 66.
35 Righteous Among the Nations is an honorific used by the State of Israel to describe non-Jews who risked their lives during the Holocaust to save Jews. To commemorate these people is one of the tasks of Yad Vashem. As of January 1, 2019, Russia has 209 Righteous Among the Nations, recognised by Yad Vashem. "Names of Righteous by Country," *Yad Vashem official website*, https://www.yadvashem.org/righteous/statistics.html (accessed March 11, 2020).
36 "Statement by Anna Shlaen from Soldato-Aleksandrovskoe, Stavropol Krai, undated," in *GARF*, f. R-7021, op. 17, d. 11, l. 116.
37 See the most famous study on the issue of disloyalty of neighbours towards Jews: Jan T. Gross, *Neighbours: The Destruction of the Jewish Community in Jedwabne, Poland* (Princeton: Princeton University Press, 2011).
38 Statement by Raisa Kogan from Essentuki of June 29, 1943, in *GARF*, f. R-7021, op. 17, d. 4, l. 14.

Thus, survivor testimonies collected by the ChGK represent the earliest personal evidence of the Holocaust in the USSR in general and in the region in particular. It is not entirely clear why Holocaust survivors wanted to share their stories in the form of a statement with representatives of an official state institution such as the ChGK. The initiative could have come from the survivors themselves, who might have wished to testify to Nazi crimes and their suffering in the first person. Simultaneously under the latent antisemitism in the USSR, which already existed in the pre-war period,[39] they might also have wanted to demonstrate their loyalty to the authorities and therefore expressed joy of liberation and their hope for a bright Soviet future. For them, after having experienced murderous German antisemitism, the USSR represented the lesser of two evils. On the other hand, ChGK members could have sought such evidence, even in keeping with the non-recognition of ethnicities, in order to testify to the gravity of Nazi crimes, since no economic damage can equal the mass murder of people (and the absolute majority of Soviet citizens killed in the occupied territory were Jews). Notably, the files of the Krasnodar Commission do not contain statements by Holocaust survivors: either the ChGK members were themselves antisemitic or those few who survived the Holocaust did not trust the ChGK and chose to keep a low profile.

– Statements given by eyewitnesses, relatives or close friends of killed Jews
The ChGK received many such statements. They are stylistically diverse and often written emotionally. Villagers usually became eyewitnesses to the Holocaust out of curiosity. One witness wrote: "For me, the rounding up of Jews was mysterious and I decided to observe a little what would happen to them and see how many Jews there were."[40] Others became accidental witnesses: Akilina Medvedeva, for instance, wanted to find out "when the mill workers will grind the grain,"[41] and Vera Dubar' "was trying to hide our cow from the Germans."[42] When witnesses happened to come across killing actions they described not only how Jews were rounded up, led, and put into gas vans, but also that they heard gun shots and people screaming. Furthermore, they incorporated into their statements additional

---

39 Antisemitism as a state policy of the USSR appeared in the late 1930s, during the establishment of Stalin's rule and reached its peak in the late 1940s and early 1950s. Kostyrchenko, *Tainaya politika Stalina*, 703–709.
40 "Statement by Mikhailov (the full name is unknown) from Taganrog, undated," in *GARF*, f. R-7021, op. 40, d. 11, l. 55.
41 "Statement by Akilina Medvedeva from the village of Kurgannaya, Krasnodar Krai, undated," in *GARF*, f. R-7021, op. 16, d. 463, l. 4.
42 "Statement by Vera Dudar' from Novorossiysk of October 13, 1943," in *GARF*, f. R-7021, op. 16, d. 11, l. 57.

pieces of information that they had learned about the fate of the Jews later on ("every single Jewish person was shot").[43] Holocaust eyewitnesses stated what they were doing at the time of the Nazi crimes as a kind of excuse, in order to explain why they had shown up near the killing site and remained alive. They did not always mention the ethnicity of Nazi victims in their witness testimony, using the depersonalised term "Soviet citizens."[44] This can be attributed to the illiteracy of the applicants and/or the process through which the statement was compiled by ChGK members.

Statements by non-Jewish relatives and close friends of Holocaust victims are characterised by a high level of emotionality and a detailed description of the Nazi crimes. Here is an example of one such statement by Dultseva (the full name is unknown) who lived in the village Divnoye (Stavropol Krai):

> Tatyana Davidovna Kantor [was] my son's wife. She worked as a schoolteacher for over eight years. Children loved her. But now the Germans came to our village, they killed Tanya Kantor's old father David Moiseevich right in the shed near the house. Poor Tanya's heart cringed at the thought of the fate of her kids and her own still young life. On August 10 [1942], Tanya lost her father, who used to admire his grandchildren. And on September 23, Tanya was taken to the police together with her two kids Inna, aged 10, and Lyudmila, aged 5. Tanya's mother Ida Bensionovna was also taken with them. At first, they were told they would be sent somewhere and ordered to get ready. But Tanya's heart was beating anxiously, having a presentiment of death. Surely these beasts are not able to understand a mother's love. So, the butchers shot Tanya, aged 26, together with her two daughters and her mother in a woodland belt on September 23 [1942]. The children's lips were smeared with some poison and the adults were shot. It is impossible to remember the little sweethearts without pain in the heart [when I saw them] on a date secretly arranged through cracks of the prison. How they begged me to save them, crying, calling for help! They are gone, my little sweethearts, along with Tanya and their grandmother, but I ask to take revenge on the German monsters for them, to help more quickly liberate such children as my grandchildren, still groaning under the heel of the German occupiers.[45]

Despite the highly emotional description of the tragedy of her family, the author of the statement did not dwell on her relationship to those who died. It is also noteworthy that this text, as well as many others, does not clarify that the German motive was to kill Jews. Only the victims' Jewish surnames explain what had happened. Possibly, the author's distanced attitude towards the Jewishness of her killed relatives reveals the deep layers of both state and societal antisemitism.

---

43 "Statement by Mikhailov."
44 "Statement by Vera Dudar'."
45 "Statement by Dultseva from the village of Divnoe, Stavropol Krai of July 7, 1943," in *GARF*, f. R-7021, op. 17, d. 9, l. 23.

The other statement was written by Dariya Kovalenko who lived in the village Novominskaya (Krasnodar Krai) and who had a tenant with a Jewish surname. The witness described the circumstances of Leib Antonovich Gavt's death in all possible details, emphasising that he was a worker and a not a party member, that he was "shot and buried in a grave along with other killed Soviet citizens."[46] Kovalenko in her statement displayed a kind and warm attitude towards her tenant, thus the hypothesis of latent antisemitism is not always the determining factor in identifying the reasons for not mentioning the ethnicity of Holocaust victims. Jews mostly turned up in the villages and farms of Krasnodar Krai only during the evacuation. Until 1941, villagers often had no idea who Jews were nor had they exposure to antisemitic ideas. Hence, many locals regarded Jewish victims if not as "friends," then certainly not as "hostiles,"[47] they were just "Soviet citizens," and many villagers may simply have lacked the education to reflect on the reasons why the Germans killed certain citizens and left others alive.

The regional commissions also collected witness testimonies from friends and colleagues of Holocaust victims, particularly from employees of medical institutions and secondary schools that had been evacuated to the North Caucasus in the autumn of 1941. A rather large proportion of teachers and medical workers was Jewish. Since re-evacuation in the summer of 1942 was impossible, the North Caucasus became the last address for the majority of Soviet doctors and scientists. Statements by colleagues described in detail how the Nazis treated the Jewish intelligentsia:

> They [Germans] began by marking Jews with distinctive star signs, that is, humiliated them morally. They posted announcements in a number of shops that Jews were not allowed to buy there. Then they began to use them as forced labourers, [. . .] forced them to do back-breaking work, beat them unmercifully for 'misbehaviour,' not to mention a great many other forms of humiliation and abuse.[48]

The authors of such statements were educated, literate people. The language of their narrative is full of epithets and descriptive adjectives. They strengthen the impression of the humiliation and abuse, which Jews suffered before they were shot.

---

46 "Statement by Dariya Kovalenko from the village of Novominskaya, Krasnodar Krai of September 7, 1943," in *GARF*, f. R-7021, op. 16, d. 434a, l. 113.
47 For more details about the categories "friend," "hostile,"and "other" see: Maya Dubossarskaya, "Svoi-chuzhoi-drugoi: k postanovke problemi," *Vestnik Stavropol'skogo gosudarstvennogo universiteta 54* (2008): 167–174.
48 "Statement by Associate Professor Virabov (the full name is unknown) from Essentuki, undated," in *GARF*, f. R-7021, op. 17, d. 4, l. 37–38.

The image of decent, kind, and sympathetic Jews who became Holocaust victims appeared in the witness testimonies written by their relatives and friends. The victim's ethnicity disappeared in many such statements and was replaced by the term "peaceful civilians" who were killed, which Soviet authorities would later use when generalising the original statements about the Nazi crimes.

– Claims by close relatives regarding the loss of the head of the family

Such statements always aimed to obtain compensation for material damages suffered under the occupation. This is the largest group of statements and the least interesting by virtue of the poverty of their language. In order to get state financial support for the loss of the head of the family, the fact of his death had to be documented. However, when a relative "left for Jewish registration" or "was seized" by the German authorities and his or her body was not found or identified during the exhumation of killing sites, the ChGK report of a relative's death under the occupation would serve as an official death certificate. That is why the surviving family members provided the ChGK with rather short statements about the circumstances of a relative's "disappearance." They obtained the signatures of two witnesses and got a ChGK report, which made it possible to receive a cash benefit or reimbursement paid in recognition of the property losses that must have occurred under the occupation to survive after the losing the head of the family.[49] As a rule, claims statements were received from the surviving members of "mixed marriages." These statements did not register mourning on the part of relatives but only asked for compensation. The passive form "was seized" [*byl iz'yat*] occurred frequently, but usually implies the loss of property or an inanimate object. This makes sense in the context of claiming compensation for the loss of a household head. The man was the breadwinner in the patriarchal household and, consequently, his "being seized" caused his "family's deprivation of its main source of subsistence, and my child was deprived of his father and his financial support at the age of 9," as Kseniya Dement'eva wrote.[50] The use of such phrases can provide evidence either of the applicant's sense of shame for their spouse's ethnicity, or of the fact that applicants simply understood the function of the ChGK statement as being to describe and possibly obtain compensation for damages caused by the Nazis under the occupation. Thus, the loss of the family member could be measured as a material loss to the Soviet society.

---

49 See: "Statements for Krasnodar and Stavropol Krais," in *GARF*, f. R-7021, op. 17, d. 1, l. 24, 36; op. 16, d. 5, l. 55, 69.
50 "Statement by Kseniya Dement'eva from Krasnodar of June 19, 1943," in *GARF*, f. R-7021, op. 16, d. 5, l. 69.

Obviously, the purposes of the different groups of statements were heterogeneous and varied from pursuing one's own interests (compensation for the loss of the family's breadwinner) to wanting to give account of what happened, to seeking "revenge on and punishment of the Nazis for what they had done."[51] Only a few statements aimed to tell about the suffering and tragic fates of Jews as a specific group under German occupation. They were written in contrast to descriptions of the deeds of the Red Army, "which liberated the Soviet people from misery" (though literally the suffering of most Soviet Jews had already finished long before the Red Army liberated the North Caucasus). The majority of the ChGK records are based on this contrast. As it was a local representative of a state institution, the sovietisation of language was not to be avoided even in the texts of personal statements. This produced a shift of emphasis from Holocaust history to the glorification of the Red Army and calls for revenge. But most important is that fact that whether or not the victim's ethnicity is mentioned in the text, the facts of mass killings of people are mentioned as a given, requiring no further explanation. Only a few statements provide an explanatory phrase, such as "was shot for being Jewish."[52] The author of one statement wrote, "according to Germans, the reason for shooting Jews is that they instigated the war, thus they were guilty."[53] Within these documents, at least, most applicants did not ask themselves why certain people (primarily Jews) were killed and others survived the German occupation. The question was not expanded upon in treatment of Jewish issues in Soviet mass media of the pre-war period; and news about Jewish organisations or the establishment of a Jewish Autonomous Oblast in the Russian Far East hardly interested any southern resident. Decrees and announcements of discriminatory measures against Jews, the establishment of Jewish Committees, and mandatory registration of Jews were put up by the Nazi authorities as soon as the occupation started. If ethnicity is mentioned in the ChGK records it is possibly unconscious, a simple use of the term "Jewish" to indicate newcomers, especially since the majority of the Jews were non-locals. The ethnic identification does not have a negative connotation in comparison with descriptions of occupiers or "hostiles," yet it points to the strangeness or "otherness" of Jews in a particular North Caucasian

---

[51] At the end of numerous statements there are common, clichéd appeals to the Commission: "I ask to bring the bastards to justice . . .," "I ask to investigate and judge Hitler's butchers . . .," "I ask to investigate the facts of guileful atrocities . . . .." See: "Statements for Krasnodar Krai and Stavropol Krai," in *GARF*, f. R-7021, op. 16, d. 4, l. 23; op. 17, d. 1, l. 11, 12.
[52] "Statement by Valentina Merosidi from Krasnodar of July 8, 1943," in *GARF*, f. R-7021, op. 16, d. 5, l. 32.
[53] "Statement by Mikhailov."

locality. Following this logic, it is possible to explain why non-Jews could refer to their Jewish relatives using terms for inanimate objects ("was seized") or that they excluded this ethnonym in emotional descriptions of the fates of Jews. Skipping a victim's ethnicity or using impersonal terms like "Soviet people" and "peaceful citizens" in personal statements makes a distinction between "hostiles" and "friends" and identifies these Holocaust victims as belonging to the latter group.

Legal Documents

– Records of interrogation with Holocaust eyewitnesses, survivors, and perpetrators

An interrogation is "a procedural act establishing a witness's or the accused person's testimony, which is drawn up using a formal language in accordance with established convention."[54] Local ChGK members interrogated eyewitnesses, Holocaust survivors, and perpetrators – both German POWs and Soviet collaborators – concerning mass killings of civilians in the occupied regions. These were conducted either independently or with the involvement of other competent authorities, including civilian or military investigators of district or urban prosecutor's offices or NKVD officers.[55] These data were immediately prepared for criminal proceedings. Most interview/interrogation[56] records were drawn up on

---

54 "Protokol doprosa," in *Sovetskii yuridicheskii slovar'*, http://soviet_legal.academic.ru/1495 (accessed March 23, 2020).
55 Aleksandr Epifanov, *Otvetstvennost' za voennye prestupleniya, sovershennye na territorii SSSR v gody Velikoi Otechestvennoi voiny: 1941–1956 gg.* (Volgograd: Volgogradskaya akademiya MVD Rossii, 2005), 44.
56 ChGK members conducted "interviews" in some regions and "interrogations" of eyewitnesses, Holocaust survivors, and/or perpetrators in the others. These two words differ in Russian in one letter: interview is "*opros*" and interrogation is "*dopros*." As for the content, these documents are identical. In a few cases is it possible to perceive a difference between them: interviews were conducted with eyewitnesses and interrogations with perpetrators. However, such a division of records is not always observed. The most probable reasons for this could be staff (shortage of professional investigators, especially in remote areas) and technical shortcomings (non-observance of instructions, lack of standard forms) of the work of local Commissions, as well as independent initiative of ChGK members. A small number of "eyewitness testimonies" [*svidetel'skoe pokazanie*] can be included in the group of interviews/interrogations, which are authors' accounts of the events they witnessed under the occupation. Eyewitness testimonies differ from personal statements in the way that they were written according to witness's words in a sufficiently formalised language and always in an office. See: "Eyewitness's testimonies for Krasnodar Krai and Stavropol Krai," in *GARF*, f. R-7021, op. 16, d. 435, l. 291–292; op. 17, d. 10, l. 201–203.

special standard forms according to prescribed rules for processing such texts.[57] Each record had an interviewer's signature and the whole file was signed by the interviewer/interrogator and the subject of interview/interrogation. Sometimes records were composed in the form of a witness report,[58] in other cases in a dialogue (question-and-answer) form.[59]

After recording personal data, a respondent was made to swear: "I have been informed of article 95 of the Criminal Code of the USSR on responsibility for giving false testimony and have signed accordingly" or, "The witness is informed about giving true testimony and that for false testimony he or she will be liable under Soviet law."[60] The fact that a witness's interview or interrogation record was sworn is crucial. On the negative side, the NKVD's involvement in the ChGK work indicates the Soviet government's determination to control information.[61] When ChGK members conducted an interrogation alone, there were no such explanations in the records. There is often a note at the end of a questionnaire "my words were recorded accurately and this is what I sign" or "the record of my words is true and checked by me."[62]

The fact that interviewees were, so to say, "on the hook" of investigating authorities is confirmed by the way an interrogation was conducted and the kinds of question asked. Even Holocaust survivors could be suspected. They were to answer such questions as "why did they survive," "did they have any connection to the occupiers," "how did they manage to conceal their ethnicity?" Consequently, their suffering did not finish with the liberation of the region. The story of Boris Kamenko is illustrative. A native of Dnepropetrovsk (Ukrainian SSR), he was evacuated together with his parents to Stavropol, which subsequently fell under occupation. As a strong young man, he was

---

[57] See an example of the interview/interrogation record form, which was used in Krasnodar, in *GARF*, f. R-7021, op. 16, d. 435, l. 135–136. Records were written by hand in many areas of the North Caucasus due to the lack of forms. See interview/interrogation records for Krasnodar Krai, in *GARF*, f. R-7021, op. 16, d. 7, l. 184–189; d. 8, l. 48; d. 434a, l. 163–164.

[58] For example, "Record of interview with Marfa Derganova from Rostov-on-Don of November 24, 1943," in *GARF*, f. R-7021, op. 40, d. 853, l. 32–33.

[59] "Record of interrogation with Mikhail Fingerut from Kislovodsk of June 23, 1943," in *GARF*, f. R-7021, op. 17, d. 5, l. 29–32.

[60] See: "Records of interrogations for Krasnodar Krai," in *GARF*, f. R-7021, op. 16, d. 7, l. 184–189; d. 11, l. 24; d. 434a, l. 182.

[61] For more details on the involvement of the NKVD in the ChGK's work in Berdichev, Ukraine, see: John Garrard, "The Nazi Holocaust in the Soviet Union: Interpreting Newly Opened Russian Archives," *East European Jewish Affairs* 25, no. 2 (1995): 16–18.

[62] See: "Records of interrogations for Krasnodar Krai and Stavropol Krai," in *GARF*, f. R-7021, op. 16, d. 8, l. 10; d. 11, l. 46–50; op. 17, d. 5, l. 29–32.

chosen among 30 other Jews to bury the killed Jews after the killing action in Stavropol.[63] After that, members of EK 12 imprisoned 20 Jewish survivors in the former NKVD building: "So, I was in that prison from August 15 [1942]. I managed to escape on December 6, 1942."[64] After the Red Army liberated Stavropol, Kamenko gave testimony to the ChGK. He recounted how the Jewish population of the city was killed and how he was able to survive.[65] During the next 20 years, NKVD authorities regularly called in Kamenko for questioning. He bore witness not only against Nazi perpetrators and their accomplices (he knew their names and could recognise them on photographs), but also had to answer the question *why* he survived, "not how, but why."[66] It is not surprising that many Holocaust survivors in later interviews asked rhetorically under which regime it was easier for a Jew to survive: under Nazi rule, constantly hiding and fearing detection and death, or in the post-war Soviet period, being suspected by Soviet authorities for having survived the occupation.[67]

Many interview/interrogation records look similar to the eyewitness testimonies discussed above. The formalities beginning and ending the document, the use of standardised expressions such as "regarding the question of brutal mass shootings of Soviet civilians by German occupants I report/inform"[68] gave official weight to these recorded facts. The actual content of the report remained the same. Both types of records consist of description of a particular Nazi crime, pointing to the sites of mass graves of Nazi victims. In terms of

---

63 According to the ChGK report, under the command of members of EK 12 3,500 Jews were shot on August 12, 1942 in the outskirts of Stavropol near the aerodrome. The second group of 500 local Jews was killed on August 14–15, 1942 near the psychological clinic. "ChGK report for Stavropol of July 16, 1943," in *GARF*, f. R-7021, op. 17, d. 1, l. 48–49.

64 Interview with Boris Kamenko, born in 1923, in *VHA USCSF*, code 30577, April 18, 1997, Dzerzhinsk, Russia.

65 Kamenko's and another six eyewitness testimonies on the case of the mass killings of Jews in Stavropol were incorporated into a single ChGK report. As for its content, this report resembles the interview/interrogation record. Here once again we face the problem of negligence in the ChGK's paperwork. "Report for Stavropol of July 3, 1943," in *GARF*, f. R-7021, op. 17, d. 1, l. 65–74.

66 Interview with Kamenko.

67 See interviews recorded for the USCSF in the late 1990s, in which some Holocaust survivors said that Committee for State Security of the USSR [*Komitet gosudarstvennoi bezopasnosti*], hereafter KGB, officers repeatedly harassed them after the war. They particularly "were interested, how we, Jews, remained alive." Interview with Dina Il'shtein, born in 1923, in *VHA USCSF*, code 33778, July 4, 1997, Zaporozhye, Ukraine; Interview with Inna Vaisburd, born in 1922, in *VHA USCSF*, code 46991–13, June 30, 1998, Haifa, Israel.

68 See: "Records of interrogation for Rostov Oblast," in *GARF*, f. R-7021, op. 40, d. 853, l. 28–29, 38–40.

style and language the formal introduction and conclusion of a record differ greatly from the arbitrary content of the main body of the text. One can find Soviet clichés only in the formal parts of the record when describing perpetrators and victims. In the body of the text eyewitnesses identify the main Nazi victims as Jews instead of "peaceful Soviet citizens," and perpetrators as the Germans, Romanians, and/or Soviet collaborators.[69] Despite the desire by party officials to emphasise the importance of ideological conformity in ChGK proceedings, it was impossible to actually write instructions about what should be reported in the testimonies themselves.

– Conversation transcripts

The ChGK member Valentina Grizodubova made an official trip to Krasnodar Krai in July 1943. She talked to eyewitnesses of Nazi crimes and to German POWs[70] in the village of Labinskaya,[71] where "1,310 peaceful citizens were killed, including 540 Jews" according to the ChGK report.[72] The central topic of her conversations was the mass killing of Jews, even though there was evidence of the killing of partisans and communists. This data is stored in typed form and was stylistically altered to make it comply with formal Soviet language. However, Jews were mentioned there as the main victim group. Such testimonial evidence is sometimes shocking. A witness, Anna Timoshkina, stated: "when Germans were shooting Jews, they forced those who were still alive to bury the bodies, even though they were already doomed to death. The last row of shot people was buried by policemen."[73] Unfortunately, the conversation transcript does not clarify how

---

69 Eyewitnesses of Nazi crimes saw what was happening, but did not often know the perpetrators' names and their positions. The indication of the perpetrators' ethnicity already meant a lot. ChGK records contain the names of Soviet collaborators much more often than those of SS members. Sometimes the ChGK members addressed Soviet intelligence to obtain the data on the Nazi military authorities in the region. For example, the Krasnodar Regional Commission made a request on July 30, 1943 to the head of the SMERSH counter-intelligence administration of the North-Caucasian Front to provide the information they possessed. The intelligence department of the headquarters of the fronts sent the requested information to the Regional Commission on August 9, 1943 with data on the German, Romanian and Slovak units and their commanders stationed in the region and responsible for the crimes along with members of Einsatzgruppe D. "Letter of the intelligence department of the North-Caucasian Front headquarters No. 4127 of August 9, 1943," in *GAKK*, f. R-897, op.1, d. 9, l. 45–47.
70 "Records of interrogations with German POWs," in *GARF*, f. R-7021, op. 16, d. 13.
71 "Transcript of the conversation between Valentina Grizodubova and eyewitnesses of the German fascist atrocities against the citizens of Labinskaya of June 4, 1943," in *GARF*, f. R-7021, op. 16, d.15, l. 11–13.
72 "ChGK report for Labinskaya of March 18, 1943," in *GARF*, f. R-7021, op. 16, d.15, l. 6–10.
73 "Transcript of the conversation between Grizodubova," l. 12.

the eyewitness was able to see the killing action in such detail, given that the Nazis went to great lengths to conceal these acts by carrying them out in forests or on the outskirts of settlements. As can be seen from the legal documents described above, the established facts about the mass killing of Jews is clearly classified as the main crime of the Nazi occupation regime.

Reports

A report [*akt*] is a record made by several persons. It verifies facts or events.[74] ChGK members drew up reports to establish the facts of human and/or material losses under the occupation. Reports summarise the original data. Practically, this meant that reports were written on the basis of statements from a specific locality. Authorised ChGK members (as a rule the chairman of the local Commission and its up to five members, at times with the addition of medical experts) or a group of up to seven persons including local citizens and/or district council deputies had the right to draw up a report. Due to the shortage of qualified specialists and sometimes even literate people, it was decided that heads of district councils should be relieved of all other duties during the work of regional commissions and charged with the responsibility of making a correct account of material damages and human losses.[75] On the one hand, such reports were useful when it came to the clarification of collected data. On the other hand, further stages of document generalisation influenced the language of reports, which started to assume a more unified form. The ChGK reports can be divided into following groups:

– Reports of a single crime

Such reports are mostly very brief. As a rule, they served as the death certificate of a person who was killed under the occupation.[76] There are numerous such ChGK reports, which state a person's fate. Here is one example:

> We, the undersigned Commission of the Executive Committee of Andreevskii District Council [*Raisovet*], including Maria Ivanovna Sidorenko, Evdokiya Maksimovna Verchenko,

---

74 "Akt," in *Slovari i entsiklopedii na Akademike*, http://dic.academic.ru/searchall.php?SWord=акт (accessed March 23, 2020).
75 "The explanatory note of the Krasnodar Regional Commission of December 28, 1943 in the ChGK," in *GAKK*, f. R-897, op. 1, d. 12, l. 1.
76 As mentioned above, a ChGK report was an official document, equivalent to the death certificate of a Soviet citizen, and was of particular importance in cases where it was impossible to find or identify the body. See: "Reports for Krasnodar Krai and Rostov Oblast," in *GARF*, f. R-7021, op. 16, d. 5, l. 31; d. 12, l. 294, 314; d. 435, l. 81; op. 40, d. 852, l. 1–57; d. 775, l. 33.

and Maria Viktorovna Popova drew up the following report stating that during the German occupation of Rostov-on-Don Tsilya Yakovlevna Gol'man, registered at 18 Bauman Street, appeared by order of German authorities and was shot. It is proven by the statements of eyewitnesses and confirmed by their signatures and a seal.[77]

There are also more detailed reports of single instances of the mass killing of Jews in the ChGK materials. On the basis of a few witness reports, the Commission "determined" the chronological order of events, acknowledged the Nazi victim groups (in many cases), accused the perpetrators, and defined the grave sites. The victim's ethnicity is almost always stated in these reports.

– Reports confirming all Nazi crimes in a single locality

Such reports summarised several personal statements. The facts of the Holocaust are highlighted as the main Nazi crime. Quotations from personal statements were often included into the core report, providing vivid, emotionally coloured examples that helped to visualise the horror of the occupation. For example, the following facts were established for the village of Aleksandriiskaya, Stavropol Krai:

> 18-year-old [Neta] Shakhet was abused and tortured by German Nazi soldiers. Her mother saw it all. Shakhet and her mother were forced to work on a road grader but later were sent back to the Gestapo. They were told to arrive at the Gestapo [headquarters] by 7 a.m. on September 12, 1942. Anticipating the upcoming tortures Lidiya Mikhailovna Shakhet, aged 52, and her daughter Neta, aged 18, evacuees from Leningrad, committed suicide by poisoning.[78]

Apart from their descriptive function, such details attract the attention of the potential reader; they were aimed at fomenting hatred of and desire for revenge on the enemy. With the same goal in mind, the names of the perpetrators and those responsible for the atrocities during the occupation, according to the ChGK members, are nearly always listed in the final paragraph of the reports. Reports consist of a description of the fates of Jews during the first weeks of occupation including information about the establishment of a Jewish Committee, assembly in a specific location with baggage according to German orders, being put into gas vans or transported to killing sites, and then being killed *en masse*. The list of killed civilians is included in almost every single report for every specific

---

[77] ChGK members developed a standard form, in which it was necessary to write down the name and the address of the victim (in the given example, this data is underlined). One copy of the report was handed over to the relatives of a victim of the Nazis. "ChGK report No. 1127 for Andreevskii District of Rostov-on-Don, undated," in *GARF*, f. R-7021, op. 40, d. 852, l. 11.

[78] "ChGK report for the village of Aleksandriiskaya, Stavropol Krai of July 17, 1943," in *GARF*, f. R-7021, op. 17, d. 9, l. 14.

locality. The victim's ethnicity is indicated either under a heading in the list, or is clear from the victim's name.[79]

– Final reports for large cities or districts

Such reports contain even more general information on the killing of "peaceful Soviet citizens" with totalled numbers of human losses. The recounting of the facts in such reports ranged from general (what happened to a city under occupation, with general data on damages and human losses) to more detailed information (description of precise Nazi crimes and identification of the perpetrators). Mass killings of Jews were identified as Nazi crimes against "peaceful civilians" along with transporting Soviet citizens to the West as forced labour, mainly in Germany, and the killing and torture of Red Army POWs, partisans and communists.[80] The Jews as the main victim group morphed into "peaceful Soviet citizens" in these reports. The language of this type of ChGK report is brief and propagandistic. Photos of mass grave sites with corpses and medical examination reports are attached. When reading such reports, one has the impression that they were written for a general reader. Their authors had perfectly mastered the idioms of formal Soviet language. Despite lacking descriptive words in sentences, the propagandistic character of these reports is seen by the use of sharp-cut metaphors in order to describe those guilty for the "peaceful civilians' pain" ("bandits," "butchers," "cannibals," "barbarians," "beasts," etc.).

– Medical reports following the exhumation the corpses

Nikolai Burdenko, the chief surgeon of the Red Army, and Vladimir Makarov, the ChGK member responsible for assessing damage caused to cultural, scientific and medical institutions, drew up special instructions for grave examination and exhumation. According to these instructions, when there were no medical workers in a town/city who could serve as forensic medical experts, the commander of the army or the front should be addressed with the request for a detachment of relevant specialists to carry out the exhumation of corpses. The NKVD, the city/town local authorities, or the field hospital of the nearest military base usually financed this procedure. A military commissar provided the workforce from recruits or from the local population. The police guarded the exhumation sites at night. The control of the exhumation was entrusted to the

---

79 See lists of killed Soviet citizens in particular regions of the North Caucasus in *GARF*, f. R-7021, op. 16, d. 5, l. 72–73; d. 435, l. 71–73, 110–113; op. 17, d. 3, l. 36; d.5, l. 24–28; op. 40, d. 5, l. 18–19, 68–70.
80 See ChGK reports in particular regions of the North Caucasus in *GARF*, f. R-7021, op. 16, d. 8, l. 165; d. 11, l. 1–17; op. 17, d. 4, l. 1–3; op. 40, d. 11, l. 1–9.

NKVD. Forensic, chemical-forensic and spectroscopic studies were carried out by relevant experts from local laboratories and, in case of their absence, in the anatomic pathology laboratories of fronts, armies or other military formations by agreement with their command.[81] As required, representatives of local public authorities and religious communities were included alongside medical experts in "authorised commissions" on exhumation. The results of exhumations were detailed in special ChGK medical reports where the exact location of the killing site was given along with the number of the graves and their size. There is also information about the position and approximate number of corpses in the graves, including the number of men, women and children and their age. According to instructions, the process of exhumation was to be either photographed or filmed.[82]

The ChGK medical reports in the North Caucasus are of varying quality. A doctor often gave a medical certification of the cause of death, such as "violent death," "poisoning with carbon monoxide," "asphyxiation," "poisoning by lubrication of the mucous membrane." Some reports contained a detailed medical description of the organs of corpses.[83] At times medical reports made by some doctors may imply his or her professional incompetence. For example, one report states that people "were violently killed along with various tortures."[84] Yet the language of most medical reports is usually literal, factual, and full of medical terms. The ethnicity of corpses is not stated in this kind of report. However,

---

[81] "The special ChGK decree of March 23, 1944, 'On Work on the Investigation and Establishment of the Method of Killing Soviet Citizens by German Fascist Invaders,'" in *GARF*, f. R-7021, op. 116, d. 38, l. 1–11.

[82] Special attention was paid to the fact that corpses needed to be filmed by the pits from which they were exhumed, but by no means in an isolated way, since this created an impression of pretence. "The ChGK Resolution on Work on the Investigation and Determination of the Killing Method of Soviet Citizens by German Fascist Invaders of March 23, 1944," in *GARF*, f. R-7021, op. 116, d. 38, l. 5. ChGK members developed this instruction after the killing of Polish military officers in Katyn forest by the NKVD in the spring of 1940. The ChGK conducted a special investigation in the autumn of 1943 to "prove" that responsibility for that crime rested with Wehrmacht soldiers who had invaded this territory in 1941. For more information about this case, see: Epifanov, *Otvetstvennost'*, 127; Stefan Karner, "Zum Umgang mit der historischen Wahrheit in der Sowjetunion: Die 'Außerordentliche Staatliche Kommission' 1942 bis 1951," in *Kärntner Landesgeschichte und Archivwissenschaft: Festschrift für Alfred Ogris zum 60. Wissenschaften*, ed. Wilhelm Wadl (Klagenfurt, 2001), 509–523.

[83] "Medical reports for Krasnodar and Stavropol Krais," in *GARF*, f. R-7021, op. 16, d. 422, l. 9–12; op. 17, d. 3, l. 30–31.

[84] "Medical report for the village of Shkurinskaya, Krasnodar Krai of February 12, 1943," in *GARF*, f. R-7021, op. 16, d. 435, l. 295.

corpse identification and lists of killed civilians give comprehensive information about Holocaust victims.

All reports were notarised by district or city executive committees of the Soviets of Worker's Deputies and registered in a special book. Each report was made in two copies. One was sent to the Central ChGK office in Moscow and the other remained where it was drawn up.

Memoranda

The results of the ChGK activity were discussed at the levels of republics, krais and oblasts. A memorandum *[dokladnaya zapiska]* is a final document, which was carefully reviewed by Soviet functionaries. It contains a summary of the republican or regional commission's activities pertaining to the collection of evidence of Nazi crimes. Table 4 provides general information about the regional commissions' results, specifying the number of human losses. It is impossible to indicate Holocaust victims in the ChGK memorandums because all the victims were referred to as "peaceful Soviet citizens." The generalisation and homogenisation of Nazi victims thus came to its apotheosis. Figures (which are highly debated among researchers[85]) replaced the details of the Holocaust and other Nazi crimes. Holocaust history was totally removed from final ChGK records. Undoubtedly, the low level of legal and economic training of the local population and party members directly affected the quality and objectivity of the ChGK records. At the same time, party elites in the liberated Soviet territories, accustomed to acting on the basis of directives issued from the "top," could not understand which figures of human losses and damage should be sent to Moscow. Relatively small numbers of victims and destruction could be read as a manifestation of collaboration and the lack of an active underground or partisan movement. Large numbers could imply unsatisfactory work on evacuation and the "manifestation of demoralising sentiments among the population." Documents delivered from the regions to the central ChGK demonstrate the confusion and lack of coordination of local authorities in resolving this

---

[85] For the conscious overstatement and faking of final figures in ChGK reports for the Ukrainian SSR, see: Niels Bo Poulsen, "Rozsliduvaniya voennikh zlochiniv 'po-sovets'ki': kritichnii analiz materialiv nadzvichainoi derzhavnoi komisii," *Golokost i suchasnist'* 1, no. 5 (2009): 31–37; Kiril Feferman, "Soviet Investigation of Nazi Crimes in the USSR: Documenting the Holocaust," *Journal of Genocide Research* 5, no. 4. (2003): 587–602.

dilemma.[86] For this reason, the primary ChGK records, mainly personal statements and records of interrogations are of particular value. They appear to be internal ChGK records and therefore were not widely publicised except for some quotations, which were included in the final reports and published in the Soviet press.

3.3 The Place of Official Wartime Sources in the Holocaust Memory Dispositif

Detailed analysis of wartime Holocaust-related sources in the North Caucasus emphasises their originality and high historical importance for the further (re)construction of Holocaust memory in the USSR. Political workers used Red Army and ChGK records for propaganda reasons: information about economic damages and human losses became the major theme in numerous newsletters, which were distributed under the rubric "In Support of an Agitator and Propagandist."[87] They were often published in *Krasnaya zvezda* [Red Star], the official newspaper of the Ministry of Defence of the USSR. For example, the second page of *Krasnaya zvezda* on December 2, 1941 was devoted to the liberation of Rostov-on-Don. All articles on this page were published under a general rubric, which noted that "the foul German invaders fired machine guns and sub-machine guns and tore to pieces many hundreds of civilians in the streets of Rostov. Our hearts are boiling with anger. There is no mercy for monsters and Hitler's murderers. Without mercy, exterminate every one of them!"[88] Red Army soldiers were the target audience for such articles. The detailed reports about infernal Nazi cruelty toward

---

[86] Marina Sorokina, "'Svideteli Nyurnberga': ot ankety k biografii," in *Pravo na imya: biografiya kak paradigma istoricheskogo protsessa: vtorye chteniya pamyati V. Iofe 16–18 aprelya 2004*, ed. Irina Flige (Saint Petersburg: "Memorial," 2005), 61.
[87] For example, Notebooks of an Agitator of the Red Army and of an Agitator of the Navy of the USSR were published between one (in 1942) and three times a month (in 1945). These publications propagated the heroism of the soldiers of the Army and Navy, covered the domestic and foreign policies of the state, generalised and disseminated the experience of propagandists and agitators. Information on mass killings of civilians in the occupied regions was published in the special column "Facts and Figures," under the general heading "The Atrocities of the Fascists." These articles devoted close attention to the fates of Red Army POWs. Mass killings of "peaceful civilians" in different Soviet regions were described generally without indicating the victims' ethnicity. See: *Bloknot agitatora Krasnoi armii* (Moscow: Voenizdat, 1942–1945); *Bloknot agitatora voenno-morskogo flota SSSR* (Moscow: Voenmorizdat, 1942–1945).
[88] Vsevolod Kochetov, "Chto uvideli nashi boitsy v Rostove," Il'ya Erenburg, "Posle Rostova," Vladimir Kozlov, "Rasstrely, ubiistva, pytki," *Krasnaya zvezda*, December 2, 1941.

Soviet citizens aimed to raise combat fervour and inculcate hatred of the enemy. The identity of victims was not so important for such a purpose. Following the end of the war, primary Red Army reports were sent to the Central Archive of USSR Armed Forces (now TsAMO) where they remained classified until 2012.[89]

ChGK records were more fortunate. Articles entitled "ChGK Messages [*soobshchenie*] on the Establishment and Investigation of the Crimes of the Fascist German Invaders and their Accomplices" were regularly published in the central Soviet press during the period of ChGK activity in different liberated regions of the USSR.[90] These articles were also reprinted in regional newspapers. The mass killing of Jews as a main Nazi victim group became one of the top news stories in published ChGK reports, especially in 1943. Despite the tendency to homogenise all Nazi victims in final ChGK records, information about Jews as "innocent Nazi victims" persisted in most volumes of a 15-volume document collection specially prepared by the GlavPU KA between 1942 and 1945.[91] Entitled "The Atrocities of the German Fascist Aggressors" [*Zverstva nemetsko-fashistskikh zakhvatchikov*], the collection was designed to expose the cruelty of Nazi policies on occupied Soviet territories. A separate volume was devoted to Nazi crimes in the North Caucasus, where the mass killing of Jews was reported along with crimes against the Peoples of the Caucasus, including the rape of women and beating of elderly people.[92] Thus, the Holocaust was submerged under the sufferings experienced by other citizens of the multi-ethnic Soviet state.

---

89 Over 200 archival files under the title "Reports about the atrocities of the fascist invaders" (TsAMO, f. 32, op. 11302) were classified until 2010–2012. Il'ya Al'tman, "Krasnaya armiya i Kholokost: k postanovke problem," in *Kholokost na territorii SSSR: materialy XIX mezhdunarodnoi yezhegodnoi konferentsii po iudaike*, ed. Arkadi Zeltser et al., vol. 1 (Moscow: Sefer, 2012), 87–88. Red Army materials on the liberation of the North Caucasian regions were also stored in the collection of the Military Council of the Field Administration of the North-Caucasian Front of the 2nd Formation. See: *TsAMO*, f. 51, op. 958, d. 52 (access: in *USHMMA*, RG-22.016).

90 Altogether, about 30 Commission reports were published in the party newspaper *Pravda* during the war, two of which concern the region under study: "Soobshhenie ChGK o zverstvakh nemetsko-fashistskikh zakhvatchikov v Krasnodare i Krasnodarskom krae," *Pravda*, July 14, 1942; Idem, *Bol'shevik*, July 13, 1942; "Soobshhenie ChGK o zverstvakh nemetsko-fashistskikh zakhvatchikov v Stavropol'skom krae," *Pravda*, August 5, 1942. Also, more than ten ChGK "Messages" were published as separate brochures. See the ChGK Message on the results of a Krasnodar trial in 1943: *Sudebnyi protsess po dely o zverstvakh*.

91 Nazi crimes in the North Caucasus presented in three volumes: Vladimir Veselov, ed., *Zverstva nemetsko-fashistskikh zakhvatchikov: dokumenty*, vol. 2 (Moscow: Voenizdat, 1942), 62–78; Idem, vol. 6 (Moscow: Voenizdat, 1943); Idem, vol. 7 (Moscow: Voenizdat, 1943), 13–32.

92 *Idem*, vol. 6, p. 6–7, 32, 38, 40.

3.3 The Place of Official Wartime Sources in the Holocaust Memory Dispositif — 85

The same information was given in a few published collections of ChGK records during and after the war.[93] Jews were mentioned as one of the Nazi victim groups but the emphasis of these publications was distinctly Soviet: the abuse of Peoples of the Caucasus under Nazi occupation was on a par with the mass killing of Jews. However, the crimes of the Stalin era, foremost the deportation of some entire Peoples of the Caucasus during the war, were completely ignored. ChGK reports published during the Soviet period highlighted the cruelty of the occupation as a whole, which led to huge economic damage and human losses. The ChGK did not seek to reveal the reasons for these human losses. It was the intentional policy of the USSR to hide evidence of the Holocaust by shifting emphasis to the sufferings of all Soviet citizens, and in this way to strengthen the spirits of Red Army soldiers. On the basis of a close reading of ChGK reports on Babi Yar and Auschwitz, the Soviet writer Lev Bezymenskii indicates that there were no signs of the ethnicity of Holocaust victims in them either. However, he affirms: "Soviet readers who had critical reading skills and were willing to decode this kind of article perfectly understood that those who had been killed in Auschwitz were Jews. But they also understood from the ChGK reports that it should not be talked about."[94] So, published records met all the requirements of the Soviet period; they were an excellent example of Soviet propaganda and had little in common with the primary data obtained from the country's regions.

ChGK records were originally aimed at presenting the prosecution case against Nazis in the International Court of Justice. The Soviets brought 520 cases at the Nuremberg trials, using mainly ChGK records. Only four of those records geographically concerned the North Caucasus. They are the ChGK

---

93 *Dokumenty obvinyayut: sbornik dokumentov o chudovishchnykh zverstvakh germanskikh vlastei na vremenno zakhvachennykh imi sovetskikh territoriyakh*, 2 vols. (Moscow: Gospolitizdat, 1943, 1945); SSSR. *Chrezvychainaya gosudarstvennaya komissiya po ustanovleniyu i rassledovaniyu zlodeyanii nemetsko-fashistskikh zakhvatchikov: sbornik soobshchenii Chrezvychainoi gosudarstvennoi komissii o zlodeyaniyakh nemetsko-fashistskikh zakhvatchikov* (Moscow: Gospolitizdat, 1946). In part, ChGK reports were published in collections of documents and materials about the Great Patriotic war for particular regions of the country. See: Boiko, *Stavropol'e v Velikoi Otechestvennoi voine*, 115–127; Mukhamed Shekikhachev, preface to *Kabardino-Balkariya v gody Velikoi Otechestvennoi voiny 1941–1945 gg.: sbornik dokumentov i materialov* (Nalchik: Elbrus, 1975), 3–19; Chermen Kulayev, ed., *Narody Karachaevo-Cherkessii v gody Velikoi Otechestvennoi voiny (1941–1945 gg.): sbornik dokumentov i materialov* (Cherkessk: Stavropol'skoe knizhnoe izdatel'stvo, Karachaevo-Cherkesskoe otdelenie, 1990), 134–177.
94 Lev Bezymenskii, "Informatsiya po-sovetski," *Znamya* 5 (1998), https://magazines.gorky.media/znamia/1998/5/informacziya-po-sovetski.html (accessed March 23, 2020).

report on the atrocities in Kislovodsk, Stavropol Krai on July 5, 1943 (USSR-1), the ChGK report on the atrocities in Teberda (USSR-63/8), ChGK reports on atrocities in Krasnodar and Krasnodar Krai on July 13, 1943 (USSR-42) and records of the Krasnodar Trial (USSR-55). Only the first two of these were used as evidence of the Holocaust in the region.[95] Already during the war ChGK records were being used as evidence of the guilt of Soviet collaborators and German POWs in public trials in Krasnodar and Kharkov in the summer and autumn of 1943. The first internationally known trial of "the atrocities of the Fascist German invaders and their accomplices" took place in Krasnodar between July 14 and 18, 1943, six months after the liberation of the city. It received wide publicity within the country and worldwide.[96] This may explain why only "peaceful Soviet citizens" were mentioned in all the published reports and investigation records.[97] The public prosecutor was interested in proving the treason of 11 Soviet citizens[98] during the occupation of Krasnodar Krai as well as demonstrating the methods of the destruction of "peaceful civilians," including by use of gas vans. He was not interested in demonstrating the motivations of the Nazis and their accomplices for carrying out the crimes.

The majority of ChGK records were not widely studied by Soviet historians. Regulations on the ChGK archive were approved by agreement with the head of

---

[95] Materials of the Nuremberg trial were published in Russian partly and all were censored. See: Roman Rudenko, ed., *Nyurnbergskii protsess nad glavnymi nemetskimi voennymi prestupnikami: sbornik materialov*, 7 vols. (Moscow: Gosyurizdat, 1957–1961). For the index of documents on Holocaust history, presented at the Nuremberg trial, see: Jacob Robinson and Henry Sachs, ed., *The Holocaust: the Nuremberg Evidence: Digest, Index and Chronological Tables* (Jerusalem: Yad Vashem, 1976), 172–175.

[96] *Sudebnyi protsess po delu o zverstvakh*; Elena Kononenko, *Pered sudom naroda* (Moscow: OGIZ; Gospolitizdat, 1943); *The People's Verdict: A Full Report of the Proceeding at the Krasnodar and Kharkov Nazi Atrocity Trials* (London, New York, Melbourne: Hutchinson & Co, 1944).

[97] Apparently, there was an internal order to officials of the military tribunals, who directed their efforts in the liberated Soviet territory at capturing and punishing Soviet collaborators. According to the order, the ethnicity of Nazi victims should not be indicated when passing sentence on a defendant. However, these documents are stored in the former republican and regional archives of the FSB and are still classified within the Russian Federation. The Ukrainian historian Olga Radchenko managed to find a comment of the military tribunal of the 2nd Ukrainian Front on the case of policeman Stepan L. of January 25, 1945: "Despite the instructions given to you earlier that you should not write in sentences such words as 'beat the Jews', 'shot the Jews', etc., you write in the sentence on the L. case: 'participated in the arrests of citizens of Jewish ethnicity'. In this case, you should have written 'took part in the arrests of Soviet citizens.'" I thank Radchenko for providing this information.

[98] 11 defendants were Soviet citizens accused of serving in SK 10a and participating in Nazi crimes. This unit killed thousands of Jews in Krasnodar and many other places.

the State Archives Administration of the NKVD. These regulations prescribed the creation of a Central Archive in the Commission "for [the purposes of] collecting, accounting and ensuring the safety of materials concerning fascist atrocities." The Central Archive being authorised as a department of the ChGK was controlled by the State Archives Administration of the NKVD of the USSR and ensured the safety of documents until they were transferred to the state archives. The head of the Commission's Central Archive was directly subordinate to Nikolai Shvernik or his deputy.[99] There are approximately 43,000 legal cases in the ChGK collection. They are nowadays stored in GARF, with copies and some original files stored in all regional state archives.[100] This may explain why the findings of the ChGK were so unfairly underestimated by both the public and historians during the whole Soviet era.[101] Due to the restricted access to the ChGK collection and the formation of a state ideology of war remembrance by the end of the war, no research on the Commission's investigations of civilian losses and Holocaust victims on Soviet soil was conducted during the Soviet period. Those few ChGK records previously published in the Soviet press were used primarily for historical research on the "Great Patriotic War." No critical analysis was involved.[102] Soviet-era publications were mainly journalistic and propagandistic.[103] Only Soviet historian Natalia Lebedeva has given a brief description of the ChGK structure and activity in her monograph.[104]

After the collapse of the USSR, ChGK records were partially declassified and researchers and the general public got access to them. They have become an

---

[99] "Dokladnaya zapiska Upravleniya gosudarstvennymi arkhivami NKVD SSSR ob organizatsii arkhiva ChGK," in *GARF*, f. R-7021, op.149, d.14, l. 5–6.

[100] Sovinformburo staff, for example, was not allowed access to the original sources collected by the ChGK. Members of the JAC who worked on *The Black Book* were allowed access to a few ChGK files carefully selected by ChGK members. Sorokina, "People and Procedures," 6.

[101] It is a question of introducing the ChGK records about mass killings of Jews and more widely "peaceful Soviet citizens" to the scholarly community. I have found only one Soviet-era dissertation, which dealt with the ChGK data on mass killing of Jews in the Adyghe AO: Vasilii Glukhov, "Adygeya v dni Velikoi Otechestvennoi voiny" (cand. of science diss. in history, Maykop State Institute, 1948), 195–213.

[102] Petr Pospelov et al., ed., *Istoriya Velikoi Otechestvennoi voiny Sovetskogo Soyuza. 1941–1945*, 6 vols. (Moscow: Voenizdat, 1961–1965); Grechko, et al., ed., *Istoriya Vtoroi mirovoi*, 12 vols. (Moscow: Voenizdat, 1973–1982).

[103] *Bol'shaya sovetskaya entsiklopediya*, vol. 47 (Moscow: Gosudarstvennoe nauchnoe izdatel'stvo, 1957), 434; Andrei Sinitsyn, "Chrezvychainye organy sovetskogo gosudarstva v gody Velikoi Otechestvennoi voiny," *Voprosy istorii* 2 (1955): 32–43.

[104] Lebedeva, *Podgotovka Nyurnbergskogo protsessa*.

almost central source for Holocaust memory in post-Soviet countries. However, researchers still face problems obtaining unrestricted access to wartime sources. It is permitted to rewrite by hand or to type up the text of documents in some central and most regional state archives. This work may last for months or years due to the huge quantity of ChGK files. Some archives (for example, GASK) permit only partial access to the researchers. Their staff appeal to the Federal Law of the Russian Federation "On Archival Affairs in the Russian Federation" of October 22, 2004, which stipulates "a restriction for a period up to 75 years on archival documents that contain information regarding private or family secrets of a citizen, his or her private life, and any data which may have risks to his or her safety."[105] Access to the FSB archives is denied or fully closed to researchers in Russia. In the late 1980s when the declassification of ChGK records started, Yad Vashem and later the United States Holocaust Memorial Museum (USHMM) signed contracts with the Russian central archives[106] to copy Soviet sources about the Holocaust.[107] In the mid-1990s these museums got most copies of the ChGK files as well as some documents from the TsAMO and central FSB archive.[108] At

---

105 "Federal'nyi zakon ot 22 oktyabrya 2004 goda No. 125-FZ 'Ob arkhivnom dele v Rossiyskoi Federatsii,' st. 25, p. 3," *Rossiyskaya gazeta*, October 27, 2004.
106 USHMM staff developed projects of agreements with Russian regional state archives in the late 2000s, but many archives, for example GASK, refused to cooperate. Interview with Vadim Altskan, in *author's archive*, code 16-MEM-US01, August 4, 2016, Washington, USA.
107 Yad Vashem began to acquire ChGK files in September 1989. Only records related to Jewish matters were copied. The USHMMA received the filmed collection via the United States Holocaust Museum International Archives Project in 1995, three accretions in 2015, and one accretion in 2016. In 2014 USHMM and GARF signed a new agreement under which ChGK records were scanned in full, but only those which concerned evidence of the Holocaust. See: Collection M.33, "Records of the Extraordinary State Commission to Investigate German-Fascist Crimes Committed on Soviet Territory," in YVA; RG-22.002M, "Selected Records of the Extraordinary State Commission to Investigate German-Fascist Crimes Committed on Soviet Territory from the USSR, 1941–1945," in *USHMMA*, https://collections.ushmm.org/findingaids/RG-22.002M_01_fnd_en.pdf (accessed March 20, 2020).
108 Russian archives were fully opened and their data declassified after the collapse of the USSR. Many researchers, who worked in the early 1990s, noted complete freedom in searching and obtaining archival data on any topic. On this tide of liberalism, agreements were concluded with USHMM and Yad Vashem. As a result, most of the Red Army reports, as well as records of the first trials in Krasnodar and Kharkov, were copied and transferred to the archives of the museums. See: RG-06.025 "Central Archives of the Federal Security Services (FSB, former KGB) of the Russian Federation Records Relating to War Crime Trials in the Soviet Union, 1939–1992 (bulk dates 1945–1947)," RG-22.016 "Reports and Investigative Materials Compiled by the Military Commissions of the Red (Soviet) Army Related to the Crimes Committed by the Nazis and Their Collaborators on the Occupied Territories of the Soviet Union

times foreign archives were the only place where primary Holocaust Soviet sources could be freely accessed.

When speaking about the role of ChGK records in the Holocaust memory dispositif in the North Caucasus, one can specify a few domains which they have influenced. Firstly, their influence is found in the research findings of regional, Russian and foreign historians. ChGK records are mostly used as a primary source to estimate the scale of the Holocaust within Russia.[109] For example, since the mid-2000s members of the French organisation Yahad-In Unum have made field trips to rural areas of the former USSR to conduct interviews with Holocaust eyewitnesses, but only after detailed study of ChGK records along with German sources.[110] Some researchers investigate the activity of the ChGK from the inside, applying critical[111] and/or linguistic[112] analysis to these sources.[113] These studies are a first step in Holocaust research in the USSR. It

---

and Eastern Europe during WWII, 1942–1945," in *USHMMA*; Collection M.40 "Military Archive in Podolsk," in YVA.

[109] See, for example, Aleksei Shevyakov, "Gitlerovskii genotsid na territoriyakh SSSR," *Sotsiologicheskie issledovaniya* 12 (1991): 3–11; Al'tman, *Zhertvy nenavisti;* Idem, ed., *Kholokost na territorii SSSR: entsiklopediya* (Moscow: ROSSPEN, Nauchno-prosvetitel'nyi Tsentr "Kholokost," 2009); Polyan, *Mezhdu Aushvitsem*; Winkler, "Rostov-on-Don 1942."

[110] Patrick Desbois and Edouard Husson, "Neue Ergebnisse zur Geschichte des Holocaust in der Ukraine: Das 'Oral History' Projekt von Yahad-In Unum und seine Wissenschaftliche Bewertung," in *Besatzung, Kollaboration, Holocaust. Neue Studien zur Verfolgung und Ermordung der Europäischen Juden*, ed. Johannes Hürter und Jürgen Zarusky (München: Oldenbourg Verlag, 2008), 178.

[111] A number of researchers have concluded that the main aim of the ChGK was propaganda, which is illustrated by examples of falsification and data juggling by ChGK members. See, for example, Sorokina, "People and Procedures," 1–35; Marian R. Sanders, "Extraordinary Crimes in Ukraine: An Examination of Evidence Collection by the Extraordinary State Commission of the USSR, 1942–1946" (PhD thesis, Ohio University, 1995); Garrard, "The Nazi Holocaust," 3–40; Poulsen, "Rozsliduvaniya voennikh zlochinyv," 25–47. Accusatory by nature is the monograph "The Circassian Tragedy", in which evidence of the falsification of facts in Karachay-Cherkessia was first published. Local authorities and regional ChGK members passed off as Nazi crimes the punitive actions of the NKVD and the material losses of the local civilians when supplying the 37th Soviet Army. Kamil' Azamatov et al., *Cherkesskaya tragediya* (Nalchik: El'brus, 1994).

[112] Irina Rebrova, "'Evreiskii vopros' na Kubani v ofitsial'nykh dokumentakh voennogo vremeni: analiz ideologicheskikh, smyslovykh i vremennykh shtampov," in *Natsistskaya politika genotsida na okkupirovannykh territoriyakh SSSR*, ed. Evgenii Rozenblat (Brest: Brestskii gosudarstvennyi universitet, 2014), 127–132.

[113] Aleksandr Epifanov, "Organizatsiya i deyatel'nost' Chrezvychainoi gosudarstvennoi komissii po ustanovleniyu i rassledovaniyu natsistskikh zlodeyanii" (cand. of science diss. in law, Academy of management of the Ministry of Interior Affairs, 1996); Stepanenko, "Deyatel'nost'

would be impossible to make an advanced analysis of the wartime sources without them. However, there is not a single published collection of wartime Holocaust sources for any region in Russia. Some regional archives have published ChGK reports in general collections of wartime sources, where Holocaust history is not a central subject.[114] A selection of Red Army reports about the Holocaust in the USSR with special attention to evidence in modern Belarus, Ukraine, and Western Russia was first published in 1996 in the "Library of the Holocaust" series of the Holocaust Center.[115] Publication of primary sources would broaden the potential readership, make the fieldwork of local historians and activists easier, and could bring the full extent of the Holocaust to light for ordinary Russian citizens.

ChGK records are also the primary source for commemorations of Holocaust victims in each Russian region. The ChGK lists of Holocaust victims are in much demand following the collapse of the Soviet Union. Revived Jewish communities and local historians try to obtain approval from local authorities to install new monuments or memorial plaques with the names of Holocaust victims on already existing Soviet monuments. Over the past few decades, regional archives have received ever more frequent inquiries from the relatives

---

Chrezvychainoi gosudarstvennoi komissii"; Pavlova, "Dokumenty Chrezvychainoi," 61–71; Al'tman, "Krasnaya armiya i Kholokost," 86–93.

114 Valeria Vodolazhskaya, et al., ed., *Stavropol'e v period nemetsko-fashistskoi okkupatsii (avgust 1942 – yanvar' 1943 gg.): dokumenty i materialy* (Stavropol: Knizhnoe izdatel'stvo, 2000), 86–93; Violetta Belokon', et al., ed., *Golosa iz provintsii: zhiteli Stavropol'ya v 1941–1964 godakh: sbornik dokumentov* (Stavropol: Komitet Stavropol'skogo kraya po delam arkhivov, 2011), 172–201; *Kuban', opalennaya voinoi (o zhertvakh i zlodeyaniyakh zakhvatchikov na territorii Krasnodarskogo kraya, vremenno okkupirovannoi v 1942–1943 gg.)* (Krasnodar: Periodika Kubani, 2005). In the early 1990s Yad Vashem published a collection of Soviet documents on the Holocaust, but only a few pertain to the region. Yitzhak Arad, ed., *Unichtozhenie evreev SSSR v gody nemetskoy okkupatsii (1941– 1944): sbornik dokumentov i materialov* (Jerusalem: Yad Vashem, 1992), 217–241. In Germany, various research centers have been carrying out the project "The persecution and killing of European Jews by Nazi Germany 1933–1945" for several years, whose objective is a comprehensive scholarly edition of original sources about the persecution and mass killing of European Jews. Volume 7 partly contains wartime sources on the Holocaust in the North Caucasus. See: *Die Verfolgung und Ermordung der Europäischen Juden durch das Nationalsozialistische Deutschland 1933–1945*, vol. 7, *Sowjetunion mit Annektierten Gebieten I. Besetzte Sowjetische Gebiete unter Deutscher Militärverwaltung, Baltikum und Transnistrien*, ed. Bert Hoppe und Hildrun Glass (München: Oldenbourg Verlag, 2011), 387, 482–499.

115 Only one published document, the Red Army report of November 30, 1941 noted "the mass killing of 60 Rostov-on-Don civilians" by the Nazis. Fedor Sverdlov, *Dokumenty obvinyayut. Kholokost: svidetel'stva Krasnoi armii* (Moscow: Center "Holocaust," 1996), 81.

of Holocaust victims.[116] State regional and some school museums curated permanent and temporary exhibitions dedicated to the Great Patriotic War, in which copies of the ChGK records are presented. ChGK reports are also cited in documentaries about the Holocaust in the USSR and reprinted in the regional press on International Holocaust Remembrance Day or anniversaries of the liberation of particular cities, towns or regions. In the following chapters, I will turn to the detailed analysis of these elements of the regional dimension of the Holocaust memory dispositif.

---

116 Archivists from GAKK developed a geographic index to the collection R-897 "Krasnodar Regional Commission for the Establishment and Investigation of Atrocities Committed by the German Fascist Invaders and Their Accomplices" to facilitate the search for the requested information. The database "Victims among peaceful civilians in the period of fascist occupation of Stavropol Krai 1942–1943" has been available since May 2017 on the website of GASK. It is created on the basis of ChGK reports within the framework of the regional movement "Let the memory live" ["*Pust' pamyat' zhivyet*"]. The database contains 8,250 records. "Pust' pamyat' zhivyet," *GASK official website*, http://zhertvy-fashizma.stavarhiv.ru/web/index.php (accessed March 20, 2020).

4 Memorial Sites: Re-framing Holocaust Memory

> While Auschwitz and its railway tracks, ramps, ruins of gas chambers, and kilometres of barbed wire *per se* has become over the past two decades a symbol of destruction, the killing sites beyond the Polish eastern border seem to remain hidden behind an invisible curtain of historical perception.
>
> Henning Langenheim, German photographer[1]

I carried out field research in the North Caucasus in the spring of 2015. It was the 70th anniversary of the Victory in the Great Patriotic War, with large-scale celebrations all over Russia.[2] On May 14, I attended the official opening of a memorial plaque to Holocaust victims on the Soviet monument in the area of Mount Koltso on the outskirts of Kislovodsk.[3] It had taken over three years for local activists to enlist the support of the city administration to commemorate the memory of Holocaust victims. Members of the Jewish community and pupils of the city's Secondary School No.2 organised the event. Andrei Kulik, the mayor of Kislovodsk, opened the ceremony, characterising it as "one of numerous small gatherings dedicated to the 70th anniversary of the Victory."[4]

In 1949, local historians the Vinogradov brothers initiated the erection of a modest obelisk on this site, in the form of a pyramid with a red five-pointed star on top (Figure 4). In honour of the 40th anniversary of the Victory, a new monument was unveiled in its place. Its composition is centred on a stone rectangle, on the front of which a bas-relief depicts a group of people made according to

---

[1] Henning Langenheim, "Mjakotino," in *Mordfelder. Orte der Vernichtung im Krieg gegen die Sowjetunion: Ausstellungskatalog*, ed. Peter Jahn (Berlin: Deutsch-Russisches Museum Berlin-Karlshorst, 1999), 12.

[2] Decree of President of the Russian Federation Vladimir Putin "On Preparing and Holding the Celebration of the 70th Anniversary of the Victory in the Great Patriotic War of 1941–1945," April 25, 2013, http://kremlin.ru/events/president/news/17977 (accessed March 1, 2020).

[3] The majority of the city's Jews (about 2,000 people) were rounded up by the Germans on September 9, 1942. They were transported to the glass factory on the outskirts of Mineralnye Vody, where they were killed. A small group of Jews (women, children, and the elderly) that were hunted down after this action, together with Soviet POWs, were killed in two ravines near Mount Koltso some hundred metres away from the Kislovodsk-Karachaevsk highway in the autumn of 1942. One more group of people were killed in the same place in January 1943 shortly before the Red Army liberated the region. In the summer of 1943, ChGK members exhumed 322 corpses in the ravines. "ChGK records for Kislovodsk," in *GARF*, f. R-7021, op. 17, d. 5.

[4] Video recording of the unveiling ceremony of a memorial plaque on the "To the Victims of Fascism" ["*Zhertvam fashisma*"] monument near Mount Koltso in Kislovodsk, Stavropol Krai, May 14, 2015, in *author's archive*, code 15-MEM-SK02.

https://doi.org/10.1515/9783110688993-004

Figure 4: A modest obelisk near Mount Koltso in Kislovodsk, 1949 (Archive of Kislovodsk "Fortress" Local History Museum, f. 108, d. 3).

design of Kislovodsk sculptor Gurgen Kurgenyan. These are figures of a woman holding a girl tightly and covering her face with her hand, and two adult men holding a third, half-dead person. Their facial expressions convey their intention to fight the enemy and protect the weaker ones. The bas-relief seems to say that all the innocents who were killed in that war will be avenged by their friends and relatives.[5] The inscription on this monument repeats the previous one. It states: "322 Soviet citizens who were awfully tortured in 1942 during the

---

5 "Pamyatnye mesta Stavropol'ya – stranitsy istorii Velikoi Otechestvennoi," *Ekimovka.ru*, https://www.war.ekimovka.ru/index.php?task=content&action=view&id=94 (accessed March 23, 2020).

temporary occupation of the city of Kislovodsk are buried here." Any mention of the ethnicity of victims was avoided throughout the Soviet era, so the majority of monuments erected at Holocaust sites were devoted to "peaceful Soviet citizens tortured or shot under the occupation." Since the dissolution of the USSR, members of revived Jewish communities, local activists, and many historians have sought to restore some kind of historical justice by installing memorial plaques indicating the ethnicity of victims of the Nazis on existing Soviet monuments, or by erecting new monuments. It has become an important task for them, since far from all Holocaust sites are marked in today's Russia.

The leaders of Kislovodsk's administration, Jewish community members, Holocaust eyewitnesses, relatives of Holocaust victims, and schoolchildren were present at the unveiling of a marble memorial plaque with the Star of David and the inscription "To the memory of the victims of the Catastrophe. Never again" in Hebrew and Russian on the Soviet monument in the area of Mount Koltso (Figure 5). The ceremony was recognised as a secular social event and the presence of the authorities demonstrated its importance to the citizens of Kislovodsk.[6] Almost all speakers emphasised the significance of this gathering and called upon young people "to remember the horrors of the war." The audience listened to tragic Soviet poetry, the recollections of eyewitnesses of the Holocaust and those of relatives of its victims. The director of Kislovodsk's "Fortress" [*Krepost'*] local history museum, Sergei Luzin, reconstructed the chronology of the 148-day occupation of the city according to archival data, including ChGK records. Spring rain set in during the middle of the ceremony, as if nature itself was mourning alongside those gathered.

Although the ultimate purpose of the gathering was to unveil a memorial plaque dedicated to Holocaust victims, emphasis was given to the "ethnic diversity of peaceful Soviet citizens," "innocently murdered brother-residents of Kislovodsk."[7] The formal Soviet language of describing Nazi victim groups is still in use in modern Russia. Moreover, the two schoolgirls who were tasked with hosting the ceremony, dressed in the traditional Soviet 'white top, dark bottom' school uniform

---

[6] Remarkably, the ceremony was not mentioned in the official city press nor on the website of the city's administration. The news was published only on the websites of Jewish organisations. See: "V Kislovodske otkryli pamyatnuyu dosku zhertvam Kholokosta," *Jewish.ru*, May 15, 2015, http://jewish.ru/ru/news/articles/171282/ (accessed March 1, 2020).

[7] According to ChGK records, the identity documents of 15 of the victims were found during the exhumation of corpses in the summer of 1943. Most of them belonged to Jewish evacuees rather than locals. "ChGK report for Kislovodsk of July 7, 1943," in *GARF*, f. R-7021, op. 17, d. 5, l. 56–58.

Figure 5: Monument "To the Victims of Fascism" near Mount Koltso in Kislovodsk, 1985 and a memorial plaque, 2015 (author's photograph).

with the Ribbon of St. George [*georgievskaya lenta*] on their chests[8] (Figure 6) did not use any of the recognised terms for the mass killing of Jews, such as "Holocaust," "Shoah," or "Catastrophe" throughout the gathering. At the end of the event, the participants lit candles and placed small stones and flowers on the monument.

According to the head of the local Jewish National and Cultural Autonomy, Viktoriya Lanovaya, it was "a festive day" for a lot of Jews, "as too long has been our journey to opening this modest plaque in memory of the Jews who died here under the occupation."[9] As a matter of fact, remembrance of Holocaust victims is

---

[8] The orange and black St. George ribbon is one of the symbols of the Victory in modern Russia. It is traditionally worn on the chest in the form of a bow as "a sign of respect, remembrance, and solidarity with heroic Russian soldiers who defended the country's freedom back in the 1940s." Russians often hang St. George ribbons on cars and handbags and wear them all the year round. The action of distributing St. George ribbons was pioneered in 2005 on the initiative of the Russian international news agency RIA Novosti and has been held annually at the expense of state and regional budgets since that time. Medium and big businesses, as well as various media, support the action. "Georgievskaya lenta – simvol dnya Pobedy," *Marsiada.ru*, http://marsiada.ru/624/lica/718/5022/ (accessed March 21, 2020).

[9] Video recording of the unveiling ceremony of a memorial plaque.

Figure 6: Official unveiling ceremony of a memorial plaque to Holocaust victims on the Soviet monument near Mount Koltso in Kislovodsk, 2015 (author's photograph).

a very long process in post-Soviet Russia. It is sometimes impossible considering the absence of a state programme, inactivity of local authorities, and/or passivity of local citizens. Local authorities often regard installing memorial plaques to particular ethnic groups as a threat, potentially leading to ethnic conflicts in ethnically mixed regions like the North Caucasus. Although Russian authorities have nominally recognised the Holocaust as the extreme manifestation of Nazi racial policy, it is not openly emphasised.[10] Such a policy is sometimes deemed to be antisemitic. Those instances in which local authorities support the initiatives of Jewish organisations to commemorate Holocaust victims are therefore important milestones in the gradual revision of Soviet ideology and its suppression of the ethnic character of many Nazi crimes.

This chapter studies non-discursive practices of Holocaust memorialisation over the entire post-war and post-Soviet periods in the North Caucasus. Symbolic and material objectifications of memory in the form of monuments

---

10 According to Feferman, the rise of Russian nationalism is the main reason for the wariness of the Russian authorities about officially recognising the uniqueness of the Holocaust. Feferman, "Pamyat' o voine i Kholokoste," 84.

and memorials,[11] memorial plaques, and Avenues of the Righteous Among the Nations are an integral element of the Holocaust memory dispositif. Depending on where and by whom these memorials are erected, these sites mark the past according to a variety of national myths, ideals, and political needs.[12] I will chronologically distinguish individual, collective, and state practices of memorialisation of Nazi victims and examine the semantics of typical Soviet and post-Soviet monuments to Holocaust victims, as well as those erected by members of the Jewish community. I am interested in how a monument to Holocaust victims (or "peaceful Soviet citizens") fits into the landscape and how local citizens interact with such Holocaust sites on a day-to-day basis. With the steady urbanisation of the post-war period many of these sites, formerly in the outskirts of settlements or deep in the countryside, have become incorporated into urban or rural districts with a developed infrastructure. I will study discursive practices of commemorative acts at Holocaust memorial sites: who visits them, when and why, how memorial gatherings are held, and what speeches are given. I will also discuss problems of Holocaust remembrance in Russia in general and in the North Caucasus in particular. Due to the ideological importance historically attached to the term "peaceful Soviet citizens" on the one hand, and the glorification of the memory of the "Great Victory of the Soviet People" on the other, commemoration of the Holocaust is still impeded by the conflicting interests of the authorities and Jewish communities, especially in ethnically diverse regions like the North Caucasus. In this context, non-Jewish local historians and activists become important actors in Holocaust remembrance, and this is a principal feature of the Holocaust memory dispositif in this region. At the same time, 'memory wars' are an integral part of Holocaust memorialisation in contemporary Russia, and the unveiling of the memorial plaque to Holocaust victims on the monument in Kislovodsk is one of a very few happy exceptions.

---

[11] Some researchers distinguish a memorial from a monument in a broader sense. For James Young, there are memorial books, memorial activities, memorial days, memorial festivals, and memorial sculptures, which are mournful or celebratory, and monuments refer to a subset of memorials: the material objects, sculptures, and installations used to memorialise a person or thing. Harold Marcuse uses the term monument to reflect objects that may be more heroic versus those that are more contemplative (memorials). Young, *The Texture of Memory*, 4; Harold Marcuse, "Holocaust Memorials: The Emergence of a Genre," *American Historical Review* 115, no. 1 (2010): 53–89. I would treat these two terms as synonyms: they both refer to a piece of stone or other material, which represents the memory of Holocaust victims/peaceful Soviet citizens. The distinction between the terms appears to be only in the size of the objects: a monument is a single object, while a memorial consists of several objects set out in a broader landscape.
[12] Young, *The Texture of Memory*, 1.

4.1 "Memory in Stone"? The Commemoration of Holocaust Victims in the North Caucasus in Historical Perspective

A monument or a memorial in a broad sense is an object (an obelisk, statue, building, etc.) erected in commemoration of a person or event. The main purpose of the erection of a monument is to remember and impede forgetting.[13] Memory and monument are to each other as process and product, although not necessarily as cause and effect.[14] Monuments and memorial complexes retain information about the past (the historical event they commemorate and its epoch, justifying the creation of such a monument according to such-and-such a design and with such-and-such a textual inscription), about the present (how a monument is incorporated into the urban or rural landscape and interacts with the society; whether it is the product of a state memory policy or a private initiative), and about the future (its physical condition and tidiness signal the extent to which contemporaries are willing to preserve the memory of a particular historical event for the next generation, what story 'in stone' they are prepared to transmit). Such a bonding of past, present, and future is a fundamental characteristic of a monument, and is especially important when studying Holocaust remembrance in the Soviet and post-Soviet context.

While Holocaust monuments erected in Western countries and in Israel have long been a subject of academic interest,[15] this has not been the case in the USSR.[16] The well-known memorial complexes erected in the USSR in the 1960s

---

13 Young, *The Texture of Memory*, 4.
14 Robert S. Nelson and Margaret Olin, ed., Introduction to *Monuments and Memory, Made and Unmade* (Chicago: University of Chicago Press, 2003), 4.
15 Adolf Rieth, *Den Opfern der Gewalt: KZ-Opfermale der Europäischen Völker* (Tübingen: Wasmuth, 1968); Young, *The Texture of Memory*; Stefanie Endlich, *Wege zur Erinnerung: Gedenkstätten und -orte für die Opfer des Nationalsozialismus in Berlin und Brandenburg* (Berlin: Metropol, 2007); Marcuse, "Holocaust Memorials," 53–89.
16 Altshuler is the only one who has extensively studied Jewish Holocaust commemorative activity in the Soviet Union. Mordechai Altshuler, "Jewish Holocaust Commemoration Activity in the USSR under Stalin," *Yad Vashem Studies* 30 (2002): 271–295. Researchers predominantly of Soviet background have recently begun studying practices of Holocaust commemoration in the (post-)Soviet space, primarily in Ukraine, Belarus, and the Baltic States. Rebecca Golbert, "Holocaust Memorialisation in Ukraine," *Polin: Studies in Polish Jewry* 20 (2008): 222–243; Arkadi Zeltser, "Pamyatniki Kholokostu v SSSR: na puti k sovremennoi pamyati," in *Kholokost: 70 let spustya: materialy mezhdunarodnogo foruma i 9-i mezhdunarodnoi konferentsii "Uroki Kholokosta i sovremennaya Rossiya,"* ed. Il'ya Al'tman, Igor' Kotler, and Jürgen Zarusky (Moscow: Tsentr "Kholokost," 2015), 141–145; Boris Kovalev, "Memorializatsiya zhertv Kholokosta na territorii Novgorodchiny," in *Ibid*, 162–168; Aleksandra Tcherkasski, "Mesto Kholokosta v sovetskom memorial'nom landshafte," in *Materialy XX ezhegodnoi mezhdunarodnoi mezhdistsiplinarnoi*

and 1970s were dedicated to all "victims of fascism" without mentioning their ethnicity. Furthermore, the Holocaust on Soviet soil occurred in pits and ditches on the outskirts of rural or urban centers. These places were mostly abandoned after the war and were then gradually incorporated into expanding urban or rural areas. Many of the bodies in these unmarked mass graves were, in effect, buried twice: first, during the mass killings, as tens and hundreds of people were gassed, shot, poisoned and buried during the war, and second, as sites without memory in the post-war period. Meanwhile, local commemorative acts at many Holocaust sites have been conducted since the liberation of a particular locality until the present day. Initiated mainly by Jewish communities or local historians, this work was sometimes supported but more often prevented by municipal, republican, and/or state authorities.

A common approach to the study of Holocaust monuments seeks to identify the presence of Jewish symbolism on monuments to Holocaust victims – images of the Star of David, the Menorah, the tables of the Ten Commandments, or Hebrew letters; concepts associated with Jewish history and tradition – and the names of victims, as well as direct indication in Russian, or the local language, that Jews are buried there.[17] Such an approach leads to a rather prosaic style of analysis, in effect doing little more than differentiating between Jewish and non-Jewish monuments. I propose instead to study the process of Holocaust memorialisation in the North Caucasus "from below" and "from above." This approach attaches importance to the activity of remembering communities (Jewish and non-Jewish), the authorities who made efforts to create a monument to "peaceful Soviet citizens" and/or specially to Jews, and the monument's further incorporation into the social life of a town, city or village at a particular historical time.[18] The categorisation of Nazi victims is a key variable in the evolution of the Soviet and post-Soviet memorial landscape, since its development was influenced both by official state interpretation of the history of the "Great Patriotic War" and the increasing generalisation of the term "peaceful Soviet citizens," used to identify all groups of Nazi victims. At the same time, there were isolated though not always

---

konferentsii po iudaike, ed. Viktoriya Mochalova, vol. IV (Moscow: Tsentr "Sefer," 2013), 85–103; Ekaterina Makhotina, "Between 'Suffered' Memory and 'Learned' Memory: The Holocaust and Jewish History in Lithuanian Museums and Memorials after 1990," *Yad Vashem Studies* 44, no. 2 (2016): 207–246; Ilja Lenskis, ed., *Holocaust Commemoration in Latvia in the Course of Time, 1945–2015: Exhibition Catalogue* (Riga: Muzejs "Ebreji Latvijā", 2017).
17 Arkadi Zeltser's monograph is devoted to an analysis of Jewish monuments to Holocaust victims in the post-Soviet space. Arkadi Zeltser, *Unwelcome Memory: Holocaust Monuments in the Soviet Union* (Jerusalem: Yad Vashem studies, 2018).
18 Idem, 41.

successful attempts to preserve the memory of particular groups of Nazi victims, primarily Soviet Jews, throughout the Soviet period. The North Caucasus was not an exception. With regards to initiatives "from below" by representatives of local remembering communities and "from above" by representatives of the Soviet and post-Soviet Russian authorities, three major stages of Holocaust memorialisation in the North Caucasus can be highlighted.

1. *The First Post-War Decade (1940s to Mid-1950s)*

Nazi authorities did not openly talk about the killing of Jews in the Military Administration zones. On the contrary, they often issued special orders for civilians to leave the localities close to the future killing sites for a few days.[19] Despite this, the majority of the local population found out almost immediately about the mass killings and the killing sites. Furthermore, locals sometimes happened to witness these killings involuntarily. They subsequently gave testimony to the ChGK and identified the grave sites. The ChGK records thus became the main source for reconstructing the geography of the Holocaust in the USSR. Nevertheless, the memorialisation of Nazi victims was not among the tasks of the ChGK.

Administratively, monuments and memorials were under the competence of the public authorities. They fell under the supervision of the Ministry of Culture and its departments in the Union Republics after their establishment in March 1953. According to a 1946 decree of the Council of Ministers of the USSR, the erection of monuments costing more than 50,000 roubles had to be approved by it.[20] Thus, decisions to erect expensive monuments dedicated to the memory of the Great Patriotic War were made primarily by state authorities at the all-Union level. At the same time, the priority of the state in that first post-war decade was to restore the country's economy. The trend across the Soviet Union at that period was to erect temporary commemorative signs and modest obelisks at the mass grave sites of Red Army soldiers and killing sites of civilians. The creation of the first monuments in the North Caucasus was initiated either by relatives and friends of war victims, or by proactive citizens, including local historians, as happened at the killing site

---

[19] For example, the day before the killing of Jews in Zmievskaya Balka, Rostov-on-Don "on August 10, 1942, the Germans told the villagers of Vtoraya Zmeevka to take products for one-two days and to leave the village on August 11 at 7 a.m." They were not allowed to return before 7 p.m. on August 11. The Germans motivated this eviction by claiming that they were conducting "large-scale firing exercises." "Memorandum to the Secretary of the Rostov Regional Committee of the VKP(b) Comrade Dvinskii of February 18, 1944," in *GARF*, f. R-7021, op. 40, d. 14, l. 5.
[20] Alexandra Tcherkasski, "Vneshnie i vnutrennie vliyaniya na razvitie sovetskogo memorial'nogo landshafta o Vtoroi mirovoi voine," *Bylye gody* 25, no. 3 (2012): 82.

near Mount Koltso. The initiative came more rarely from members of Jewish communities, because they had a very limited presence in the region, and little influence. Post-war initiatives from below happened across the Soviet Union, but the active presence of non-Jewish local activists and historians is a unique feature of the regional dimension of Holocaust memorialisation in the North Caucasus.

The first monuments to Holocaust victims in the North Caucasus did not contain Jewish symbols or inscriptions in Hebrew or Yiddish. This is in contrast to monuments in other Soviet regions, primarily on the territory of modern-day Belarus, Ukraine, and Lithuania; that is, the area of the former Jewish Pale of Settlement.[21] Post-war monuments had classical forms – such as an obelisk, a tall pylon or a truncated pyramid between 1.5 and 5 metres high – were made of stone, metal or concrete, painted white, and crowned with a five-pointed red star.[22] The same monuments were erected for fallen Red Army soldiers; therefore there was initially no difference in architectural representations in memory of war heroes and victims. The first monuments were used to mark a site as meaningful without any specification of that meaning. The inscriptions on the first monuments had two parts: informational and ideological. The informational part varied little and, as a rule, consisted of an indication of the approximate number of victims – "peaceful" or "Soviet" citizens – and the approximate date of their killing. The dates during which the city/town or village had been occupied were more frequently found on the inscriptions than the exact dates of the Nazi killing action. These inscriptions usually ended with a call for revenge on the enemy and/or a promise to remember the victims. An obelisk at the killing site of Jews near the glass factory beyond Mineralnye Vody in Stavropol Krai is a typical

---

[21] Many of these monuments (for example, in Ponary, Lithuania, 1945) were subsequently pulled down and the activists who erected them were repressed as a result of the campaign against "rootless cosmopolitans" between 1948 and 1953. Inna Gerasimova, "Novaya istoriya starogo pamyatnika," *Mishpokha* 22 (2008): 90–97; Altshuler, "Jewish Holocaust Commemoration," 279–294; Genrikh Agranovskii and Irina Gusenberg, *Po sledam litovskogo Ierusalima: pamyatnye mesta evreiskoi istorii i kul'tury: putevoditel'* (Vilnius: Pavilniai, 2011), 601–602; Zeltser, *Unwelcome Memory*, 122–141.

[22] Such monuments were erected at killing sites in Stavropol Krai: the village of Novoselitskoe, 1943; Pyatigorsk, near Proval, 1943; Stavropol, Urochishche Stolbik, 1944; the village of Letyaya Stavka, 1945 (?); the glass factory beyond Mineralnye Vody, 1948; the village of Staromarevka, 1944; Kislovodsk, near Mount Koltso, 1949; the town of Ipatovo, 1952; in Krasnodar Krai: Krasnodar, 1944 (?); Ust-Labinsk, 1945 (?); Armavir, 1945 (?); in Rostov Oblast: Zmievskaya Balka in Rostov-on-Don, 1943; Petrushina Balka in Taganrog, 1945; and in the Cherkess AO: the city of Cherkessk, before 1952. The precise dating and locating of these first post-war monuments is often impossible, since most of them were not preserved and state/regional records of monuments were not kept in the 1940s.

example of the post-war memorial architecture. Constructed from ashlar blocks in 1948 (the sculptor is unknown), the obelisk bears a carved inscription, which reads: "7,500 Soviet citizens perished here, violently killed by German fascist invaders on September 6–9, 1942. The motherland will not forget you"[23] (Figure 7).

Figure 7: Monument "To the Victims of Fascism" near the glass factory in Mineralnye Vody, 1948. It is now part of the memorial complex together with the first memorial stone (1943) and granite slabs with the names of 1,034 Holocaust victims (2019) (author's photograph, 2015).

---

[23] "Monument Certificate 'To the Victims of Fascism,' Mineralnye Vody, 1948," in SGMZ, f. 104, d. 63, l. 1.

The memory of perished relatives, acquaintances and citizens was still fresh; calls for revenge and remembrance bore not only the stamp of Soviet ideology, but also embodied the feelings and thoughts of the survivors.

A monument erected by victims' relatives always had a simpler form. As a rule, it was a single stone up to one metre high. It had more personal inscription, which also ended with ideological calls addressed to the future. The form of such monuments resembles gravestones in Jewish cemeteries; the inscription is addressed to all the victims buried in this place. The 1948 monument near the glass factory beyond Mineralnye Vody was erected at the site where victims' relatives – possibly local Jews who had returned from the evacuation or relatives of the Jewish evacuees – had erected a modest stone back in 1943 with the inscription "To relatives, close ones, and friends killed by German-fascist monsters on September 9, 1942. The motherland has avenged you. We will not forget the bloody atrocities of the enemy"[24] (Figure 8).

Figure 8: The first obelisk at the killing site near the glass factory in Mineralnye Vody, 1943 (GASK, O-04278).

---

24 Photo of the obelisk at the killing site near the glass factory beyond Mineralnye Vody, 1943, in *GASK*, O-04278. This first monument was knocked over when the official monument was erected in 1948. After years of neglect, it was finally reset and whitewashed in 2019, becoming a part of the new memorial complex erected there.

Some Russian researchers and local historians have identified Jewish symbols in the preserved monuments of the post-war decade. A granite obelisk was erected in 1944 near Urochishche Stolbik along the Stavropol-Sengileevskoe road, where 700 local Jews were killed[25] (Figure 9): "Soon the Star of David was replaced with a red one,[26] and the word 'Jews' with 'Soviet citizens':[27] the Holocaust was a taboo subject in the Soviets."[28] Soviet authorities replaced the Star of David on post-war monuments with the five-pointed one[29] and erased Jewish inscriptions[30] in some western regions of the USSR. This happened, for example, in Nevel, a town in Pskov Oblast. However, I failed to find any documented proof or photos of any first monument (including the one near Urochishche Stolbik) with Jewish symbols in the North Caucasus.

Thus, the first stage of the commemoration of Holocaust victims was initiated from below. Local citizens, local historians, and less frequently Jewish survivors erected architecturally modest monuments. This period did not mark the memory of the Great Patriotic War and its victims as a 'historical memory' because the war was still vivid in everyday experience. Almost immediately throughout the North Caucasus, monuments to war victims, following the example of monuments to its heroes, obtained a Soviet style both in their form (obelisks or truncated pyramids with a five-pointed star on top or on the sides of an obelisk) and in their content (standardised inscriptions "to the victims of

---

25 "List of monuments of history and culture of Stavropol Krai," in *GASK*, f. R-1852, op. 16, d. 2682a, l. 8.
26 In the form in which the monument is preserved, the five-pointed star does not tower over the monument, but is fastened to the top part of the obelisk; it is in fact four stars, one on each side, and the stars are made of the same material as the monument itself: grey granite, painted red.
27 The following is carved on the plaque: "Buried here are Soviet citizens brutally tortured by Nazi butchers during the occupation of Stavropol in 1942–1943."
28 Fatima Magulaeva, "Zemlei zabity nashi rty," *Otkrytaya gazeta*, May 16–23, 2012, http://www.opengaz.ru/stat/zemley-zabity-nashi-rty (accessed March 2, 2020).
29 Magen Davids were replaced with five-pointed stars on both monuments at the killing site of Jews in Nevel. One monument was erected at the killing site of men, the other one at the killing site of women and children. Both monuments were erected within several hundred metres of each other back in the 1940s. On one of the monuments, the modification was most likely made during its making. Mikhail Ryvkin and Aleksandr Frenkel', "Pamyatniki i pamyat'," *VEK* 7, no. 8 (1989): 24; Zeltser, *Unwelcome Memory*, 177–178.
30 Zeltser provides examples of changes to the inscriptions on monuments at the killing sites of Jews of Pyryatyn (Ukraine) and Zhlobin (Belarus). Zeltser, *Unwelcome Memory*, 165–166. At the same time, the monuments erected in Soviet times at Yama [lit. "pit"] in Minsk, in the cities of Uzda, Dzerzhinsk, and in the town of Uzlyany (in modern-day Belarus) had inscriptions in Yiddish, which continued to exist throughout the Soviet period, as Inna Gerasimova has indicated. Gerasimova, "Novaya istoriya," 94–97.

Figure 9: Granite obelisk near Urochishche Stolbik along the Stavropol-Sengileevskoe road, 1944 (author's photograph, 2019).

fascism" and ideological calls for revenge and eternal remembrance). An important feature of these first monuments is their direct link to the killing sites. The majority of the monuments were registered in the 1960s, yet not all of them were kept in good repair, or the surrounding area was cleaned up, which caused their natural destruction. The exact number of monuments erected in the first post-war decade is not possible to determine. I found data on dozens of early monuments to "peaceful citizens" in each region of the North Caucasus, but I managed to visit only some of them; those that are still looked after and to which there is a path.

2. Mid-1950s to 1991[31]

The USSR became a member of the United Nations Educational, Scientific and Cultural Organisation (UNESCO) in 1954,[32] which led international authorities to have influence on the process of memorialisation. The commemoration of the Great Patriotic War thus became the state's concern and a point of honour. Regional departments of the Ministry of Culture gradually inventoried all existing monuments; schools and enterprises were then obliged to care for most of them, most frequently the Tombs of the Unknown Soldier. The Soviet memorial culture of war heroes and war victims was elaborated in this period. The construction of grandiose memorial complexes became a common practice throughout the country and reached its peak in the mid-1960 and 1970s. The Soviet leaders considered the victory in the war as a logical consequence of the Russian Revolution of 1917 and the building of socialism in the USSR. Fascism (the term "national socialism" was used much more rarely, as socialism could not be considered national)[33] was deemed to be an extreme form of capitalism. That is why the commemoration of war heroes and victims "who died for socialism in the struggle against barbarous capitalism" was the only possible option. By the 20th anniversary in 1965, when Victory Day became a holiday again,[34] many monuments in honour of the heroism and glory of the Red Army and the Soviet people were being erected. New

---

[31] Tcherkasski researches memorial activities in the former western republics of the USSR and divides this period into two stages (mid-1950s to late 1970s and 1980s to 1991). Tcherkasski, "Vneshnie i vnutrennie," 83–86. The further division of Holocaust memorialisation in the North Caucasus into more segmented periods is not appropriate because of the absence of significant changes throughout this historical period.

[32] "UNESCO in brief – Mission and Mandate," *UNESCO official website*, http://www.unesco.org/new/en/unesco/about-us/who-we-are/history/ (accessed March 23, 2020).

[33] Georgi Dimitrov, "The Fascist Offensive and the Tasks of the Communist International in the Struggle of the Working Class against Fascism: Main Report Delivered at the Seventh World Congress of the Communist International, August 2, 1935," in Georgi Dimitrov, *Selected Works*, vol. 2 (Sofia: Sofia Press, 1972), https://www.marxists.org/reference/archive/dimitrov/works/1935/08_02.htm (accessed March 2, 2020).

[34] May 9 had symbolic significance in the USSR, because it was an official public holiday from 1945 to 1947. It was downgraded to a working holiday between 1948 and 1964 and was upgraded again to a public holiday in 1965. Carmen Scheide, "Kollektive und Individuelle Erinnerungsmuster und der Großen Vaterländischen Krieg (1941–1945)," in *Stalinistische Subjekte: Individuum und System in der Sowjetunion und der Komintern, 1929–1953*, ed. Brigitte Studer und Heiko Haumann (Zürich: Chronos, 2006), 438–440; Guido Hausmann, "Die unfriedliche Zeit: Politische Totenkult im 20. Jahrhundert," in *Gefallenengedenken im globalen Vergleich: Nationale Tradition, politische Legitimation und Individualisierung der Erinnerung*, ed. Manfred Hettling (München: Oldenbourg Verlag, 2013), 424.

memorial complexes had to show that the war was won by all Soviet people, and through this form of remembrance, to consolidate and reinforce loyalty to the Communist regime.[35] The tendency to universalise the "victims of fascist terror" was an integral aspect of Soviet memory politics.[36] Monuments were erected to commemorate war heroes (soldiers and partisans) and victims or peaceful citizens. The commemoration of the later was included into the official memorial culture since the mid-1960s, in contrast to the post-war period.[37]

From the late 1950s on, memorial complexes began to be erected incorporating both architectural (stelae, triangular or rectangular granite blocks) and sculptural (figures of soldiers, mourning mothers, unbowed civilians destined for death) elements. The largest memorial complexes to "peaceful citizens" in the North Caucasus were erected to celebrate the 30th Victory anniversary (1975) in Krasnodar (the "Memorial to 13,000 People of Krasnodar – Victims of Fascist Terror," Figure 10), in Zmievskaya Balka, Rostov-on-Don ("In Memory of the Victims of Fascism," Figure 11), and "Kholodnyi Rodnik" [lit. Cold Spring] in Stavropol (Figure 12). An important element of any memorial complex to war heroes (less so for Nazi victims) was the Eternal Flame, as a symbol of the eternal remembrance of the war.[38] The monuments obtained some new features and meanings in this period. Since they were now projects of the state, and not private ones, the 'expressing subject' became the city or a town as an administrative unit. The 'expression' was directed outside, its content alienated and impersonal and now with a governmental, ideological character. Such 'administrative' monuments were constructed for show (i.e., to show them to higher authorities).[39] It is significant that the memorial complex in Petrushina Balka on the outskirts of Taganrog took almost ten years to complete. The main goal of party officials was to give an air of ambitiousness and pomposity to the remembrance of the killing

---

[35] Peter Jahn, "Niemand Vergessen – Nichts Vergessen?," in Jahn, *Mordfelder*, 28.
[36] Zvi Gitelman, "The Soviet Union," in *The World Reacts to the Holocaust*, ed. David S. Wyman (Baltimore: Johns Hopkins University Press, 1996), 295–323.
[37] Tcherkasski, "Vneshnie i vnutrennie," 86.
[38] See: Reinhart Koselleck, *Zur politischen Ikonologie des gewaltsamen Todes: ein deutschfranzösischer Vergleich* (Basel: Schwabe, 1998). For the symbolism of the Eternal Flame in Soviet memorial culture, see: Ekaterina Makhotina, "Symbole der Macht, Orte der Trauer: Die Entwicklung der rituellen und symbolischen Ausgestaltung von Ehrenmalen des Zweiten Weltkriegs in Russland," in *Medien zwischen Fiction-Making und Realitätsanspruch: Konstruktionen historischer Erinnerungen*, ed. Monika Heinemann (München: Oldenbourg Verlag, 2011), 293–295, Anna Yudkina, "'Pamyatnik bez pamyati': pervyi vechnyi ogon' v SSSR," *Neprikosnovennyi zapas*, 3 (2015): 112–134.
[39] Natalya Konradova and Anna Ryleeva, "Helden und Opfer Denkmäler in Russland und Deutschland," *Osteuropa* 4–6 (2005): 349–350.

Figure 10: Memorial "To the Victims of Fascist Terror" in Pervomaiskaya Roshcha, Krasnodar, 1975 (author's photograph, 2015).

Figure 11: "Memorial to the Memory of the Victims of Fascism (Zmievskaya Balka)," Rostov-on-Don, 1975 (author's photograph, 2013).

4.1 "Memory in Stone"? The Commemoration of Holocaust Victims — 109

Figure 12: "Kholodnyi Rodnik" memorial complex in Stavropol, 1975 (author's photograph, 2019).

site, but the memorial's construction was restricted by financing. A project to erect a sculptural composition following the example of Babi Yar in Kyiv[40] was discussed; however, the idea was never realised.[41]

Aside from the giant forms of these memorial complexes dedicated to peaceful Soviet citizens, another distinctive feature was their geographical remoteness from the killing sites. In the places where new monuments were erected (as a

---

[40] For many Soviet Jews, Babi Yar (in Kyiv, Ukraine), the main example of Holocaust concealment and concurrently the struggle for memory, became a central symbol of the Jewish tragedy in the USSR. For the history of the erection of the monument in Soviet times, see: Tatyana Evstaf'eva, "Babii Yar: poslevoennaya istoriya mestnosti," in *Babyn Yar: masove ubyvstvo i pam"yat' pro n'oho*, ed. Vitalii Nakhmanovych (Kyiv: Ukrainskii tsentr vyvchennya istoriyi Holokostu, 2012), 21–31; Iryna Klimova, "Babyn Yar in Sculpture and Painting," in Hrynevych, *Babyn Yar*, 259–274.

[41] Viktor Voloshin and Valentina Ratnik, *Vchera byla voina: Taganrog v gody nemetsko-fashistkoi okkupatsii (oktyabr' 1941 – avgust 1943 gg.)* (Taganrog: OOO "Lukomor'e," 2008), 366–373; "Iz informatsii otdela kul'tury Rostovskogo obkoma KPSS o sostoyanii raboty nad pamyatnikami v gorodakh Rostove, Taganroge, Shakhty," in *Zaveshchano pomnit' . . . Donskie arkhivy – 70-letiyu velikoi pobedy*, ed. Lyudmila Levendorskaya (Rostov-on-Don: Al'tair, 2015), 238.

rule, city parks),[42] a symbolic reburial of the remains might be conducted to give the place historical significance. That happened, for example, in Krasnodar in Pervomaiskaya (now Chistyakovskaya) Roshcha during the erection of the "Victims of Fascist Terror" memorial complex in 1975.[43] The modest post-war obelisks at the killing sites could be preserved, as happened in Stavropol. The early post-war monument near Urochishche Stolbik was retained after the erection of the "Kholodnyi Rodnik" memorial complex. Sometimes memorial complexes were planned and designed for sites where early war monuments were already erected. Thus, the "To the Victims of Fascism" memorial complex was opened in Zmievskaya Balka in Rostov-on-Don in 1975. It replaced modest post-war monuments initially erected close to the killing site. It was already the third official monument in Zmievskaya Balka (in addition to two private obelisks that had also been erected on the site). The first 2.8-metre-high concrete monument was erected to commemorate over 10,000 Nazi victims by order of Rostov City Council on May 25, 1943 and had the inscription: "To victims of the Nazi terror of 1942–1943." In 1958 a new monument was erected in the form of a monumental sculpture of two soldiers: one, bareheaded, clinging to the staff of a red banner, the other frozen in mournful kneeling, clasping a weapon of revenge in his hands. The inscription on the pedestal read: "1941–1945. Eternal glory to the heroes who fell fighting for the freedom and independence of our Motherland. From the workers of Zheleznodorozhnyi District of the City of Rostov. May 9, 1958."[44] With each new monument, the total number of victims given out by the authorities increased – from 10,000 to 27,000 – but the reluctance of the region's authorities to recognise this place as the largest Holocaust site within modern Russia resulted in a lawsuit in 2012, as I discuss below.

There were instances when post-war monuments at the killing site were demolished after the erection of new memorial complexes. This happened in Krasnodar. One of the monuments – a six-metre-high stele in commemoration of 500 killed Jews at the crossroads of Solnechnaya and Moskovskaya streets – was dismantled because it was a hindrance to paving a highway and "cars

---

[42] For example, the "Kholodnyi Rodnik" and "Victims of Fascist Terror" memorial complexes in Stavropol and Krasnodar are situated in parks. In Petrushina Balka, trees were specially planted and the area currently resembles a forest rather more than a park.
[43] Georgii Panyutin, "Pamyatnik nepokorennym," *Sovetskaya Kuban'*, May 9, 1975.
[44] "Monument Certificate at Vtoraya Zmeevaya Balka, Zheleznodorozhnii District," in *GARO*, f. R-4096, op. 1, d. 93, l. 42–42a; "Ob otkrytii pamyatnika v poselke Vtoraya Zmeevka," *Molot*, May 13, 1958; Gennadii Belen'kii and Nikolai Red'kov, "Memorial'nyi kompleks 'Pamyati zhertv fashizma' v Zmievskoi balke," in *Pamyatniki monumental'nogo iskusstva goroda Rostova-na-Donu* (Rostov-on-Don: Donskoi izdatel'skii dom, 2016), 36–37, 64–65.

were forced to go around the monument at the crossroads." The second monument – a two-metre metal stele further along Moskovskaya Street, at the site of the former anti-tank ditch where about 7,000 people were buried – was demolished to make way for the building of the Kaskad factory in 1971. In 1982, a new monument to "Soviet citizens" was erected on the territory of the Kaskad factory, access to which was limited by the security guards at the factory. In the late 2000s, it was decided to move the monument outside the factory. The monument in the form of a two-metre brick wall with the inscription "Eternal memory to the victims of fascism" is currently located near the central entrance to the factory building on Moskovskaya Street. A small informational plaque says "Here perished 7,000 people, including aged people and even babies at the hands of the Nazi invaders in the period from August 1942 to February 10, 1943." Thus, the erected monument does not reflect Holocaust history and is located away from the killing site itself.[45]

In the 1960s and 1970s the canonisation of the Soviet soldier, his triumph and its meaning, was taking shape according to the state ideology. War victims were also remembered as fighters who died for their Motherland. Memorial complexes and monuments to "victims of fascism" were often given the highly patriotic title "To the Unconquered" [*Nepokorennym*].[46] There is almost always an image of an elderly person, a woman,[47] embracing a child and "a doomed but undefeated man" in sculptural representations of Nazi victims.[48] Their faces are "severe:"[49]

---

45 For actual grave sites in Krasnodar, see: "ChGK report for Krasnodar, June 30, 1943," in *GARF*, f. R-7021, op. 16, 5, l. 11–15. For problems of preserving the memory of Holocaust victims in Krasnodar, see: "Appeal of Leonid Ivanov to the chairman of the Krasnodar Committee for the Protection of Historical Assets Alla Achkasova, February 6, 2001," in *GAKK*, f. P-807, op. 1, d. 181, l. 32–35; Valentina Kayuk, "Zabytye rvy," *Krasnodarskie izvestiya*, February 10, 2001; Vadim Ivanov "Eto nuzhno ne mertvym, eto NUZHNO? Zhivym. . .: O zhertvakh fashizma v Krasnodare i dal'neyshikh sud'bakh massovykh zakhoronenii okkupatsionnogo perioda," *project website of the society "Memorial," "Uroki istorii XX Vek,"* http://urokiistorii.ru/node/259 (accessed March 2, 2020); "Pamyatnik zhertvam fashizma ustanovlen v Krasnodare," *Informatsionnyi portal "Yuga,"* May 7, 2009, https://www.yuga.ru/news/154060/ (accessed March 2, 2020); Interview with Yurii Teitelbaum, in *author's archive*, code 16-MEM-KK-01, June 1, 2016, Bat Yam, Israel.
46 For example, the sculptural composition of the memorial complex in Krasnodar is called "To the Unconquered." The monument in Novorossiysk has the same name.
47 The figure of a woman as a symbolic image of the Motherland is typical for monuments to war heroes (for example, the most famous "Rodina-mat'" [lit. Motherland] monument is in Volgograd, former Stalingrad). It can also be found in the sculptural composition "To the Victims of Fascism" in Zmievskaya Balka, Rostov-on-Don.
48 "Monument Certificate of the history of the memorial complex 'Kholodnyi Rodnik' in Stavropol, 1975," in SGMZ, f. 104, d. 50, l. 3.
49 Panyutin, "Pamyatnik nepokorennym."

"They reflect both contempt for death and boundless faith in the Victory, deep hatred for the enemies who have not broken the freedom-loving spirit of the Soviet people, who spared neither blood nor life for the glory of their dear Motherland."[50]

Most memorials are figurative in the style of socialist realism and they often depict groups of people symbolising solidarity and anti-fascist resistance. These figures tower over the visitor, which again emphasises the "greatness and invincibility of the Soviet people." The image of heroism predominates in the iconography of state memorials. 'Ready-to-consume' visual images of Nazi victims were also formed in this period. These are archetypical, non-national, and depersonalised; like the Unknown Soldier, whose monument was erected in almost every town and city of the USSR. Memorial inscriptions were even more homogenous. They told of "Soviet citizens," townspeople or "peaceful citizens – victims of fascism" and appealed to future generations "to remember and honour their feat." Artistic representations and inscriptions on Soviet monuments were often designed both to honour the war heroes (fallen soldiers, partisans who fought the enemy) and remember the victims ("peaceful citizens") who were heroised in this way. The pedestal of the monument "To the Unconquered" in Novorossiysk, for instance, bears the following inscription: "To the people of Novorossiysk and Soviet Army paratroopers tortured and shot here in 1943 by fascist invaders. Eternal glory to the fallen for the freedom and independence of our beloved Motherland." One of the plaques at the memorial complex in Pervomaiskaya Roshcha in Krasnodar bears the carved inscription: "To the citizens of Krasnodar, tortured and killed in gas vans, massacred by Hitler's butchers in August 1942 – February 1943." Soviet memorial culture turned "peaceful Soviet citizens" into war heroes, leading people to forget the real reasons for their death, as well as the fact of the Holocaust itself.

Along with large-scale state projects to construct memorial complexes, the tradition of erecting small obelisks and monuments at killing sites was continued.[51] In the North Caucasus, many killing sites, especially in rural areas, were situated in wasteland, so it was much easier to erect a private monument in these sites. At the same time, the architectural expression of a monument had to meet Soviet standards. As a rule, their erection was still initiated by local

---

50 "Monument Certificate of the monument 'To the Unconquered,' Novorossiysk, 1963." Provided by the Department of State Protection of Cultural Heritage of Krasnodar Krai.
51 After commemoration of the war came under the aegis of the Ministry of Culture in the USSR, private monuments could be erected only in cemeteries and with the permission of the authorities. Aleksandra Tcherkasski and Leonid Terushkin, "Strategii sovetskoi evreiskoi obshchestvennosti po uvekovechivaniyu pamyati na primere ustanovki pamyatnikov evreyam-zhertvam fashizma," *Bylye gody* 29 (2013): 71.

historians and relatives of Holocaust victims.[52] For example, in 1973, a Soviet Army colonel came to a sorrowful place "in the area of Ladozhskaya village, where more than 130 Jews, mainly evacuees from today's Ukraine, Moldova, Belarus and Lithuania were killed in the anti-tank ditch in 1942. He erected a modest obelisk there"[53] (Figure 13). In the 1980s a Holocaust survivor Semen Dikenshtein being already retired began to look for the place where he survived the killing of all his relatives. At that time, he was 12 years old and did not remember the exact place where his family had been evacuated. Years later he met his rescuer, who told him that the tragedy had taken place in the village of Balabanovka, Rostov Oblast. However, the village was not revived after the war and the killing site of more than 100 Jews turned into a wasteland. Dikenshtein published a story in *Literaturnaya gazeta*, a weekly cultural and political newspaper published in the Soviet Union and in Russia. As a result, the authorities of Ol'khovyi Rog, the town closest to the killing site, erected a modest tombstone at the grave site.[54]

Again, it is impossible to define the total number of monuments to Holocaust victims ("peaceful Soviet citizens") erected in the North Caucasus in the Soviet period, since not all monuments have been listed as objects of cultural heritage and thus subject to protection and preservation. Members of Krasnodar's Jewish community searched for Soviet monuments at Holocaust killing sites in Krasnodar Krai. They found seven such monuments. These are the monument to Musya Pinkenson in Ust-Labinsk (1972), monuments in the town of Gulkevichi (1967), in the villages of Novominskaya (at the local cemetery, date unknown), Belaya Glina (date unknown), Ladozhskaya (1973), Udobnaya (1976) and Mostovskoi (1965).

---

[52] Nevertheless, the activity of local historians in commemorations of Holocaust victims in the North Caucasus in the Soviet period was not very noticeable as compared with other regions of the country. For example, the activity of the local historian Elena Ivanova to commemorate Holocaust victims was widely known in Bryansk and Smolensk Oblasts. In 1979, Arthur Khavkin, a Moscow researcher of Holocaust history who was collecting Holocaust evidence for *The Black Book*, prepared an album of monuments at the killing sites, which counted more than 20 photographs of monuments from various places of the USSR. As for the North Caucasus, he included only the monument in Kislovodsk in his album. Elena Ivanova, *Vyzyvaya ogon' na sebya: polozhenie evreev pri "novom poryadke" gitlerovskikh okkupantov v 1941–1943 godakh.* (Moscow: Tsentr "Kholokost," Rostov-on-Don: Feniks, 2011), 130–133; "Arthur Khavkin's personal collection," in *Holocaust Center Archive*, f. 46.
[53] He was obviously the only surviving boy, whom the villagers of Ladozhskaya rescued. Victor Bensman, "I put' ikh pereseksya . . .," *Vozrozhdenie: gazeta evreev Kubani,* January, 2015.
[54] Interview with Semen Dikenshtein, born in 1929, in *VHA USCSF*, code 7334, December 7, 1995, Holon, Israel.

Figure 13: Modest obelisk to the victims of fascism in the village of Ladozhskaya, Krasnodar Krai, 1973 (author's photograph, 2015).

None of them had an indication of the ethnicity of victims or Jewish symbols on them.[55] As for the other regions, such data are either missing or incomplete.

Thus, the commemoration of the war in the Soviet era memorialised its heroes and victims. It was almost completely monopolised by the Soviet state since the 1950s, which formed the canons of this form of memory in stone. The commemoration of the war was officially based on glorification of its heroes and not on mourning its victims. It was also influenced by the philosophy of the workers' movement: "Don't mourn – fight!"[56] Though there were universalised images of heroes (soldiers as courageous warrior-defenders) and victims (mother, child, elderly man) in sculptural compositions, the memory of them became even more impersonal and remote. This remoteness was also expressed in the geographical distance between the places of memorials erected just 'for show' and the killing/grave sites, which resulted in the desecration and ultimate loss of the latter. Israeli historian Arkadi Zeltser argues that monuments erected by the authorities from the 1960s to 1980s

---

[55] Yurii Teitelbaum, "Memorializatsiya Kholokosta v Krasnodarskom krae: opyt i problemy," in *Pamyat' o Kholokoste: problemy memorializatsii: materialy 6-i mezhdunarodnoi konferentsii "Uroki Kholokosta i sovremennaya Rossiya,"* ed. Il'ya Al'tman (Moscow: Tsentr i Fond "Kholokost," 2012), 54.
[56] Makhotina, "Symbole der Macht," 295–296.

should be considered a state alternative to the initiatives of the Jews (as well as of the non-Jewish dissident intelligentsia) and the result of foreign pressure.[57] In the North Caucasus, the sovietisation of the memory of "victims of fascism" without their ethnic identification repeated common trends throughout the country, and because of the lack of, or relative passivity, of Jewish communities or intelligentsia it was accepted as a given. The memorial complex in Zmievskaya Balka was erected at the largest grave site in the North Caucasus. However, unlike Babi Yar in Kyiv, Yama in Minsk, Rumbula in Riga, or Ponary in Vilnius, the history of Rostow-on-Don's Zmievskaya Balka was relatively little known among the liberal intelligentsia and neo-Zionist activists, who assumed the role of restoring Jewish institutions in the USSR since the late 1960s, and did not obtain wide coverage in the uncensored literature [*samizdat*] of the time.[58] Such a silent acceptance of the official Soviet politics of commemoration, on the one hand, and an appeal to the ethnic diversity of the North Caucasus, on the other hand, would significantly impede initiatives "from below" from restoring historical justice and properly identifying Holocaust victims in the region after the fall of the USSR.

3. *1991 to the Present Day*

The Perestroika of the mid to late 1980s ushered in a slight liberalisation of Soviet life, leading to a more vocal presence of Jewish communities, among others, in public life and the declassification of some archival files. In the 1990s researchers turned to the study of ChGK records. Representatives of Jewish communities and local historians used them as the main source when proposing and erecting new monuments at the killing sites and for identifying the names and numbers of Holocaust victims. With the collapse of the USSR, the administrative apparatus for the preservation of historical and cultural objects dedicated to the Great Patriotic War also broke down. In the early 1990s, following economic crisis and general political instability, many Soviet-era monuments were abandoned. Then, once again, the erection of monuments fell to the initiative of Jewish communities and local historians, but only by agreement with local authorities.

---

57 Zeltser, *Unwelcome Memory*, 37.
58 Semen Charnyi, "Rol' evreiskikh obshchin yuga Rossii v sokhranenii i memorializatsii pamyati o Kholokoste (na primere Rostova-na-Donu)," in *Istoriya Kholokosta na Severnom Kavkaze i sud'by evreiskoi intelligentsii v gody Vtoroi mirovoi voiny: materialy 7-i mezhdunarodnoi konferentsii "Uroki Kholokosta i sovremennaya Rossiya,"* ed. Kiril Feferman (Moscow: Tsentr i Fond "Kholokost," 2013), 116.

This period is characterised by the return of a direct link between the actual Holocaust killing/grave site and the monument, which almost always sports Jewish symbols – such as the Star of David and/or the Menorah – and has inscriptions in Hebrew.[59] The monument's shape often makes reference to Jewish culture too: either a stone or stones looking like the gravestones of Jewish cemeteries or a marble/brick composition in the form of a Sefer Torah [Torah scroll] or an open book. The 2014 monument to Holocaust victims in the village of Stepnoe, Stavropol Krai, for instance, is made in the form of a Sefer Torah; the 2006 monument in the village of Kuzhorskaya, Republic of Adygea is in the form of a stone; and the 2013 monument in Petrushina Balka, Taganrog is in the form of an open book (Figure 14). The inscriptions on the monuments are brief, giving the dates of the Nazi killing action, the number of Holocaust victims, and often state that it is a Holocaust site. Interestingly, the very first monument in the North Caucasus to be marked explicitly as a Holocaust site was erected in 2006 in Kuzhorskaya on Pervomaiskii Kurgan, Republic of Adygea at the private expense of citizens of Maykop and Krasnodar (including non-Jews) in memory of 36 Jewish evacuees killed under the command of SK 11b in August 1942 (Figure 15).[60] So far, it is the only official monument to Holocaust victims in the territory of the Republic of Adygea and the surrounding Krasnodar Krai. Sometimes the ethnicity and the number of all victims buried in a particular site are indicated on the monument's inscription. In 2012, a new inscription was carved on the Soviet monument Arzgirskaya Balka, Stavropol Krai thanks to dedicated archival research and fieldwork by the local historian Anatolii Karnaukh. It reads: "In September 1942 in Arzgirskaya Balka, the Nazis and their allies shot 695 people, mainly elderly men, women and children: 675 Jews, 15 Russians, including 4 partisans, 15 Moldovans. Eternal memory to the innocent victims." Indication of the victims' ethnicities is one way of replacing the Soviet euphemism "Soviet citizens."[61] However, the terms "Soviet citizens" or "peaceful

---

59 For example, monuments to Holocaust victims in Stavropol Krai: the village of Kurskaya, 2015, Novoromanovskii, 2015, and the memorial inscription on the monument near Mount Koltso, Kislovodsk, 2015.
60 Yakov Frenkel', "Ekho Kholokosta pod Maykopom," *Vozrozhdenie: gazeta evreev Kubani*, October, 1–31, 2006.
61 At the same time, the indication of the victims' ethnicities on monuments is a common practice in Europe, primarily at the sites of former concentration camps (for example in Flossenbürg, Germany). Peter Heigl, *Konzentrationslager Flossenbürg in Geschichte und Gegenwart: Bilder und Dokumente gegen das Zweite Vergessen* (Regensburg: Buchverlag der Mittelbayerischen Zeitung, 1989), 63.

4.1 "Memory in Stone"? The Commemoration of Holocaust Victims — 117

Figure 14: "Monument to Jewish children – Holocaust victims" in Petrushina Balka in the form of an open book, Taganrog, 2013 (author's photograph, 2016).

Figure 15: Monument to Holocaust victims in the village of Kuzhorskaya, Republic of Adygea, 2006 (author's photograph, 2013).

citizens"[62] are also often read in the inscriptions. There are a few explanations for this. First, the term "Soviet citizens" is understood literally by many people of the older generation: they were all Soviet citizens living in the Soviet Union, so that the killed Jews were also *ipso facto* Soviet citizens. Secondly, Russia, being a legal successor state of the USSR, continued the tradition of honouring the veterans and celebrating the Great Victory; Russians honour the heroes and remember "peaceful Soviet citizens." Thirdly, in those regions of the North Caucasus where there was antisemitism (particularly Krasnodar Krai) the creators of the monuments consciously used Soviet terminology, avoiding explicit reference to a Jewish presence. The erection of Orthodox crosses at the killing sites of Jews in Novopokrovskii district, Krasnodar Krai, which was funded by local sponsors, is an extreme example of such an activity. Ivan Boiko, a local historian, wanted first of all to mark all the grave sites but he rejected the idea of erecting a monument with Jewish symbols on it, as he feared acts of vandalism.[63]

Indeed, acts of vandalism at memorial sites with Jewish symbols or which make reference to Jews being buried there have been recently recorded in different regions of Russia and in other post-Soviet countries.[64] During discussions of what to write on a new monument near Urochishche Stolbik in Stavropol, the following inscription was decided upon: "Remember! Here is the ash of Soviet citizens, including 3,500 Jews, tortured and shot during the Fascist occupation of Stavropol in 1942–1943." Local authorities warned, "if we write 'Jews' on the monument today, tomorrow there will not be a single stone left." This fear was rapidly confirmed. The plaque was first broken several times then, when it was replaced with a thick shatterproof slab the word "Jews" was painted over[65] (Figure 16). In 1998, a year after a modest stone (a monu-

---

[62] For example, the inscriptions on monuments in the village of Chelbasskaya, Krasnodar Krai and Kuzhorskaya, Republic of Adygea commemorate the memory of "peaceful citizens shot in 1942" and "36 Soviet citizens" respectively.

[63] Monuments in the form of an Orthodox cross were erected in 2005 in the Budenovskaya woodland belt and on the farm of Za Mir in the village of Novopokrovskaya. These monuments have not been registered as objects of cultural heritage yet, which, according to Boiko, is caused by the lack of a targeted State policy to commemorate Nazi victims. Ivan Boiko, "Opyt i problemy memorializatsii zhertv Kholokosta na raionnom uroine," in Al'tman, ed., *Pamyat' o Kholokoste*, 62; Idem, *Etapy bol'shogo puti: k 90-letiyu Novopokrovskogo raiona* (Krasnodar: Kniga, 2015), 83–112.

[64] For example, the web portal jewish.ru, reported acts of vandalism in Ternopil and Lviv (Ukraine), and in Pskov Oblast (Russia) between 2015 and 2017.

[65] Magulaeva, "Zemlei zabity nashi rty"; Interview with Efim Fainer, in *author's archive*, code 15-MEM-SK-10, May 20, 2015, Stavropol, Russia.

Figure 16: Vandalism to the monument erected in 2005 near Urochishche Stolbik in Stavropol (photograph by Efim Fainer, 2012).

ment with the Star of David, the Menorah, the Russian inscription "To Jews – victims of the fascist genocide" and the Hebrew inscription "Every person has a name" on it) was erected in Petrushina Balka, unknown vandals tore off all the inscriptions and Jewish symbols, which were made of non-ferrous metals.[66] In 2003 a similar situation occurred in Novorossiysk: vandals poured paint over the monument's memorial plaque "To the Unconquered" and tore off a wreath

---

[66] Tankha Otershtein, "Peredaite eto detyam vashim . . .," *Vekhi Taganroga* 4 (2000): 39. It is possible that this incident was not an act of vandalism, but instead an instance of the plundering of non-ferrous metals and other valuable items during the economic crisis of the 1990s.

from the Jewish religious community.[67] Such actions demonstrate the lack of educational programmes about the Holocaust in Russia and unwillingness to tolerate any symbolic presence of Jews in the landscape by many contemporary Russians.

A distinctive feature of post-Soviet Holocaust remembrance, particularly in the North Caucasus, is its evolution from a unified, impersonal memory of war heroes and war victims to attempts at remembering people as persons and individuals. In his poem "Requiem" [*Rekviem*], the well-known Soviet poet Robert Rozhdestvenskii appealed: "Let us remember everyone by their name" [*vspomnim vsekh poimenno*].[68] This phrase has become a leitmotif for celebrations of Victory Day[69] and lately for the commemoration of Holocaust victims. It has become possible to identify the names of many Jews killed in the North Caucasus using archival data, to inscribe their names on plaques, and to find and contact their relatives. The search for names is complicated significantly by the lack of both Soviet and German sources on the Holocaust in the region. Not all refugees and evacuees to the North Caucasus were registered, their migration was at times not recorded, there were almost no German lists of registered Jews, and not all the surnames of murdered Jews were mentioned in the ChGK records. Yet, after more than 60 years local activists and members of Jewish communities are restoring bit-by-bit the Holocaust history of particular localities. For example, of the

---

[67] Svetlana Ladozhskaya, "Antisemity oskvernili pamyatnik – reportazh iz Novorossiiska," *Agentstvo evreiskikh novostei*, July 30, 2003, http://aen.ru/index.php?page=article&category=anti&article_id=156 (accessed December 9, 2019).

[68] Robert Rozhdestvenskii, "Rekviem (Vechnaya slava geroyam . . .): Poema," *website of classical and contemporary Russian poetry*, https://45parallel.net/robert_rozhdestvenskiy/rekviem_vechnaya_slava_geroyam.html (accessed March 29, 2020).

[69] A clear example of the personification and personalisation of the memory of the war is the annual all-Russian Immortal Regiment [*Bessmertnyi polk*] campaign, "during which the participants march in a column and carry banners with photographs of their relatives who participated in World War II, as well as share family stories of their heroes who defeated fascism in the People's Chronicle on the website of the campaign." In 2015, the civil initiative "Immortal Regiment" was included in the federal programme to prepare the celebration of the 70th anniversary of the Victory, which over 4 million people all over Russia attended. "Letopis' Polka," *Moypolk.ru*, https://www.moypolk.ru/letopis-polka (accessed March 3, 2020). The personalisation of World War II memory in the public sphere is due to the continued interest of young generations. Some sociologists regard this process not only as an attempt by the authorities to preserve Victory Day as "the main holiday of the Soviet people's triumph," but also as having a future potential for the memory of Stalinist victims to leave the private sphere for the public one. Mischa Gabowitsch, Cordula Gdaniec, and Ekaterina Makhotina, ed., introduction to *Kriegsgedenken als Event: Der 9. Mai 2015 im postsozialistischen Europa* (Paderborn: Ferdinand Schöningh, 2017), 45–48.

600 Holocaust victims in the small village of Stepnoe, Stavropol Krai, only 32 names were identified.[70] As a rule, the identified names are carved on a new monument or a new plaque that is installed on an already existing monument.[71] In 2004, due to the return of war veteran Aleksandr Falko to his childhood village of Chelbasskaya, Krasnodar Krai, some work was done to restore an abandoned killing site of evacuated Jews (mostly from Leningrad) in Chelbasskaya Roshcha. As a teenager Falko became a Holocaust eyewitness. When the war ended, he lived in Ukraine for a long time and after coming back to his native Chelbasskaya he made every effort to restore the memory of Holocaust victims. As a result, two monuments were erected in 2005 with the support of members of a local veteran organisation and the village administration. The one is erected on the grave site itself in Chelbasskaya Roshcha, and takes the form of two pillars stuck in the ground. There are two plaques on the pillars: the first reports the occurrence of a tragic event here in 1942 and the second indicates the names of the victims: "V.A. Bobchii, D.T. Nemchenko, Chernobrovaya, T.A. Shevchenko, the Rubelshkezer family – father, daughter Dora, granddaughter – and 6 more Jews."[72] Not far from this place, on the edge of a forest (the grove [*roshcha*] has become a forest during the last half a century), the second monument was erected with the inscription, "May the memory of peaceful citizens shot in 1942 live forever. From descendants," carved on a marble plaque.[73]

The only monument in the whole of the North Caucasus dedicated to child victims of the Holocaust is that in Petrushina Balka erected in 2003 to more than 1,500 Jewish children who died at this site.[74] The inscription indicates the names and ages of 60 young Holocaust victims. Tankha Otershtein, a young prisoner of fascism [*maloletnii uznik fashizma*], devoted his entire life to the study of Holocaust history in Taganrog. The monument is a direct result of his research.[75] Members of Rostov-on-Don's Jewish community have been searching

---

70 "Tridtsat' dva imeni iz shestisot," *Novosti RJC*, September 10, 2014, http://www.rjc.ru/rus/site.aspx?IID=2644956&SECTIONID=85646 (accessed December 4, 2019).
71 For example, the monument to Mountain Jews in the village of Bogdanovka, Stavropol Krai; monuments in the village of Novominskaya, Krasnodar Krai, in the villages of Stepnoe and Arzgir, Stavropol Krai, and in Petrushina Balka, Rostov Oblast.
72 Valentin Tsvetkov, *Na zemle predkov: stranitsy istorii ZAOPZ "Volya" i stanitsy Chelbasskoi* (Maykop: "Poligraf-Yug," 2012), 441–442.
73 The second monument was erected because the village administration wanted a monument in a site with access, so that commemorative events could be held there in the future. Idem, 442.
74 Lev Shkol'nik, "Detyam – zhertvam Kholokosta," *Agentstvo Evreiskikh Novostei*, October 31, 2013 http://www.aen.ru/?page=articleoftheday&article_id=3424 (accessed December 4, 2019); Dar'ya Vseglyadova, "Pamyat' o Kholokoste," *Taganrogskaya pravda*, December 17, 2014.
75 Otershtein, "Peredaite," 37–40; Interview with Otershtein, in *author's archive*.

for and identifying the names of Jews killed in Zmievskaya Balka for more than five years. As a result of archival research by Vladimir Rashka, a Jewish archivist in Rostov and with the support of Yurii Dombrovskii, a member of the board of trustees on the Russian Jewish Congress's (RJC) "Holocaust Remembrance" programme, the names of 3,361 victims were identified. They were recorded in the *Martyrology*, a commemorative book of Holocaust victims.[76] A similar book is planned to commemorate the names of Jews, buried near the glass factory beyond Mineralnye Vody in Stavropol Krai.[77] It is notable that Holocaust remembrance usually began with compiling memorial books – *yizker-bikher* in European countries[78] – whereas such martyrologies are one of the final stages of memorial work in Russia, and particularly in the North Caucasus. This has only become possible after a partial revision of the Soviet commemorative culture of World War II.

The installation of commemorative plaques on houses where Holocaust victims, mainly well-known Jewish cultural and scientific figures, lived and at the locations where Jews are instructed to assemble also express the increasing tendency to commemorate Holocaust victims. Several plaques were installed in Rostov-on-Don in 2002. One was mounted on the façade of the house on Pushkinskaya Street 83, the address of Sabina Spielrein-Scheftel, a well-known psychoanalyst and student of Carl Gustav Jung.[79] Another commemorated the as-

---

[76] Inna Shvartsman, ed., *Kniga pamyati: martirolog zhertv Kholokosta: Rostov-na-Donu, Zmievskaya Balka, 1942 god* (Rostov-on-Don: Rostovskaya evreiskaya obshchina, 2014). As of March 2017, the Rostov list has 4,007 names, out of which 3,601 were confirmed by researchers of the Hall of Names in Yad Vashem. These names were planned to be added to the plates of the memorial complex in Zmievskaya Balka, which has been under restoration since the spring of 2017. This intention has not been realised as of spring 2020.

[77] Interview with Mikhail Akopyan, in *author's archive*, code 15-MEM-SK-01, May 13, 2015, Mineralnye Vody, Russia. In 2019 there was opened the new memorial complex. The names of more than 1,000 Holocaust victims are carved in granite. "V Minvodakh otkryli memorial zhertvam Kholokosta," *RIA Novosti*, June 18, 2019, https://ria.ru/20190618/1555676843.html (accessed March 4, 2020).

[78] Yizker-bikher, or Yizkor books remember the life and destruction of European Jewish communities according to the most ancient of Jewish memorial media: words on paper. For a murdered people without graves, without even corpses to inter, these memorial books often served as symbolic tombstones. Young, *The Texture of Memory*, 7.

[79] Memorial plaque on the house at Pushkinskaya Street 83 states: "In this house lived the famous student of C. G. Jung and S. Freud, psychoanalyst Sabina Spielrein. 1885–1942." An oak was planted in her memory at the killing site in Zmievskaya Balka in 2012, just as she wished in a will written during her youth in Switzerland. Il'ya Karpenko, Evgenii Movshovich, and Inna Shvartsman, "Za gran'yu psikhoanaliza," *Lekhaim* 11 (2005), http://www.lechaim.ru/

sembly point of Jews and was installed at Bolshaya Sadovaya Street 23.[80] However, large-scale work to commemorate the last address not only of famous people, but also of ordinary Jews – like the Stolpersteine project, which was initiated in 1992 and has spread through Germany and other European countries[81] – is still to be done in Russia.

The commemoration of Holocaust victims is the main task of the principal Russian Jewish organisations, such as the Holocaust Center and Foundation, the RJC, and the Federation of Jewish Communities in Russia (FJCR). Most of these organisations were established in the 1990s. Thus, the RJC initiated the Restore Dignity [*Vernut' dostoinstvo*] project in 2009. Its main objectives are to identify Holocaust killing/grave sites, erect new monuments to commemorate Holocaust victims or to bring existing monuments in conformity with *halakha*, traditional Jewish law, and to find data about all those murdered by the Nazis.[82] Thus, for example, seven monuments in Stavropol Krai were erected in 2016 due to local historian Anatolii Karnaukh's activity with support from the RJC, the Holocaust Center, "Even Ha'ezer" Social Assistance and Support Foundation, and the Association of Protestant Communities in Russia. Among them are the Arzgirskaya Balka memorial in Arzgir, monuments to Holocaust victims in the villages of Novoromanovskoe and Serafimovskoe in Arzgir District, in the villages of Spasskoe and Shishkino in Blagodarnenskii District and in the farm of Nevdakhin in Trunovskii District.[83] The Restore Dignity project can be implemented only with the help of local partners and volunteers, such as local historians, war veterans, employees of local and regional museums, schoolteachers, and members of Jewish

---

ARHIV/163/VZR/020.htm (accessed March 3, 2020); Aleksandr Olenev, "Molodoi dub v Zmievskoi Balke v pamyat' o zhenshchine-legende," *Vechernii Rostov*, April 27, 2012.
80 "Decrees of the Mayor of Rostov-on-Don No. 1614 of August 19, 2002 and No. 2163 of October 16, 2002," in *Archive of the Rostov department of the All-Russian Society for Protection of Historical and Cultural Monuments* [*Vserossiiskoe obshchestvo okhrany pamyatnikov istorii i kul'tury*], hereafter *VOOPIiK*.
81 The artist Gunter Demnig remembers the victims of National Socialism by installing commemorative brass plaques in the pavement in front of their last address. Stolpersteine [lit. "stumbling stones"] are now installed in 1,099 locations in Germany and in 20 other European countries. See: *Stolpersteine.eu*, http://www.stolpersteine.eu/start/ (accessed March 3, 2020).
82 "The Restore Dignity project," *RJC official website*, https://rjc.ru/ru/projects/project-42 (accessed March 3, 2020).
83 Nadezhda Klimchenko, "Budem pomnit'! Pomnit' vsegda!," *Zarya: obshchestvenno-politicheskaya gazeta Arzgirskogo raiona Stavropol'skogo kraya*, September 13, 2016; Interview with Karnaukh.

communities. They conduct research in their native localities and help obtain legal permission from the regional authorities to erect or renew the existing monuments. As Yurii Kanner, the RJC's president and Il'ya Al'tman have repeatedly emphasised, the commemoration of Holocaust victims could be carried out only with the assistance of local authorities.[84] Yurii Teitelbaum, chairman of the Jewish community in Krasnodar was engaged in Holocaust commemoration in Krasnodar Krai between 1996 and 2011.[85] He travelled all over the krai, met local historians and activists, researched in the regional archives, made a list of existing monuments at Holocaust sites and initiated the erection a few monuments. One of them he planned to erect at the crossroads of Moskovskaya and Solnechnaya streets in Krasnodar, at the first killing site of Jews in the city, where previously a post-war modest obelisk had been located.[86] However, protracted negotiations, the lack of sufficient financing, and the passivity of local authorities were an obstacle to his plans.[87] Sometimes members of Jewish communities prefer to obtain permission from local authorities by themselves without support of a central Jewish organisations. It can be easier to install a commemorative plaque indicating a Holocaust site on existing Soviet monuments rather than seeking to implement a large-scale project. The work of the Kislovodsk Jewish community described at the beginning of this chapter is a good example of such a local commemorative act. The rule "the fewer organisations participate in the commemoration, the more effective is its outcome" has been proven repeatedly.

The commemoration of not only Holocaust victims but also of Righteous Among the Nations has become another significant trend of the past decade. Thus, in 2010 a monument to the Circassian Mother was erected in the mountain

---

[84] "Zayavlenie Rossiiskogo evreiskogo kongressa i tsentra 'Kholokost' v svyazi s obnaruzheniem v Stavropole massovykh zakhoronenii zhertv Velikoi Otechestvennoi voiny," *RJC official website*, August 26, 2009, http://www.rjc.ru/rus/site.aspx?SECTIONID=91208&IID=621204 (accessed December 3, 2019).

[85] For the results of this activity, see: Teitelbaum, "Memorializatsiya Kholokosta v Krasnodarskom krae," 53–59.

[86] I am grateful to Teitelbaum that during my stay at Bat Yam, Israel, he not only found the time to give me a detailed interview about his activity as the chairman of the Jewish community of Krasnodar, but also shared his archive, which contains correspondence with the RJC, copies from regional archives, sketches of several monuments he designed, all issues of the newspaper "Revival: Newspaper of the Kuban Jews" [*"Vozrozhdenie: gazeta evreev Kubani"*], and other valuable sources.

[87] In 2012, Teitelbaum and his family moved to Israel, where they currently reside. According to him, the resistance of local authorities to commemorating the Holocaust in Krasnodar Krai and the indifference of Russian Jewish organisations, including the RJC and the Holocaust Center were the main reasons for his decision to emigrate. Interview with Teitelbaum.

village [*aul*] of Beslenei in the Karachay-Cherkess Republic. During the war the aul's residents adopted 32 orphans from Leningrad, who had fled through Krasnodar Krai into the North Caucasus. After the war, many of them remained in Circassian families and were brought up as full members of the Circassian community. Being widely known in the Soviet era, this story resulted in the erection of a post-Soviet monument dedicated to the villagers who saved the children. It is a bronze sculpture of a mother embracing a child[88] (Figure 17).

Figure 17: Monument to the Circassian Mother in Beslenei aul, the Karachay-Cherkess Republic, 2010 (author's photograph, 2015).

It was done thanks to the efforts of Aleksandr Okhtov, the director of the National Cultural Autonomy of Russian Circassians, and rescued children themselves with the support of Boris Ebzeev, the president of the Karachay-Cherkess Republic during 2008–2011. The Avenue of Righteous Among the Nations from Stavropol Krai was planted to honour local citizens near Mount Koltso in 2006. By 2020, seven citizens of Stavropol Krai had received this award from Yad

---

[88] "V aule Beslenei otkryt pamyatnik zhitelyam aula, priyutivshim detei iz blokadnogo Leningrada," *official website of the Administration of the Karachay-Cherkess Republic*, May 7, 2010, http://kchr.ru/news/detailed/4686/ (accessed March 23, 2020). More about this deed see: Yair Auron and Aleksandr Okhtov, *Podvig milisediya: spasenie detei iz blokadnogo Leningrada v cherkesskom aule Beslenei* (Moscow-Cherkessk-Jerusalem, 2018).

Vashem. They are Sergei Metreveli, Anna Popova, Ol'ga Pylneva, Aleksandra Zholtaya, Evdokiya Pal'onaya and Akim Pal'onyi, and Varvara Tsvileneva. All of them come from Kislovodsk.[89] There are such Avenues of Righteous Among the Nations in Israel and in other countries of the world, being a new and very promise experience for Russia and the North Caucasus.

The transnational scope of Holocaust commemoration[90] characteristic of contemporary society is also seen in the North Caucasus. The names of Holocaust victims from the region are recorded in the Yad Vashem Hall of Names. There are also examples of the erection of monuments to North Caucasian Holocaust victims abroad. After the USSR collapsed, many Jews emigrated mainly to Germany, Israel, and the USA. A community of Mountain Jews – former citizens of the Republic of Kabardino-Balkaria – settled in New York. One of these migrants, Aleksandr Kishiev, initiated the erection of a memorial stone in honour of 473 people (his relatives among them) who became victims of the Holocaust in the village of Bogdanovka in 1942.[91] This memorial stone is placed among more than a hundred other such memorial stones dedicated to the memory of individual Jews, groups of Jews, and other victim groups killed in different places in Europe at a memorial complex erected in the early 2010s in Manhattan Beach Park, Brooklyn. The inscription of this memorial stone is carved in both English and Russian and reads: "In memory of hundreds of relatives and friends, residents of the village of Bogdanovka in Stavropol region, killed by the Nazis and their collaborators in September, 1942" (Figure 18). The individual memory of the Holocaust as private pain and sorrow transcends geographical borders along with the expatriates themselves. Thus, as Holocaust memorials address transnational audiences, they become an important element in international Holocaust remembrance.

With the collapse of the USSR, commemoration of Holocaust victims by Jewish communities and local historians received a new impulse. Many Holocaust

---

[89] Afanasii Krzhizhanovskii, "Na Stavropol'e vspominali zhertv Kholokosta," *Otkrytaya gazeta*, 36, September 14–23, 2016, http://www.opengaz.ru/stat/vernyom-svoyo-dostoinstvo (accessed March 3, 2020); the Righteous Among the Nations Database for Kislovodsk, *Yad Vashem official website*, https://righteous.yadvashem.org/?search=%D0%9A%D0%B8%D1%81%D0%BB%D0%BE%D0%B2%D0%BE%D0%B4%D1%81%D0%BA&searchType=righteous_only&language=ru (accessed March 3, 2020).

[90] According to Marcuse, Holocaust memorials, as a new genre of commemorative art distinct from older forms, are addressed to transnational audiences. Marcuse, "Holocaust Memorials," 54.

[91] Lyuba Yusufova, "Prazdniki v N'yu Yorke: den' materi i den' pamyati," *Stmegi Novosti*, May 14, 2014, https://stmegi.com/posts/17672/prazdniki_v_nyu_yorke_den_materi_i_den_pamyati_9669/ (accessed March 23, 2020); Svetlana Danilova, "Kholokost na Severnom Kavkaze i iskazhenie faktov v kontekste sovremennykh sobytii v dannom regione," *Zametki po evreiskoi istorii* 2 (2012), http://berkovich-zametki.com/2012/Zametki/Nomer2/Danilova1.php (accessed March 23, 2020).

Figure 18: Memorial stone commemorating the Holocaust victims (Mountain Jews) of Bogdanovka village, a part of memorial complex in Manhattan Beach Park, Brooklyn, New York, USA (author's photograph, 2019).

killing/grave sites have been marked with commemorative plaques and monuments from the Soviet period have been given new meaning in the context of Holocaust remembrance. At the same time, contemporary activists have to deal with the legacy of Soviet ideology. Regional authorities sometimes do not consider the erection of memorials to Holocaust victims to be important, stating that people of various ethnicities became Nazi victims in the USSR. However, such claims reduce the historical significance of the Holocaust. The recognition of the Holocaust does not minimise the fact that there were other Nazi victim groups who were persecuted for political and other reasons, including ethnic or racial ones. Acts of vandalism and local activists' fear of possible desecration of monuments with Jewish symbols are the consequence not only of the Soviet ideological attitude towards Holocaust remembrance, but also of the current position of the Russian Federation that recognises the Holocaust nominally yet does not actively contribute to its inclusion in the official memory politics of the "Great Patriotic War." This explains the tendency of local authorities to seek not to mention the Holocaust and Jews "in order not to inflame ethnic discord in the ethnically diverse North Caucasus:" it is the consequence of the state's reluctance to act and the lack of large-scale educational programmes, which could make Russians more tolerant. The events of 2012 in Rostov-on-Don represented an extreme example of disagreement between city authorities and the Jewish community concerning the recognition of the Holocaust.

4.2 Between Official Ideology and Private Memory: Is Zmievskaya Balka the Largest Holocaust Site in Russia?

The relation between a state and its memorials is not one-sided. On the one hand, official agencies are in a position to shape memory explicitly. On the other hand, once created, memorials take on lives of their own, often stubbornly resistant to the state's original intentions.[92] Using the example of the historical evolution of the memorial site in Zmievskaya Balka, I will trace the conflict of interests between remembering communities and the official state position in the commemoration of Nazi victims.

Almost immediately after the mass killing in Zmievskaya Balka the site became a place where Jews commemorated their murdered relatives. But the Soviet government, in the person of Aleksandr Baikov, the Representative of the Council for the Affairs of Religious Cults in Rostov-on-Don under the Council of Ministers of the USSR prevented members of the Jewish community from carrying out any acts to commemorate the Holocaust victims. In the first post-war years the Jewish community managed to organise a commemorative ceremony for victims in Zmievskaya Balka on the Remembrance Days, August 11–12. In 1949, Baikov "warned the members of the executive organ of the Jewish synagogue that Jews and other peoples were buried in Zmievskaya Balka and among them were believers, atheists and people of other faiths." He suggested that the Jewish community "should not carry out a commemorative ceremony on the mass grave in Zmievskaya Balka."[93] The process of state control over communities of believers

---

92 Young, *The Texture of Memory*, 3.
93 According to the Soviet legislation on cults, based on the decree of the SNK and the All-Russian Central Executive Committee [*Vserossiiskii tsentral'nyi ispolnitel'nyi komitet*] "On Religious Organisations" of April 8, 1929, religious organisations had no right to organise any

occurred parallel to the developing of state ideology towards the Great Patriotic War. After 1950, according to Baikov's reports, "there were no commemorations on the mass grave," but members of the Jewish community persisted for some time in their intention to preserve the memory of Holocaust victims in Rostov-on-Don. So, in the 1950s the head of the Jewish community approached Baikov several times with a request to allow the erection of a monument to the killed Jews and to organise fundraising for this purpose among the Jews of Rostov. That permission was never received for the same formal reason of the diverse ethnicities of the victims. However, Baikov allowed members of the Jewish community to visit the mass grave on the Remembrance Days and to place a wreath in 1956. He also recommended the Rostov Executive Committee [*Ispolnitel'nyi komitet*, or *Ispolkom*] to erect a greater monument on the mass grave.[94]

From the late 1950s the membership of the Jewish community in Rostov sharply decreased, and after the death of the permanent rabbi Meir Shai in 1960, a new rabbi was not appointed.[95] When the municipal government decided to build a memorial complex in Zmievskaya Balka in the mid-1970s, the authorities completely suppressed the commemorative activity of Rostov's Jews. Soviet functionaries sought to prevent any religious activity in the USSR. They tried to control and suppress any form of public religious activity in every possible way whether it was fundraising for the erection of a monument or carrying out any ceremonies, including Holocaust remembrance. Soviet authorities were evidently not ready to distinguish the Holocaust, as a specifically Jewish tragedy, from the common fate of all Soviet people who experienced Nazi occupation.

The memorial in Zmievskaya Balka was erected on the eve of the 30th anniversary of the Victory, in May 1975. It was officially named according to the canons of Soviet memorial policy and was listed in the register of the Cultural Administration of the Rostov Municipal Executive Committee[96] as "Memorial to the Memory of the Victims of Fascism (Zmievskaya Balka)." Of course, there was no mention of the Jews as the principal Nazi victim group. The authors of

---

public campaigns outside churches, synagogues, and prayer houses. But in spite of the warning received from the controlling authority, in 1949, Jewish community members sent out 256 invitations to the Remembrance Day commemoration, which occasioned Baikov's written request to the Council for the Affairs of Religious Cults. "Materialy o rabote upolnomochennogo po Rostovskoi oblasti," in *GARF*, f. R-6991, op. 3, d. 823, l. 78–79. For more details, see: Charnyi, "Rol' ereiskikh obshchin," 111–112.

94 "Materialy o rabote upolnomochennogo po Rostovskoi oblasti, 1949–1956" in *GARF*, f. R-6991, op. 3, d. 823, l. 190; d. 825, l. 129; d. 828, l. 83–85.

95 Charnyi, "Rol' evreiskikh obshchin," 116.

96 This organisation was responsible for erecting the memorial complex.

the project defined the grave site of thousands of people as the spatial and compositional center of the memorial complex, which is preserved in the form of a deep crater. All the main elements of the memorial complex are oriented towards it.

> A pedestrian alley leads from the 'Hall of Sorrow' ['*zal skorbi*'], which is the name of the museum building, down the slope to the lower platform with the 'Eternal Flame.' This alley has a ring shape and symbolises the 'Death Road' ['*doroga smerti*'], the road along which men and women, old people and children went to the execution site. Five commemorative pylons symbolising the five years of the war, 1941–1945, were placed in front of the alley, alternating with low-rise granite headstones of mass graves. The 'Memorial Square' ['*ploshchadka pamyati*'] with a bowl of 'Eternal Flame' and a sculptural composition 'Victims of the Shooting' ['*zhertvy rasstrela*'] (Figure 19) are located in the geometric center of the memorial complex. A sculptural composition represents the last few minutes before shooting. A woman is raising her arms. A child is embracing the knees of his mother, as if trying to hide behind her. An old man is stretching out his tied hands [. . .] The leading element in the composition of the memorial complex is this sculpture made in reinforced concrete. It psychologically and emotionally influences the viewer. The cultivated elements of the landscape can be considered as an additional means of artistic expressiveness of the whole complex. Scarlet sage on mass graves,

Figure 19: Sculptural composition "Victims of the Shooting" at the "Memorial to the Memory of the Victims of Fascism (Zmievskaya Balka)," Rostov-on-Don, 1975 (author's photograph, 2013).

rows of poplars, and firs along the upper 'Alley of Memory' ['*alleya pamyati*'] symbolise the unity and memory of generations.[97]

A permanent exhibition "about the Nazi atrocities in occupied Rostov-on-Don" was displayed in the Hall of Sorrow during Soviet times. A plaque with a quotation from Rozhdestvenskii's poem was installed on the black marble wall in the hall opposite the entrance. It reads:

Remember!	*Pomnite!*
Through the ages,	*Cherez veka,*
through the years,	*cherez goda,* –
Remember!	*Pomnite!*
Those,	*O tekh,*
who will return no more –	*kto uzhe ne pridet nikogda,* –
Remember!	*pomnite!*
Do not wail!	*Ne plach'te!*
In your throat	*V gorle*
curb those moans,	*sderzhite stony,*
bitter moans.	*gor'kie stony.*
Of the memory	*Pamyati*
of the fallen	*pavshikh*
be worthy![98]	*bud'te dostoiny!*

As the years passed, eyewitnesses of the war grew old, private memories of the war gradually turned into historical memory, and the Soviet euphemism "peaceful Soviet citizens" was taken by most citizens of the USSR as a given. Holocaust memory was completely erased from the public sphere, becoming the personal and family tragedy of Soviet Jews.

After the USSR collapsed, the ownership of the memorial in Zmievskaya Balka became uncertain. Black marble from the Hall of Sorrow was plundered, the sculptural composition required restoration. The embodied "memory in stone" of this monument lay unclaimed in a context of rapid ideological and political unravelling in the transitional period of the 1990s. At the same time, the

---

[97] Anatolii Zubkov, Kolokola pamyati: o tekh, kto pal za mir, za zhizn', za schast'e nashe (Rostov-on-Don: Rostovskoe knizhnoe izdatel'stvo, 1985), 30–32; "Preservation Order No. 115–10 for the cultural heritage object of regional significance 'Memorial complex to the memory of the victims of fascism in Zmievskaya Balka 1942,' August 17, 2010." Provided by the City Cultural Administration of Rostov-on-Don.

[98] Robert Rozhdestvenskii, "Requiem," transl. Tristam Barrett.

Jewish community of Rostov-on-Don was revived[99] and many archival materials were declassified. Evgenii Movshovich, a Jewish native of Rostov-on-Don who survived the Holocaust thanks to his timely evacuation to Western Siberia,[100] devoted his entire life to researching the history of the Jews of Rostov.[101] He was the author of the text of the first commemorative plaque that appeared on the façade of the museum building in Zmievskaya Balka in 2004 in accordance with a decree of the Mayor of Rostov-on-Don. It reads: "On August 11–12, 1942, more than 27,000 Jews were killed by the Nazis here. This is the largest Holocaust memorial in Russia."[102] The commemorative plaque remained there for seven years and survived a renovation of the memorial complex in 2006–2007. Moreover, the memorial complex "In Memory of the Victims of Fascism" in Zmievskaya Balka was registered as "the largest Holocaust monument in Russia" in the registration document of the city's Cultural Administration [*upravlenie kul'tury*] in its 2010 version.[103]

However, in November 2011 the Cultural Administration unilaterally removed this memorial plaque[104] and replaced it with another one, which included data on the memorial itself and its creators:

> Here, in Zmievskaya Balka in August 1942, more than 27,000 peaceful citizens of Rostov-on-Don and Soviet POWs were killed by the Hitlerite occupiers. Among those killed were representatives of many ethnicities. Zmievskaya Balka is the largest site in the Russian Federation of the mass killing of Soviet citizens by fascist invaders during the Great Patriotic War – it is a historical monument of regional importance. The

---

99 The Jewish community was revived and registered as a legal organisation in 1991. Interview with Mikhail Potapov, in *author's archive*, code 13-MEM-RO-01, February 27, 2013, Rostov-on-Don, Russia.
100 Interview with Evgenii Movshovich, in *author's archive*, code 13-X-PO-04, February 27, 2013, Rostov-on-Don, Russia.
101 Movshovich, *Ocherki istorii evreev*.
102 Decree of the Mayor of Rostov-on-Don of September 20, 2004. For the text of the decree, see Yanina Chevelya, ed., *Zmievskaya Balka: vopreki* (Rostov-on-Don: Feniks, 2013), 232. I am grateful to the Chairman of the Rostov department of the VOOPIiK Aleksandr Kozhin, who was responsible for the approval of the text of the memorial plaque by city authorities, for providing me with information and access to the archives of the Rostov department of the VOOPIiK.
103 "Appendix No. 2 to Preservation Order No. 115–10 for the cultural heritage object of regional significance 'Memorial complex to the memory of the victims of fascism in Zmievskaya Balka 1942,'" August 17, 2010, l. 3. Provided by the City Cultural Administration of Rostov-on-Don.
104 The first plaque was initially placed in the museum of the memorial in Zmievskaya Balka, where it lay on the floor for a few years. Nobody knew exactly what it was: an artefact or an ownerless object, which no one had the heart to throw away. The plaque is currently a part of the permanent exhibition of the Museum of Jewish Heritage and the Holocaust of the Memorial Synagogue at Poklonnaya Gora in Moscow. Interview with Aleksander Kozhin, in *author's archive*, code 16-MEM-RO-02, September 7, 2016, Rostov-on-Don, Russia.

memorial complex "In Memory of the Victims of Fascism" in Zmievskaya Balka was erected in 1975 according to the project of a creative team consisting of: architects: R.I. Murad'yan, N.N. Nerses'yants, sculptors: N.V. Avedikov, E.F. Lapko, B.K. Lapko. It was renovated in 2009.

This event caused serious discussions in the press, at numerous round tables and in public debate.[105] The RJC openly expressed its position: "It was unacceptable to erase all memory of the greatest tragedy of our time from future generations. It is important to warn the descendants about the danger of Nazi ideology and its consequences."[106]

Rostov's remembering community, represented by members of the Jewish community, public figures, and intelligentsia expressed shock at the unauthorised replacement of the memorial plaque, which in fact violated the order of the municipal authorities, the complete removal of any mention of the Holocaust in the text of the second plaque, and the fact that the reference to Jews was replaced by the term "peaceful Soviet citizens."[107] Thus, the remembering community felt that Russian society was stepping backward to Soviet times, when it was not permitted to openly mention Jews as specific victims of the Holocaust.

The conflict reached its peak in the spring of 2012, when Rostov lawyer Vladimir Livshits, a representative of the second generation of Holocaust victims in Rostov-on-Don, filed a lawsuit in the Kirovskii District Court against the City's Cultural Department, the Rostov Region Ministry of Culture and the City Administration. He demanded recognition that the removal of the first plaque was illegal. One of his most important arguments, which he repeated more than once during the almost six-month-long case was the following: "The statement

---

105 See, for example, Yanina Chevelya, "Privlech' mir k Zmievskoi Balke," *Rostov ofitsial'nyi*, November 23, 2011; Idem, "Delo doshlo do suda," *Rostov ofitsial'nyi*, May 30, 2012; Il'ya Shvets, "Vokrug memoriala v Zmievskoi Balke vsplyli novye prichiny konflikta," *Vechernii Rostov*, February 9, 2012; Grigorii Reikhman, "Zmievskaya Balka: 'svobodno ot evreev?,'" *Vesti: Israil' po-russki*, December 22, 2011; Ellen Barry, "Protest by Jewish Group over Memorial," *The New York Times*, January 25, 2012.
106 "Otkrytoe pis'mo o situatsii vokrug memoriala 'pamyati zhertv fashizma v Zmievskoi Balke,'" *Agentstvo evreyskikh novostei*, February 10, 2012, http://aen.ru/?page=brief&article_id=63353 (accessed December 4, 2019).
107 Al'tman considered the fact that the new text of the memorial plaque refers to "peaceful citizens" an act of "dejudification of the Holocaust," the main symbols of which were the permanent exhibition in the museum of Auschwitz until the late 1980s and the monument in Babi Yar of the mid-1970s. Il'ya Al'tman, "Mesto Kholokosta v rossiiskoi istoricheskoi pamyati," in Al'tman, *Pamyat' o Kholokoste*, 19. For more about the phenomenon of dejudification of the Holocaust in Poland and other countries of the former Soviet Bloc, see: Ronald J. Berger, *The Holocaust, Religion, and Politics of Collective Memory* (New Brunswick: Transaction Publishers, 2012), 163.

about the mass killing of peaceful citizens in Zmievskaya Balka without mentioning that they were Jews is a sophism, i.e., a well-considered and dissembled logical error based on a certain ideological approach. I would consider this kind of ideological approach to constitute a partial denial of the Holocaust through its understatement."[108]

It is notable that both sides of the lawsuit appealed to the same sources, primarily the ChGK records, but they interpreted them each in their own way. The defendants, recognising the fact that Jews were buried in Zmievskaya Balka appealed to the historical name of the memorial complex ("Victims of Fascism"), in other words, the burial place of "people of different ethnicities."[109] Another piece of evidence that the defendants brought to bear was the inaccuracy of the total numbers of the Jews killed. Indeed, the figure of 27,000 Jews mentioned on the first plague was too high. A ChGK report for Rostov-on-Don of February 17, 1943 says: "According to preliminary data, the number of Jews shot, poisoned, and tortured in Rostov-on-Don in the period from July 23, 1942 to February 13, 1943 amounts to 15,000–18,000 people."[110] ChGK Report No. 1231 for Zheleznodorozhnyi District of Rostov-on-Don of November 23, 1943 states:

> "[. . .] in Peschano-kamennyi kar'er [lit. Sand-Stone Quarry] more than 15,000 people were shot and buried in three graves [. . .] More than 10,000 Jews – men, women and children – were shot and buried in a grave in Balka on the edge of the grove which is 500 metres east of the Vtoraya Zmeevka village [. . .] More than 27,000 peaceful citizens of Rostov all together – men, women, children and POWs – were shot and buried near Vtoraya Zmeevka village.[111]

---

108 From Vladimir Livshits' statement of claim. Quoted after: Chevelya, *Zmievskaya Balka*, 245. This thesis was repeated many times by the claimant during the legal proceedings. I am grateful to Aleksandr Kozhin for providing a video of the proceedings.
109 According to Aleksandr Kharchenko, a Rostov historian, candidate of historical sciences, whom the City Cultural Administration hired to compile a historical background of the memorial complex, "what happened on the territory of Balka in 1942 is a tragedy for all the people of Rostov-on-Don. The memory of this tragedy is our collective memory. This is what the text of the information plaque installed in 2011 says. Trying to segregate any part of the victims from all the others, to distinguish it from others seems to me blasphemous. And it is doubly blasphemous to profit from our ancestors' blood in the year of the 70th Remembrance Day in Zmievskaya Balka, one of the most tragic pages of the Great Patriotic War." Aleksandr Kharchenko, "Vopros o doskakh stal delom printsipa," *Vechernii Rostov*, May 25, 2012.
110 "ChGK report for Rostov-on-Don of February 17, 1943," in *GARO*, f. 3613, op. 1, d. 450, l. 8.
111 "ChGK report No. 1231 for Zheleznodorozhnyi district of Rostov-on-Don of November 23, 1943," in *GARO*, f. 3613, op. 1, d. 30, l. 7–10. For an analysis of the existing quantitative data on Holocaust victims in Zmievskaya Balka, see Winkler, "Rostov-on-Don 1942," 105–130.

However, Livshits's claims were different. His main point was not only about the illegal dismantling of the memorial plaque, but also about the substitution of historical concepts, which he considered to be a manifestation of antisemitism and covert denial of the Holocaust at the official level in modern Russia. This produced a vicious circle, in which unresolved problems of the Soviet era became a sticking point in the development of tolerance in modern Russian society. Furthermore, the lawsuit and its heated discussion led to Jewish organisations being blamed for inciting ethnic hatred. Thus, in January of 2012 a round-table discussion of the status of the Zmievskaya Balka memorial took place in the press center of the Interfaks-Yug news agency with the participation of Cossacks, war veterans, representatives of patriotic, ethnic and other non-governmental organisations, researchers and local historians. Many participants were against "changing the status of the 'In Memory of the Victims of Fascism' memorial complex in Zmievskaya Balka as one of the biggest memorials of Russia dedicated to the tragic events of the Great Patriotic War."[112] Neo-Nazi organisations who were also involved in the conflict reacted. They demanded that the Prosecutor's Office investigate whether the RJC had ever incited ethnic hatred since "considering the Zmievskaya Balka memorial as a Holocaust site could lead to destabilisation of the society, to different uncontrolled destructive actions, and even to riots."[113]

The lawsuit was conducted exactly 70 years after the mass killings in Zmievskaya Balka. Many delegations from all over Russia and abroad came to commemorate the Holocaust victims instead of attending the on-going lawsuit. The Rostov Jewish community organised a March of the Living [*marsh zhivykh*],[114] a new documentary film *Suffering of Memory* [*Stradanie Pamyati*] about the history of the Holocaust in 1942 in Rostov-on-Don was shown, and the international conference "Lessons of the Holocaust and modern Russia," organised by the Holocaust Center took place during the commemorative week in early August 2012.[115]

---

112 Quoted in: Chevelya, *Zmievskaya Balka*, 281; Elena Bondarenko, "Kakova ona, istoricheskaya pravda 'Zmievskoi Balki'?," *Molot*, January 27, 2012.
113 The statement to the Prosecutor's Office of Rostov-on-Don was compiled by members of the public organisation "Russian Image" *["Russkii Obraz"]* and the Russian national cultural center "We are Russians" *["My russkie"]*. Chevelya, *Zmievskaya Balka*, 282.
114 The international campaign March of the Living was first held in 1988. Taking place annually on Yom Hashoah – Holocaust Remembrance Day – the March of the Living itself is a three-kilometre walk from Auschwitz to Birkenau as a tribute to all victims of the Holocaust. *International March of the Living official website*, https://motl.org/ (accessed March 23, 2020).
115 For information in the media on commemorative events in Zmievskaya Balka in August 2012 see: Yana Fomina, "Vkhod na Zmievskuyu Balku 70 let spustya," *Moskovskii komsomolets na Donu*, August 15–22, 2012.

Vladimir Livshits lost the lawsuit. On June 22, 2012, during the proceedings, Decree No. 471 of the Administration of Rostov-on-Don was issued, which cancelled the previous decree on installing first plaque. Thus, Livshits's claims could not be satisfied because the first plaque had lost its legal status, and the judgement was made in favour of the defendants. The judgement stated: "Nobody calls in question the fact that the memorial in Zmievskaya Balka is a place where peaceful citizens including Jews were annihilated. The memorial in Zmievskaya Balka commemorates all peaceful citizens, POWs and other categories of Soviet people.[116] In this connection, the plaintiff's arguments that the removal of information about the predominance of Jews among Nazi victims in Zmievskaya Balka is tantamount to Holocaust denial by the defendants, are untenable."[117]

However, a compromise was found.[118] On April 28, 2014 a third plaque was installed on the front wall of the Hall of Sorrow at the memorial complex. The Rostov-on-Don Special Commission for Naming Socially Significant Sites, Preserving the Names of Famous Persons and Memorable Events affirmed its text, which reads: "Here, in Zmievskaya Balka, in August of 1942 more than 27,000 peaceful citizens of Rostov-on-Don were killed including POWs. Among them there are representatives of many ethnicities. Zmievskaya Balka is the largest place of mass killing of Jews on the territory of Russia during the Great Patriotic War." Chief Rabbi of Russia Berl Lazar, President of the FJCR Aleksandr Boroda, leaders of the Rostov Jewish community, representatives of regional and city administrations, and thousands of Rostov Jews including relatives of Holocaust victims attended the unveiling ceremony of the third plaque.[119]

Unfortunately, the principal aim of Livshits, members of the community in Rostov, and many public figures "not to underline Jewish victims but to restore the ethnicity of Holocaust victims and to remind contemporary Russian society of Nazi crimes when the object of genocide could become any people"[120] was

---

116 In the 1960s, after Evgenii Evtushenko's poem *Babi Yar* was published, the main motif of public debates in the USSR was almost the same: "not only Jews were killed here." See: Edith W. Clowes, "Constructing the Memory of the Holocaust: The Ambiguous Treatment of Babii Yar in Soviet Literature," *Partial Answers* 3, no. 2 (2005): 153–182.
117 "The judgement of the Kirovskii District Court of Rostov-on-Don on case no. 2-3201/12 of October 15, 2012." I am grateful to Aleksandr Kozhin for providing a copy of this document.
118 The judgement of the Kirovskii court was appealed at a higher judicial level, which apparently led to some consensus. Interview with Kozhin.
119 "Kholokost: Rostov: Fakty," in Shvartsman, *Kniga Pamyati*, 24–25; Diana Volkova, "V Zmievke 'Primerili' Dosku," *Rostov ofitsial'nyi*, April 23, 2014.
120 These are the head of the RJC press service Mikhail Savin's words. Quoted in: "Memorialu Kholokosta ne khvatilo nuzhnykh spravok," *project website of the Society "Memorial"–"Uroki istorii XX vek,"* January 24, 2012, https://urokiistorii.ru/article/2874 (accessed March 23, 2020).

not achieved. Holocaust memory remains the private memory of remembering communities within Russian society. The passive position of most Rostov citizens – the legal proceedings were open for all citizens and was widely broadcasted, but no protest action was registered either in Rostov-on-Don or in other cities in Russia – is a sign of indifference and implicit acceptance of the official concept of the "Great Patriotic War," which does not include the Holocaust.

4.3 Monuments as Social Agents in Translating the Memory of Holocaust Victims

Any monument has several meanings. This is especially true for those commemorating war. Initially it may be constructed to carry out a historical task, that is, to mark a site of glory (victory in a battle, feats of soldiers, etc.) or mourning (of killed soldiers or civilians). But then a monument assumes its place within the landscape of a city or village. Local citizens and tourists interact with it and around it. As Halbwachs has argued, it is primarily through membership in religions, national, or class groups that people are able to acquire and then recall their memories at all.[121] That is, both the reasons for and the forms of memory are always socially mandated. It is part of a system of socialisation, whereby fellow citizens gain common history through the vicarious memory of their forbears' experiences. In this sense monuments become social agents. They become mediators, translating historical memory of the event to future generations. Bearers of social or collective memory use them as a matter of deference and the central object of commemorative acts. Memorials provide sites where groups of people together create a common past for themselves and retell the constitutive narratives and 'shared' stories of their past.[122] Monuments at Holocaust killing/grave sites have been used in various commemorative acts since the end of the Second World War. According to the form and content of such practices three basic models of commemoration can be defined.

First is the traditional religious Jewish commemorative model. It includes the reading of Jewish prayers, *El Malei Rachamim* [God full of Mercy] and *kaddish* (hymns in praise of God) on Remembrance Day (as a rule on the *Yahrtzeit*, which in Yiddish means "anniversary" and refers to the day on which a person passed away) in synagogues and also at killing/grave sites, and the lighting of

---

[121] See: Maurice Halbwachs, *Das Gedächtnis und seine sozialen Bedingungen* (*Les cadres sociaux de la mémoire*) (Frankfurt am Main: Suhrkamp, 2012).
[122] Young, *The Texture of Memory*, 6–7.

memorial candles at sites and at homes.¹²³ The main agents of this commemorative model are individuals, mainly members of Jewish communities, who have come together because of emotional bonds between community members and a killing site. It is not necessary that participants at the commemoration be relatives, more likely they can be thought of, in Jay Winter's term, as "fictive kinship groups." Such people were the key agents of Holocaust remembrance in the USSR.¹²⁴ However, as discussed above, no traditional religious Jewish commemorative acts were allowed in the post-war North Caucasus. In modern-day Russia, this commemorative model is not often seen, as religious ceremonies are accompanied by secular practices in most cases.¹²⁵

Second is the Soviet commemorative model, according to which mass gatherings at memorial sites took place on Victory Days all over the USSR and in modern-day Russia. The activities related to monuments for heroes and victims of the war were strictly ritualised. The presence of state representatives was obligatory. They usually gave a speech, addressed to the past (about the "cruelty of the war," "the importance of the Soviet Soldier's Feat," "the sorrow of peaceful citizens") and also with a nod to the future ("it is a great joy to live in the country which defeated fascism"). The higher the rank of the functionary giving the speech, the more important the gathering was. A Guard of Honour comprised of Pioneers, who kept vigil over the Eternal Flame, was an obligatory element of any commemorative event. At the end of the ceremony, a minute of silence was held in honour of the dead and wreaths and flowers were placed on the monument. Soviet memorial culture remembered all victims of the war in the context of the heroic struggle of every fallen person for the Soviet state. The absolute domination of this patriotic conception of the war did not leave room for other facts – such as the Holocaust, the fates of POWs, Stalin's deportation of peoples – to be incorporated into official remembrance of the war. Such annual mass gatherings were carried out near grand memorial complexes. The

---

123 Zeltser, "Pamyatniki Kholokostu v SSSR," 143.
124 Jay Winter, "Remembrance and Redemption"; Idem, *Sites of Memory, Sites of Mourning: The Great War in European Cultural History* (Cambridge: Cambridge University Press, 1995).
125 The recital of the mourner's *Kaddish* prayer by a rabbi is followed by a secular gathering with children (address to the future) and authorities (a symbol of the importance of the event) in attendance. For example, a public gathering was held on May 8, 2015, the day before Victory Day in the village of Novominskaya, Krasnodar Krai. The reason was the unveiling of a renovated Soviet monument, where Holocaust victims' names had been added. Members of the Jewish religious community of Krasnodar took part in the gathering. They read a prayer, and then participants listened to a speech of the headman of the village. At the end of the event, the participants placed flowers and wreaths on the monument. "Sobytiya. Vremya mestnoe," *Kanevskaya TV*, aired Mai 13, 2015.

actual killing/grave sites were excluded from official Soviet ceremonies because of the difficulties of accessing them.[126] Urban centers and rural hinterlands lived independently from each other. Whereas monuments created by the Soviet government to commemorate war heroes were not only incorporated into official memorial culture but also appeared in life of the average Soviet citizen as sites of memory, pride and grief, the actual sites of the Holocaust were forgotten and abandoned. Sometimes they were erased entirely, as urbanisation led wartime grave sites to be covered by the asphalt of new streets and roads.

Third is the mixed commemorative model, typical for modern Russian society. Rabbis can read prayers,[127] light candles, and carry out a religious ceremony according to Jewish tradition while the gathering as a whole is secular, with the state officials, honoured guests, and schoolchildren, the observation of a minute's silence and the placing flowers, wreaths, and small stones on the monument.[128] Mass gatherings have important patriotic and educational meanings. The speeches given reflect contemporary state ideology and the official positions of the government and different social groups on remembrance of the past. It is notable that one can hear mention of the killing of Jews more often nowadays at such gatherings. The opportunity to speak is given to witnesses of the tragedy, who become honoured agents of Holocaust commemoration along with state representatives.[129] Thus, a permanent participant of nearly all gatherings held in Mineralnye Vody next to the monument by the glass factory is a pianist and World War II veteran, Emil Zigel', whose parents were killed at that site. Despite his advanced years, he still comes from Israel to Stavropol Krai each September to

---

126 See information about the gathering held around monuments "To the victims of Fascism" in Soviet times: Anatolii Sorochan, "Nepokorennye," *Komsomolets Kubani*, May 9, 1975; Panyutin, "Pamyatnik nepokorennym"; "Torzhestva na Stavropol'e, posvyashchennye 30-letiyu velikoi pobedy," *Stavropol'skaya pravda*, May 10, 1975.

127 When an event is held, both a rabbi and an Orthodox priest can sometimes read a memorial prayer one after the other, since the Nazi victims may not have been only Jewish. For example, representatives of mixed marriages and Soviet POWs are buried in Zmievskaya Balka. "Svet protiv t'my," *Rostov ofitsial'nyi*, August 15, 2012.

128 Viktoriya Aleksandrova, "Na Stavropol'e vspominali zhertv Kholokosta," *Stavropol'skaya pravda*, September 12, 2012; "Druzhba narodov prevyshe vsego," *Vremya: gazeta goroda Mineralnye Vody*, September 16, 2015; Anatolii Dagilov and Zalman Mataev, "Evreiskaya obshchina Nal'chika pochtila pamyat' zhertv Kholokosta v Bogdanovke," *Gazeta evreev Severnogo Kavkaza*, May 9–23, 2012.

129 As a rule, the relatives of Holocaust victims have an honorary right to open a new monument. See: "Pamyatnik zhertvam fashizma," *Maykopskie novosti*, October 26, 2006.

remember his relatives and acquaintances.[130] Such a tradition is popular among many elderly Holocaust survivors, who summon the strength to make annual visits to the grave site of their relatives. For example, Holocaust survivor Sergei Amiramov visited his relatives' grave in Bogdanovka every year on May 9, as "the authorities promoted and held gatherings there."[131] Another Holocaust survivor, Semen Dikenshtein, visited Millerovskii district of Rostov Oblast, every year as well, becoming a fixture of the village administrations' commemorations at Ol'khovyi Rog Holocaust site: "They all treated me very well there, organised a journey to this monument. Veterans, children, pioneers, schoolchildren went there, too. It was like a ritual. We all went there together with the village headmen."[132] Acting as what Latour terms "quasi-objects, quasi-subjects," monuments are examples of a premodern cultural hybrid that modernity in its most powerful phases attempted to purify and neuter. Latour found that hybrids trace and enact social networks, thus revealing the intricate relations that operate around and through the monument.[133] Annual visits of Holocaust survivors to the grave site of their relatives and the above-mentioned acts of vandalism on monuments to Holocaust victims show a wide spectrum of social practices of commemoration as well as varying roles and meanings of Holocaust monuments for the contemporary society.

There are also special forms of commemorative activity in some cities and villages, which specifically relate to Holocaust remembrance. Thus, in the village of Arzgir, a Torchlight Procession [*fakel'noe shestvie*] of senior schoolchildren to the monument, which was known among villagers as the "monument to the Jews," occurred annually even during Soviet times.[134] This tradition was revived in 2015, on the 70th anniversary of the Great Victory. As reported on the municipality's website, "schoolchildren of the Patriot club headed the torchlight procession; they were followed by more than 150 torch bearers and more than 100 procession participants. Especially symbolic was that the procession passed the same way as the Nazi victims in 1942 before their execution."[135] The March of the Living took place for the first time in August of 2012 in Rostov-on-

---

[130] Dmitrii Triandafilidi, "Bud' proklyata voina!. . .," *Vremya: gazeta goroda Mineralnye Vody*, September 12, 2012; Irina Borisova, "Blokade, Kholokostu i 'Osventsimu' posvyashchaetsya . . .," *Vremya: gazeta goroda Mineralnye Vody*, January 31, 2015.
[131] Interview with Sergei Amiramov, born in 1937, in *VHA USCSF*, code 39735, January 12, 1998, Nalchik, Russia.
[132] Interview with Dikenshtein.
[133] Bruno Latour, *We Have Never Been Modern* (New York: Harvester Wheatsheaf, 1993).
[134] Interview with Elena Okhmat, in *author's archive*, code 15-MEM-SK-06, May 17, 2016, village of Arzgir, Stavropol Krai, Russia.
[135] "Fakel'noe shestvie," *news on the website of Arzgir municipal district*, May 10, 2015, http://arzgiradmin.ru/news.htm, (accessed March 24, 2020).

4.3 Monuments as Social Agents in Translating the Memory of Holocaust Victims — 141

Don thanks to the efforts of members of the Jewish community. A procession of people with yellow stars on their sleeves walked the Road of Death from Rostov Zoo to the monument in Zmievskaya Balka. "During the walk different stories, recollections, and sorrow over those for whom this road became their last were heard."[136] Since then the March of the Living has been carried out annually in Rostov-on-Don. In 2017, it was held in other Russian cities and villages for the first time.[137] The participants of such processions regardless of their age, ethnicity, and religion "all walk together the path of sorrow and grief to commemorate victims of the Holocaust." These processions, as a form of commemoration, have a symbolic meaning. Walking the path of Jews doomed to death by other people during peacetime is simultaneously a public expression of grief and solidarity against the Holocaust and future genocides.

The dates when memorials in Russia are visited have changed over time. During the post-war period, they were visited on the anniversaries of dates when Jews had been killed (these dates were still remembered by relatives and friends of Holocaust victims). During the Soviet period, memorials were always visited on February 23 (Day of the Soviet Army and Navy), on May 9 (Victory Day), less often on November 7 (Day of the October Revolution), and hardly ever on the days of the liberation of cities/towns and villages or on June 22 (the beginning of the Great Patriotic War). In post-Soviet Russia, May 9 and less often February 23 are the most important dates for mass gatherings near monuments of war heroes and victims. Activities dedicated to the anniversary of Victory Day are carried out over several days. The so-called "memory routes" increasingly include not only official monuments but also monuments at Holocaust sites, which are usually situated far away from the city or regional center.[138] Jewish communities and sometimes local

---

136 Fomina, "Vkhod na Zmievskuyu Balku."
137 Mumin Shakorov, "'Marsh zhivykh' v Rostove-na-Donu," *Radio Svoboda*, August 10, 2017, https://www.svoboda.org/a/28672067.html (accessed March 23, 2020). The RJC applied to regional Jewish communities in Russia to organise the March of the Living. The processions took place in Arzgir, Kislovodsk, Mineralnye Vody, and Stavropol (in the North Caucasus) and in Kaliningrad, Novozybkov (Bryansk Region), Orel, and other cities. "'Marshi zhivykh' proidut po vsei Rossii k 75-letiyu tragedii v Zmievskoi Balke," *RIA Novosti*, August 8, 2017, https://ria.ru/religion/20170808/1499975998.html (accessed March 23, 2020); Nadezhda Klimchenko, "'Marsh zhivykh' v pamyat' o zhertvakh massovykh rasstrelov v Arzgirskoi Balke," *Zarya: obshchestvenno-politicheskaya gazeta Arzgirskogo raiona Stavropol'skogo kraya*, September 12, 2017.
138 See, for example, the programme of events for May 9, 2017 in Krasnodar and Stavropol: "Programma meropriyatii na 9 maya 2017 goda v Stavropole," *Komsomolskaya pravda*, April 24, 2017, https://www.stav.kp.ru/daily/26670.5/3692478/ (accessed March 23, 2020); "Krasnodar: prazdnichnye meropriyatiya s 5 po 9 maya 2017 goda – Den' Pobedy," 3 May, 2017, *website of*

authorities also organise gatherings near monuments on International Holocaust Remembrance Day (January 27),[139] on the Remembrance Days of the mass killing of Jews in a particular locality (for example, in Stavropol Krai at the beginning of September and in Rostov-on-Don on August 11–12).

Soviet monuments erected in city centers or parks have eventually become places of relaxation rather than sites of sorrow and mourning. At the same time, monuments to Holocaust victims on the outskirts of a city/village or in the fields/woods are rarely incorporated into the socially developed space; they generate different meanings from those situated in a city neighbourhood. In this sense, any memorial marker in the landscape, no matter how alien to its surroundings, is still perceived in the midst of its geography, in some relation to the other landmarks nearby.[140] In 2015–2016, I conducted an indicative online and paper survey in three big cities of the North Caucasus: Rostov-on-Don, Krasnodar, and Stavropol. The aim of the survey was to clarify what citizens know about monuments dedicated to World War II and especially about monuments to victims of the Nazis. The other goal of the study was to find out how war monuments are included into the daily life of citizens, how, when, and what kinds of interaction may have taken place. One block of questions concerned the history of the erection of three Soviet memorial complexes – Zmievskaya Balka in Rostov-on-Don, "Kholodnyi Rodnik" in Stavropol, and "To the Victims of Fascism" in Krasnodar – and their subsequent incorporation into the commemorative culture of contemporary citizens. People of different professions, gender and age took part in the survey, which received responses from a total of 500 participants. The results of the survey confirmed my hypothesis. Citizens know only the large, official monuments in their cities, which often become the topics of news in the regional media or are situated in their neighbourhood. If monuments to Holocaust victims are on the outskirts (as for example with the two monuments near Urochishche Stolbik in Stavropol or the "To the Victims of Fascism" monument in Moskovskaya Street in Krasnodar) and are not incorporated into official memory of the Great Patriotic War, the average citizen is not aware of them. The "To the Victims of Fascism" memorial complex in Pervomaiskaya Roshcha in Krasnodar and "Kholodnyi Rodnik" in Stavropol have nowadays morphed into recreation zones for young people and adults. The proximity of these parks attracts citizens keen to escape the bustle of the main roads and enjoy the greenery. There are also

---

news and events in Gelendzhik and Krasnodar Krai, http://yug-gelendzhik.ru/krasnodar-prazdnichnye-meropriyatiya-s-5-po-9-maya-2017-goda-den-pobedy/ (accessed March 23, 2020).
139 Inna Kalinicheva, "27 yanvarya – mezhdunarodnyi den' pamyati zhertv Kholokosta," *Pul's litseya g. Mineralnye Vody* 5, January 2015.
140 Young, *The Texture of Memory*, 7–8.

attractions for children and cafés in Pervomaiskaya Roshcha. Thus, the monument is no longer a remembrance site but rather a convenient place for recreation.

There is a slightly different situation in Rostov-on-Don. The memorial complex in Zmievskaya Balka is near the road leading out of the city and there is no park there, which is why it fulfils its primary function as a memorial site to Nazi victims. The 2012 lawsuit over the memorial plaque, which was widely broadcast in the city's media, also promoted the historical education of citizens. Rostov-on-Don had the highest proportion of respondents who explicitly linked their city's memorial to the history of the Holocaust. 73 per cent of respondents identified the memorial in Zmievskaya Balka as "dedicated to Jewish locals and evacuees killed by Nazis." On the contrary, respondents in Stavropol and Krasnodar associate the monuments in their cities only with the killing of "peaceful citizens." In fact, neither of the monuments in Krasnodar and Stavropol are erected at the sites of mass killing of Jews and other civilians. Nevertheless, the official Russian ideology still does not distinguish different groups of Nazi victims, using the Soviet euphemism "peaceful citizens," which also dominates the rhetoric of many Russian citizens.

The majority of respondents in all three cities visit memorials only on May 9, primarily as participants of the Victory Day events. Only 18 per cent of respondents in Rostov-on-Don visit Zmievskaya Balka on International Holocaust Remembrance Day or on the local Days of Remembrance, August 11–12. For comparison, only 2 per cent of respondents in Krasnodar visit the "To the Victims of Fascism" monument on International Holocaust Remembrance Day. Curiously, Krasnodar also has the highest percentage of respondents who do not know what the Holocaust is (18 per cent). The results reflect the regional authorities' attitude to the memory of Holocaust victims: they are still not distinguished from the general category of "peaceful citizens." Thus, Soviet monuments still have the seal of Soviet ideology; they seldom become sites of mass gatherings dedicated to the commemoration of Holocaust victims. Holocaust memory continues to be private and mostly Jewish, which is more or less noticeable in the public life of cities with a strong Jewish community (in Rostov-on-Don). The Soviet practice of not distinguishing any ethnic group from the homogenous categorisation of Nazi victims remains current in contemporary Russia, which actually leads to the falsification of historical facts. The majority of today's Russian citizens are unconscious bearers of this official state ideology concerning the Great Victory. At the same time, memorials to Holocaust victims as well as Soviet memorial complexes in the North Caucasus are oriented only to local remembering communities, and have not become transnational commemorative sites. In most cases they are still unknown to international Holocaust remembering communities and therefore excluded from tourist "memory routes." Despite being the largest site of the Holocaust within the borders of modern-day Russia, Zmievskaya Balka still has no worthy place in the national culture of Holocaust remembrance.

Scholarly definitions of certain Holocaust sites as "a negative place of memory,"[141] or a "counter-memorial"[142] reflect a special functional directionality. These are not merely memorials related to a historical trauma (sites for the work of mourning by the survivors), but also places with a peculiar moral-pedagogical function, embodying the slogan: "Never again!"[143] But such a perception of the Holocaust is a characteristic of West European, Israeli, and American cultures. Russian memorials, be they Soviet or not, have not been oriented towards this "Never again!" narrative. They fix memories of these sites not as much as those of the Holocaust, if at all, but rather as sites of the mass killing of peaceful Soviet citizens. These, according to the rhetoric of Soviet and later Russian authorities, obtain the characteristics of heroic victims, without reference to the Nazi racial ideology which led to the targeted annihilation of most of those victims. The heroisation intrinsic to post-Soviet memorial culture is directed inwards, i.e., to the glory of the Soviet warrior, citizen, and patriot, but not outwards in an appeal to universal humanity not to repeat the crimes of the Nazis.

---

141 Reinhart Koselleck, "Formen und Traditionen des Negativen Gedächtnisses," in *Verbrechen Erinnern. Die Auseinandersetzung mit Holocaust und Völkermord*, ed. Volkhart Knigge and Norbert Frei (Bonn: BpB, 2005), 21.
142 Young, *The Texture of Memory*.
143 Koselleck, "Formen und Traditionen"; Sandra Petermann, *Rituale machen Räume: Zum kollektiven Gedenken der Schlacht von Verdun und der Landung in der Normandie* (Bielefeld: Transcript, 2007), 26.

5 Depicting Jewish Fates: Artistic Responses to the Holocaust in the North Caucasus

Dear Irina,
I am pleased to receive your email! I am always glad of any attention to the poem "Occupation," because I would like it not to die with me but to teach people to confront any war, big or small.

I worked as a chemistry teacher in the village school in Bratskii, Ust-Labinsk District [Krasnodar Krai]. Once I was asked to write a poem for a Victory Day gathering. I talked to our old handicraft teacher, Petr Samuilovich Sidorenko, and, according to his story, wrote a small poem describing the liberation of our farm from the Nazis. The poem was read at the gathering and my name was mentioned as the author. After that people came up to me and told their stories about the occupation. Everybody wanted his or her story to be described in verse too. The short poem began to grow, and grew into a long one. Most dramatic was the story told by Vladimir Ivanovich Vinnikov from the Novoselovka farm, located near Bratskii and belonging to Bratskii rural settlement. He spoke of the shooting of 137 Soviet citizens of Jewish origin. I tried to retell his story almost word for word [. . .] In 2004, when the poem "Occupation" had already assumed its current form, an organiser of after-school activities and I staged it at the school. The actors were our pupils. We even made a scale model of the hut onto the roof of which the boys climbed and gazed into the audience, as if the shooting took place there. The play was shown to the district school principals during their meeting at the school. These very manly people cried, and afterwards one of them stole the script. It only made me happy: it meant that "Occupation" would be read elsewhere. I have been retired since 2005, but every year before the anniversary of the farm's liberation I am invited to the school to read the poem. I am glad to see the children listen attentively; some of them cry. There are no conclusions in the poem. But I hope each person will draw them for him or herself.[1]

I received this email in July 2015 from Svetlana Chechulina, the author of the poem "Occupation." I had to gather fictional responses to the Holocaust in the North Caucasus bit by bit. As a rule, the majority of them were issued in a limited edition or, as in the case of "Occupation," published on the Internet on special literary portals. One of the most successful ways to find these sources was to seek out regional authors and enter into correspondence with them.

Artistic responses to the Holocaust, just like its history, were first studied on the basis of Western exemplars.[2] Until recently, the acceptance of Babi Yar as a

---

[1] Svetlana Chechulina, email to the author, July 31, 2015.
[2] For literary responses to the Holocaust in general, see: Lawrence L. Langer, *The Holocaust and the Literary Imagination* (New Haven: Yale University Press, 1975); Berel Lang, ed., *Writing and the Holocaust* (New York: Holmes and Meier, 1988); Saul Friedlander, ed., *Probing the Limits of Representation: Nazism and the "Final Solution"* (Cambridge: Harvard University Press, 1992); Idem, ed., *Holocaust Literature: A Handbook of Critical, Historical, and Literary Writings* (Westport:

symbol of the Holocaust in the USSR remained in the shadow of the monstrous machine of Auschwitz. Holocaust fiction that described killings in the ravines in broad daylight and in the middle of villages did not conform to the classic model of Holocaust testimony. They were not able to destabilise the notions implicit in "Holocaust literature," the "literature of destruction," or "testimonial literature." Unlike the "concentration camp literature" which describes deportation to distant, closed and guarded places, the "literature of ravines" reflects the transformation of native and friendly landscapes into landscapes of death. It breaks the clichés of Western discourse, which associates the Holocaust with mass killings leaving neither witnesses nor traces. In contrast to the impersonal, industrialised death at Auschwitz, death in the ravines was personal and intimate.³ Likewise, neither world nor Soviet cinema (feature films and documentaries alike) has paid any attention to the history of the Holocaust on Soviet soil. The reasons are not difficult to understand. First, the perpetrators hardly ever filmed their massacres (they occasionally took snapshots, but not everywhere). Second, the Soviet cameramen who filmed the excavations of killing sites after their liberation did not focus on the Jewishness of the victims, but rather obscured their Jewish identities. Nevertheless, many acclaimed Soviet writers and poets, composers and directors, including some of Jewish origin, made significant contributions to

---

Greenwood Press, 1993); Alan Rosen, ed., *Literature of the Holocaust* (Cambridge: Cambridge University Press, 2013); Jenni Adams, ed., *The Bloomsbury Companion to Holocaust Literature* (London: Bloomsbury, 2014). For visual responses to the Holocaust, see: Barbie Zelizer, *Remembering to Forget: Holocaust Memory through the Camera's Eye* (Chicago: University of Chicago Press, 1998); Janina Struk, *Photographing the Holocaust: Interpretations of the Evidence* (London-New York: I.B. Tauris, 2004); Janet Walker, *Trauma Cinema: Documenting Incest and the Holocaust* (Berkeley: University of California Press, 2005); Lawrence Baron, *Projecting the Holocaust into the Present: The Changing Focus of Contemporary Holocaust Cinema* (Lanham: Rowman and Littlefield, 2005); Idem, "The First Wave of American 'Holocaust' Films, 1945–1959," *American Historical Review* 115, no. 1 (2010): 90–114; Aaron Kerne, *Film and the Holocaust: New Perspectives on Dramas, Documentaries, and Experimental Films* (New York: Continuum, 2011); Judith B. Kerman, ed., *The Fantastic in Holocaust Literature and Film: Critical Perspectives* (Jefferson: McFarland, 2015). For music in the Holocaust, see for example: Shirli Gilbert, *Music in the Holocaust: Confronting Life in the Nazi Ghettos and Camps* (New York: Oxford University Press, 2005); Matthew Boswell, *Holocaust Impiety in Literature, Popular Music and Film* (Basingstoke: Palgrave Macmillan, 2012); Moshe Avital, "The Role of Songs and Music during the Holocaust," *Journal of Jewish Music and Liturgy* 31 (2011–2012): 51–60.

3 Assia Kovriguina, "Le Témoignage en URSS: Qui est l'Auteur ?" *Fabula / Les Colloques, Témoigner sur la Shoah en URSS*, http://www.fabula.org/colloques/document2856.php (accessed June 29, 2017).

the artistic response to the Holocaust.⁴ Most of these pieces are about the life and death of Jews in the ghettos on the territory of modern Ukraine and Belarus or in the concentration and death camps of Eastern Europe. Published uncensored abroad, many pieces by Soviet authors reached readers in the homeland only after the dissolution of the USSR. Scholars have only recently turned to the study of Holocaust imagery of the Soviet period.⁵ Olga Gershenson, who specialises in

---

4 For the most famous Soviet literary responses to the Holocaust, see: Il'ya Selvinskii, "Ya eto videl!," *Bol'shevik*, January 23, 1942, reprint in *Krasnaya Zvezda*, February 27, 1942; Il'ya Erenburg, *Burya: Roman* (Magadan: Sovetskaya Kolyma, 1947); Evgenii Evtushenko, "Babii Yar," *Literaturnaya gazeta*, September 19, 1961; Anatolii Kuznetsov, *Babii Yar: roman-dokument* (Moscow: Molodaya gvardiya, 1967); Anatolii Rybakov, *Tyazhelyi pesok: roman* (Moscow: Sovetskii pisatel', 1979); Vasilii Grossman, *Treblinskii ad* (Moscow: Voennoe izdatel'stvo, 1945); Idem, *Zhizn' i sud'ba: roman* (Kuybyshev: Knizhnoe Izdatel'stvo, 1990). For the most important Soviet feature films and documentaries, see: *Nepokorennye*, directed by Mark Donskoi (Kievskaya kinostudiya, 1945); *Gott mit uns*, written by Grigorii Kanovich and Vitautas Zhalakyavichyus (1961): the script was banned by the censors; *Obyknovennyi fashizm*, directed by Mikhail Romm (Kinostudiya "Mosfil'm," 1965); the TV series *"Tyazhelyi pesok,"* directed by Anton and Dmitrii Barshchevskii (Kinokompaniya "Risk," 2008); *Tufel'ki*, directed by Konstantin Fam (Costa Fan Production et al., 2012); *Rai*, directed by Andrei Konchalovskii (DRIFE Productions i Prodyuserskii Tsentr Andreya Konchalovskogo, 2016). For a selected filmography of Soviet and world feature films and documentaries, see: "Metodicheskie materialy," *Holocaust Center official website*, http://holocf.ru/образовательная-программа/методические-материалы/ (accessed February 11, 2020). For the most famous musical works, see: Dmitri Klebanov, *Symphony no. 1 "Pamyati muchenikov Bab'ego Yara,"* 1945; Dmitri Shostakovich, *Symphony no. 13,* 1962; Evgenii Stankovich, *Kadish-rekviem "Babii Yar,"* 1991; Igor' Levin, *Poem "Bol' zemli for symphony orchestra, mixed and children's choruses,"* 2014.
5 For an analysis of literary responses to the Holocaust in the USSR, see: Il'ya Al'tman, Alla Gerber, and Dmitrii Prokudin, ed., *Kholokost v russkoi literature: sbornik urokov i metodicheskikh rekomendatsii* (Moscow: Tsentr i fond "Kholokost," 2006); Harriet Murav, *Music from a Speeding Train: Jewish Literature in Post-Revolution Russia* (Stanford: Stanford University Press, 2011); Maxim D. Shrayer, *I Saw it: Il'ya Selvinskii and the Legacy of Bearing Witness to the Shoah* (Boston: Academic Studies Press, 2013); Annie Epelboin and Assia Kovriguina, *La Littérature des Ravins: Écrire sur la Shoah en URSS* (Paris: Robert Laffont, 2013); Leona Toker, "The Holocaust in Russian Literature," in Rosen, *Literature of the Holocaust*, 118–127; Idem, "On the Eve of the Moratorium: The Representation of the Holocaust in Il'ya Ehrenburg's Novel The Storm," in Marat Grinberg et al., *Representation of the Holocaust in the Soviet Literature and Film,* Search and Research 19 (Jerusalem: Yad Vashem, the International Institute for Holocaust Research, 2013), 37–58; Il'ya Kukulin, "Sovetskaya poeziya o Vtoroi mirovoi voine: ritorika skrytoi amal'gamy," in *SSSR vo Vtoroi mirovoi voine: okkupatsiya: Kholokost: Stalinizm*, ed. Oleg Budnitskii and Lyudmila Novikova (Moscow: ROSSPEN, 2014), 328–351; Idem, "Russian Literature on the Shoah: New Approaches and Contexts," *Kritika: Explorations in Russian and Eurasian History* 18, no. 1 (Winter 2017): 165–175. For an analysis of Soviet and Russian feature films and documentaries on the Holocaust, see: Jeremy Hicks, *First Films of the Holocaust: Soviet Cinema and the Genocide of the Jews, 1938–1946* (Pittsburgh: University of Pittsburgh Press, 2012); Idem,

Jewish cultural studies, highlighted two principal features of Soviet Holocaust films, which can be generalised to all Soviet art. These are 1) universalisation – the Holocaust was regarded as part of a universal Soviet tragedy with Jews euphemistically labelled "peaceful Soviet citizens" – and 2) externalisation, when crimes against Jews were treated as such but the Holocaust was set outside the borders of the Soviet Union.⁶ During the last decades, on a par with professional artists, representatives of the second generation and Holocaust survivors themselves as well as talented and sympathetic citizens such as Svetlana Chechulina have striven to interpret again and again the tragedy of Soviet Jews during World War II, to tell this story to future generations, through the medium of artistic representation.

An important aspect of my research to portray the Holocaust memory dispositif in the North Caucasus, therefore, consisted in searching for elements of this artistic response and (when possible) for the authors themselves. This task was complicated by the fact that such literary works are, as a rule, published in very limited print runs and rarely spread beyond a single district, or at most a region or a North Caucasian republic. Collections of poems or stories were often published by authors at their own expense and rarely found a broad readership. Documentaries shot in post-Soviet times had a similar fate: they were rarely broadcast on local and central television channels or screened in cinemas. Only now and then (and not in all regions of the North Caucasus) a film screening

---

"Otrazhenie Kholokosta v sovetskikh dokumental'nykh fil'makh voennogo vremeni i ego vliyanie na pamyat' o zhertvakh voiny," in Al'tman et al., *Kholokost: 70 let spustya*, 178–182; Olga Gershenson, *The Phantom Holocaust: Soviet Cinema and Jewish Catastrophe* (New Brunswick: Rutgers University Press, 2013); Idem, "The Holocaust on Soviet Screens: Charting the Map," in Rosen, *Literature of the Holocaust*, 101–121; Idem, "Dobrovol'naya amneziya: Kholokost v sovremennom rossiiskom kino," in Al'tman, *Kholokost: 70 let spustya*, 183–191; Valérie Pozner, ed., *Filmer la Guerre 1941–1946: les Soviétiques face à la Shoah: Mémorial de la Shoah: à l'occasion de l'Exposition "Filmer la guerre: les Soviétiques face à la Shoah 1941–1946" 9 janvier – 27 septembre 2015, conçue par le Mémorial de la Shoah* (Paris: Mémorial de la Shoah, 2015). See also an analysis of Soviet photographers' and artists' works: David Shneer, *Through Soviet Jewish Eyes: Photography, War, and the Holocaust* (New Brunswick: Rutgers University Press, 2012); Marina Genkina, "Jewish Artists in the Soviet Union in the 1960s–1970s," in *Jews of Struggle: The Jewish National Movement in the USSR, 1967–1989*, ed. Rachel Schnold (Tel Aviv: Beit Hatefutsoth, 2007), 80–104; "Zhizn' i iskusstvo Zinoviya Tolkacheva," Yad Vashem official website, https://www.yadvashem.org/ru/education/educational-materials/lesson-plans/tolkachev.html (accessed February 11, 2020).

6 Gershenson, *The Phantom Holocaust*, 2.

with a public discussion may be organised during commemorative events.[7] At the same time, works published on the Internet now receive a broader audience than works on paper or CD.[8] Imagery becomes an integral element of the Holocaust memory dispositif: some literary responses to the Holocaust in the region become a mandatory part of a school's literature curriculum, some are quoted during after-school activities, commemorative events, and official gatherings near Holocaust monuments, and some are specially written for feature films and documentaries. Published in print and on the Internet, these works of art become subject to discussion by remembering communities, prolonging the life of Holocaust remembrance among contemporary generations. Given that the Holocaust lies outside the actual experience of today's audiences and readers, its imagery becomes a way of living it vicariously, and in the absence of actual recall, forming what memory studies scholar Alison Landsberg terms "prosthetic memories."[9]

My research identified the layers of imagery (mostly in works of fiction and documentaries) in which authors interpreted Holocaust history and/or described the fates of survivors in particular localities of the North Caucasus. Taking into account the common method of chronologically ordering literary responses to the Holocaust and identifying similarities between them in different periods of Russian history, both during the existence of the USSR and afterwards,[10] I conduct a thematic analysis of different types and genres of imagery relating to the

---

7 For example, Yurii Kalugin's film *Stradanie pamyati* about the events in Zmievskaya Balka was shown in the context of the commemorative events in Rostov-on-Don in 2012. Fomina, "Vkhod na Zmievskuyu Balku." In Taganrog and Rostov-on-Don, city public libraries often arrange special film screening, where they show feature films and documentaries about the Holocaust and then discuss them. See: "Vecher pamyati zhertv Kholokosta," *Don State Public Library official website*, September 16, 2007, http://www.dspl.ru/news/5212.html (accessed March 23, 2020); "Kinoprem'era Yuriya Kalugina," *Ibid*, August 7, 2013, http://www.dspl.ru/news/6376.html (accessed March 23, 2020); "Kholokost – klei dlya oboev," *Taganrog Central City Public Library official website,* August 29, 2013, http://www.taglib.ru/news/Holokost_-_klei_dlya_oboev.html (accessed March 23, 2020); "Prosmotr fil'ma Mumina Shakirova 'Kholokost – klei dlya oboev'," *Ibid*, 10 February 2014, http://www.taglib.ru/news/Prosmotr_filma_Mymina_Shakirova_ «Holokost_-_klei_dlya_oboev».html (accessed March 23, 2020).
8 For example, the websites Proza.ru and Stihi.ru have become widely known in Russia. Talented authors present their pieces here reflecting among others the theme of the Holocaust.
9 Alison Landsberg, *Prosthetic Memory: The Transformation of American Remembrance in the Age of Mass Culture* (New York: Columbia University Press, 2004).
10 Toker, "The Holocaust in Russian Literature"; Gershenson, *The Phantom Holocaust*.

Holocaust in the North Caucasus.[11] I am more interested here in understanding the ways in which the Holocaust is represented by the authors, readers, viewers, and critics of these pieces – that is, the actors in the Holocaust memory dispositif – than I am in tracing the attempts of the Soviet/Russian state to regulate the content of such works. Where possible, I will trace the story of a particular work after its publication or release, because paying attention not only to its production but also to its public reception permits a better understanding of the functioning of the regional dimension of the Holocaust memory dispositif.

5.1 "My Jews:" The Occupation and Holocaust History in the North Caucasus in Soviet/Russian Imagery

Various artistic responses to Holocaust history in the North Caucasus emerged after the collapse of the USSR. Since the topic of the mass killing of Soviet Jews under the Nazi German military administration was not publicly covered (except for a short period in the wartime media and then during the Thaw), and the Jewish communities of the Caucasus were weak or absent in the post-war period, military fiction and cinema developed according to the dominant Soviet ideology with careful attention to the history of military operations, and partisan and underground movements. The history of the registration of the Jews and their subsequent mass killing is presented as an episode in a series of detailed descriptions of the everyday fight against the enemy in occupied towns and villages. Special attention in such pieces is devoted to producing psychological portraits of young underground fighters and experienced party leaders. Following ideological conventions, Soviet writers aspired to portray the image of a war hero in their works, not only at the front but also in the rear, so that the Holocaust was accorded very little attention in Soviet imagery of the occupation of the North Caucasus.

---

11 Of all types of imagery, I will dwell only on the analysis of Soviet and Russian fiction and documentaries, since they give the deepest insight into the artistic responses to the Holocaust in the North Caucasus. Musical compositions are very fragmentary in the region. I have not discovered thematic works of fine art and post-war photographs, nor any thematic art installation in the region.

In his short story "Heroes of Taganrog" [*Geroi Taganroga*],[12] the Soviet writer Genrikh Gofman[13] depicted the Holocaust in the city of Taganrog through the actions of young partisans. They learned about the registration of Jews and strove to warn as many Jewish families as possible of potential massacres. "By the curfew they had warned about twenty families. The young partisans were happy to realise that their efforts were not in vain: some of those who had been warned agreed to leave the city."[14] The next episode in the sequence, the mass killing of "more than 2,000 Jews" gathered in Vladimirskaya Square, is shown in Gofman's tale through the fate of a boy rescued by the fisherman Kuz'ma Turubarov. He literally snatches the boy out of a column of Jews being escorted to a killing site.[15] This episode is narrated in one of the first chapters of the book. The author does not further touch upon the topic of Jews in the Holocaust. The partisans of Taganrog did indeed play an important role in rescuing not only this one boy (called Tolik in the story), but also other Jews: evacuees, refugees, and locals.[16] However, Gofman's intention was to show the significance and efficiency of the underground movement in a Soviet city under occupation, so the description of that fight – e.g., successful acts of sabotage, preparation of an uprising, and the rooting out traitors – overshadowed the rescuing of Jews. Remarkably, the part of this story about the fate of the Jews of Taganrog was translated into Yiddish and published in the Soviet Yiddish-language literary magazine *Soviet Homeland* [*Sovetish Heymland*].[17] This separate edition acquired a completely different meaning: it was a story of the underground resistance's struggle for the survival of Jews and their desire to prevent the death of innocent people.

There was a short period in the history of the development of the Soviet war narrative when recent Holocaust memory was still alive, and Soviet writers

---

12 Genrikh Gofman, *Geroi Taganroga: dokumental'naya povest'* (Moscow: Molodaya gvardiya, 1966). The short story had about a dozen reprints.
13 Genrikh Gofman (1922–1995) was a Soviet colonel and military pilot during World War II, a Hero of the USSR. He is the author of dozens of short stories about World War II and its heroes.
14 Gofman, *Geroi Taganroga*, 55.
15 Ibid, 55–58. The 1966 edition of the short story contains several illustrations, among which there is a drawing of the Jews (women and children) being escorted to the killing site. Gofman, *Geroi Taganroga*, 56–57.
16 See, for example, an interview with Inna Vaisburd, who was rescued by the underground fighters of Taganrog. Interview with Inna Vaisburd, born in 1922, in *VHA USCSF*, code 46991-13, June 30, 1998, Haifa, Israel.
17 Genrikh Gofman, "Dos iz geshen in Taganrog" [in Yiddish], transl. Moisei Itkovich, *Sovetish Heymland* 2 (1966): 119–124.

who visited the North Caucasus strove to provide the descendants with stories about the mass killing of Jews in various localities of Southern Russia. The archival collection of the JAC is made up of various literary responses to the Holocaust in the region. These works take the form of stories and edited literary articles for Soviet newspapers.[18] Some of them were published in Yiddish during the war, in the JAC's newspaper *Unity [Einigkeit]*, some stories were prepared for *The Black Book of Soviet Jewry* by Il'ya Erenburg and Vasilii Grossman, but most of them never reached their readers because of censorship or ignorance of Yiddish. In these first stories, Soviet authors, mostly of Jewish origin, described in detail the period of Nazi military administration, their primary focus being the fate of Jews. Thus, Viktor Shklovskii's[19] story "Kislovodsk" presents the edited memoirs of a 79-year-old survivor, Moisei Evenson, and contains a detailed description of Jewish suffering under German rule: "Jews had to sew a six-pointed white star six centimetres in diameter onto their chests, then an order was issued for their resettlement, which resulted in the mass killing of over 2,000 Jews of Kislovodsk, of whom only a few survived."[20] Written right after the liberation of Kislovodsk, the story includes some details of the last moments in the lives of Evenson's Jewish acquaintances: "Doctor Benenson's severely ill son was taken to the Jewish gathering place on a stretcher"; "according to a Gestapo officer, if you did not kill small [Jewish] children, they would grow big." The story is perceived as a common Jewish tragedy, and the appeal for the rebirth of the human conscience that was killed along with thousands of massacred Jews

---

18 See stories and articles by Soviet writers and correspondents about the Holocaust in the North Caucasus in Russian, English and Yiddish: Il'ya Vatenberg, "Kak nemtsy ubili doktora Dubinskogo v Krasnodare," in *GARF*, f. R-8114, op. 1, d. 132, l. 129–130 (access: in *USHMMA*, RG-22.028M); Lev Kvitko, "Fascist atrocities in Nalchik," in *GARF*, f. R-8114, op. 1, d. 181, l. 336; d. 234, l. 64–65; d. 406, l. 22–23 (access: in *USHMMA*, RG-22.028M); Mirra Zheleznova, "Novye dannye o nemetskikh zhertvakh v Kislovodske," in *GARF*, f. R-8114, op. 1, d. 30, l. 122–123 (access: in *USHMMA*, RG-22.028M); Rivka Rubina, "Further Details of Jewish Massacres in Krasnodar," in *GARF*, f. R-8114, op. 1, d. 161, l. 386–390 (access: in *USHMMA*, RG-22.028M); Idem, in *GARF*, f. R-8114, op. 1, d. 200, l. 226–232 (access: in *USHMMA*, RG-22.028M).
19 Viktor Shklovskii (1893–1984) was a half-Jewish Russian and Soviet literary theorist, critic, and writer. He worked as a screenwriter on numerous Soviet films. During World War II was evacuated to Alma-Ata [present-day Almaty]. He edited collected stories for *The Black Book* as a member of Literary Commission of the JAC.
20 Viktor Shklovskii, "Kislovodsk," in *GARF*, f. R-8114, op. 1, d. 432, l. 107–115; d. 62, l. 292–301 (access: in *USHMMA*, RG-22.028M). Published: Viktor Shklovskii, ed., "Nemtsy v Kislovodske," in *Chernaya Kniga*, ed. Vasilii Grossman and Il'ya Erenburg (Moscow: AST: CORPUS, 2015), 293–297.

gives Evenson the hope "of seeing a reckoning, otherwise the path we [Jewish survivors] have travelled will be useless."[21]

An address to Red Army combatants with the demand for revenge is the most common trope of Soviet wartime art. It is no coincidence that Soviet wartime newsreels recorded destroyed cities, towns, and villages, the corpses of Soviet victims, and grieving relatives, with women crying over the bodies of children, young women, and elderly victims.[22] A voice-over, commenting on the suffering of those who survived the occupation, called for revenge on the perpetrators for these crimes: "We shall not forget! We shall not forgive!" [*Ne zabudem! Ne prostim!*].[23] Holocaust victims were included in the overall depiction of "peaceful citizens." The historian of Soviet film, Jeremy Hicks, believed that this "sovietisation" of the Holocaust appeared during World War II. Soviet documentaries reinforced this sovietisation by recording how local citizens recognised the victims (as though grieving people pointed to the universalisation of the tragedy, and crying women in the frame could also refer to the symbol of the "Motherland"[24] mourning her dead children). Thus, Soviet wartime documentaries created an image of the Holocaust that corresponded to the theme of Nazi atrocities against Soviet citizens in general, with Slavic peoples foremost among them.[25]

---

21 Quoted in: Shklovskii, "Kislovodsk," in *GARF*, f. R-8114, op. 1, d. 432, l. 111–112, 115 (access: in *USHMMA*, RG-22.028M). The published version does not contain Evenson's final philosophical reflections.

22 *Soyuzkinozhurnal no. 114* (ed. by Roman Gikov and filmed by Arkadii Levitan, Georgii Popov, and Andrei Sologubov) released in December 1941, was devoted to the first liberation of Rostov-on-Don from the one-week-long occupation in November 1941. For a detailed analysis of this Soviet documentary newsreel, see: Hicks, *First Films of the Holocaust*, 49–58. See also wartime newsreels, which show destroyed cities, including the localities in the North Caucasus, and bodies of massacred "Soviet citizens": *Otomstim*, directed by Nikolai Karamzinskii (Tsentral'naya kinostudiya dokumental'nykh fil'mov, 1942), in *Russian State Film and Photo Archive* [*Rossiiskii gosudarstvennyi arkhiv kinofotodokumentov*], hereafter *RGAKFD* no. 4488; *Soyuzkinozhurnal no. 9*, directed by Irina Setkina, Z. Dembovskaya (Tsentral'naya kinostudiya dokumental'nykh fil'mov, 1943), in *RGAKFD*, no. 4866; *Zhertvy gitlerovtsev v Pyatigorske*, cameramen Andrei Sologubov and Yakov Avdeienko (Tsentral'naya kinostudiya dokumental'nykh fil'mov, 1943), in *RGAKFD*, no. 5994; *Po sledam fashistskogo zver'ya*, directed by Shalva Chagunava (Tbilisskaya kinostudiya, Tsentral'naya kinostudiya dokumental'nykh fil'mov, 1943), in *RGAKFD*, no. 14170.

23 *Soyuzkinozhurnal no. 114*, min. 6:20–6:25.

24 "Motherland is calling!" [*Rodina-mat' zovet!*] was a famous wartime poster, created by the artist Iraklii Toidze in late June 1941. For an analysis of the development of the image of the Motherland and its incorporation into a political, national, military and gender discourses of Russian society, see: Oleg Ryabov, "'Rodina-mat'': Istoriya obraza," *Zhenshchina v rossiiskom obshchestve*, 3 (2006): 33–46.

25 Hicks, "Otrazhenie Kholokosta," 180–181.

The universalisation of Holocaust victims, or the sovietisation of the Holocaust, reached its peak in the post-war period, when Soviet authorities prohibited the publication of *The Black Book*,[26] detailing the mass killing of Jews in occupied Soviet territories.[27] It consisted of eyewitness testimonies, collected by writers and correspondents of the JAC and selected by Grossman and Erenburg for publication as a separate edition.[28] The section on the RSFSR included stories about the mass killing of Jews in Rostov-on-Don and Essentuki, A. Nankin's letter of June 8, 1943 "In Stavropol" [*V Stavropole*], and Evenson's aforementioned memoirs "Germans in Kislovodsk" [*Nemtsy v Kislovodske*], edited by Viktor Shklovskii.[29] Composed in the genre of an assemblage (we can see the heterogeneity of texts, the constant change of scale: from individual stories to generalising essays and back, the description of the same event through the eyes and in the style of different people),[30] *The Black Book* was to become a literary monument to Soviet Jewish victims. Translated into many languages and printed abroad, *The Black Book* was first published in Russia only in 2015.

---

[26] The full title of the book is "The Black Book: On the Evil, Universal Murder of Jews by the German-Fascist Aggressors in the Temporarily Occupied Regions of the Soviet Union and in the Camps of Poland during the War 1941–1945" [*Chernaya kniga: o zlodeiskom povsemestnom ubiistve evreev nemetsko-fashistskimi zakhvatchikami vo vremenno okkupirovannykh raionakh Sovetskogo Soyuza i v lageryakh Pol'shi vo vremya voiny 1941–1945 gg.*]. The original title and its translation are not historically correct. The term "Camps of Poland" refers to the concentration and extermination camps that were situated on the territory of occupied Poland.

[27] For the fate of *The Black Book*, see: Il'ya Al'tman, "K istorii 'Chernoi knigi'," in *Neizvestnaya "Chernaya Kniga,"* ed. Yitzhak Arad (Moscow/Jerusalem: Tekst, 1993), 16–28; Idem, "Memorializatsiya Kholokosta v Rossii: istoriya, sovremennost', perspektivy," *Neprikosnovennyi zapas*, 2–3 (2005): 254–257; Idem, "The History and Fate of the Black Book and the Unknown Black Book," in *The Unknown Black Book: The Holocaust in the German-Occupied Soviet Territories*, ed. Joshua Rubenstein and Ilya Altman (Bloomington and Indianapolis: Indiana University Press, 2008), XIX–XXXVIII.

[28] This collection included 118 documents written in Russian or translated from Yiddish: letters and transcripts of stories of survivors, essays by writers included in the specially created Literary Commission for the preparation of a book on the history of the Holocaust in individual cities and republics of the USSR, death letters handed over by inmates of prisons and ghettos, ChGK files, testimonies of witnesses and reports of interrogations of perpetrators, protocols and reports by the Nazi authorities about mass killings, and a work by Soviet lawyer and ChGK member Il'ya Trainin on Nazi racial policy and antisemitism.

[29] Vasilii Grossman and Il'ya Erenburg, ed., *Chernaya kniga* (Moscow: AST: CORPUS, 2015), 286–288, 291–299.

[30] Il'ya Kukulin carried out a literary analysis of *The Black Book* and also traced the history of the assemblage genre in Soviet literature. Il'ya Kukulin, *Mashiny zashumevshego vremeni: kak sovetskii montazh stal metodom neofitsial'noi kul'tury* (Moscow: Novoe literaturnoe obozrenie, 2015), 237–249.

For the functioning of the Holocaust memory dispositif in modern Russia, it is notable that the funds for the editing and publication of *The Black Book* were raised using crowdfunding on the Internet.[31]

Modern authors often use the assemblage method to reconstruct the historical reality of a particular occupied North Caucasian locality. Vladislav Smirnov's[32] book, "Rostov in the Swastika's Shadow" [*Rostov pod ten'yu svastiki*], for instance, is an assemblage of fragmented memoirs of ordinary people who survived the occupation of Rostov-on-Don.[33] The author conducted over 50 interviews with eyewitnesses, which he divided into thematic blocks, from recollections of the first bombings of Rostov-on-Don in the autumn of 1941 to stories of joy at the city's liberation in February 1943. Holocaust history is presented as a period of the occupation, but since most of the storytellers were not Jews, the episodes about the rounding up of Jews for registration followed by their killing are mentioned in the book only in passing, as something that did not personally touch the citizens of occupied Rostov-on-Don. Only L. Grigoryan's (his first name is unknown) testimony conveys the tragedy of the Holocaust, since his mother was Jewish.[34]

The story of Jews under the occupation of the North Caucasus has become a key topic in many works of fiction and documentaries in recent decades. The first wave of detailed attention to Holocaust history in post-Soviet Russia emerged from the collapse of the USSR and attempts critically to revisit the country's past, including the history of World War II. In the 1990s, the first stories appeared and a documentary was filmed about the Holocaust in the North Caucasus. Boris Sergel[35] expressed the atmosphere of Jewish pre-war life in the seaside city of Novorossiysk in the story "My Jews" [*Moi evrei*]. The author de-

---

31 In 2014, the project was supported by 537 shareholders, more than 1.2 million roubles were collected. "Proekt 'Chernaya kniga' Vasiliya Grossmana i Il'i Erenburga," *Planeta.ru*, https://planeta.ru/campaigns/4631 (accessed March 23, 2020). This way, the publishing house CORPUS could publish not only *The Black Book* itself, but also an *Unknown Black Book*, which included previously unpublished materials, kept in the JAC archival collection in GARF. Grossman and Erenburg, *Chernaya Kniga*; Il'ya Al'tman, ed., *Neizvestnaya "Chernaya kniga": materialy k "Chernoi knige"* (Moscow: AST: CORPUS, 2015).
32 Vladislav Smirnov (1939–2014) was a Soviet and Russian scholar, journalist, and local historian. He is the author of a number of literary works, including historical and documentary tales about Rostov Oblast and poetic collections.
33 Vladislav Smirnov, *Rostov pod ten'yu svastiki* (Rostov-on-Don: "Rostovkniga," 2006).
34 Idem, 190, 212.
35 Boris Sergel (born in 1930) is an activist of the Regional Organisation of the Juvenile Prisoners of Concentration Camps [*Regional'naya organizatsiya "Soyuz byvshikh maloletnikh uznikov fashistskikh kontslagerei"*]. He worked as history teacher in St. Petersburg.

scribed in detail the characters of the local Jewish intelligentsia, so that the history of the mass killing of nearly all the city's Jews is conveyed with more emotion.[36] Thanks to the temporary opening of some FSB archives in the early 1990s, film director Yurii Kalugin[37] included the testimony of Holocaust eyewitnesses and ChGK reports revealing the tragic details of the mass killing of Jews in Rostov-on-Don in his documentary *Judenfrei: Free of Jews* [*Judenfrei: svobodno ot evreev*].[38] Shots from military documentaries follow scenes of modern everyday life, which strengthens the effect of becoming part of a tragedy that could happen to anyone. For the first time, the reader and viewer could experience the distant and unfamiliar tragedy of the Holocaust as something close and personal through the power of artistic expression. It was perhaps such emotional truthfulness that has hindered these artistic works from reaching an audience: Boris Sergel's story was never published, and Yurii Kalugin's film has never been screened on national or regional television.

In the 2000s, a new stage in literary responses to the Holocaust in the North Caucasus began. By this time, most Holocaust survivors among the children of the war generation had retired and many of them had migrated to Israel or the United States. The new cultural and social experience, an abundance of free time, along with the constant emotional pressure of the wartime past let many of them not only give various interviews to journalists and historians, but also discover their own literary talent. Yakov Krut, a former citizen of Rostov-on-Don, who survived the Holocaust in Krasnodar Krai, wrote a series of autobiographical stories collected in "Tale of a Gifted Life" [*Povest' o podarennoi zhizni*]. Yurii Burakovskii, a former denizen of Kyiv, who survived the Holocaust in Stavropol Krai, composed a poem called "Pravokumka" [the name of a village on the right bank of the river Kuma in Stavropol Krai].[39] In their works, these authors depicted various aspects of everyday life among "good" and "bad" neighbours under German rule, mass killings of Jews including their relatives and acquaintances, the unlikely survival of their families, and the happiness of liberation. The first-person narration increases the readers' empathy for the main characters, and the attention paid to details fixed in the minds of children brings the

---

36 Boris Sergel, "Moi evrei," 1997, in *personal archive of Nadezhda Suvorova*.
37 Yurii Kalugin (born in 1938) is a Russian documentary filmmaker and professor at the St. Peterburg State University of Film and Television. He was born in and studied in Rostov-on-Don.
38 *Judenfrei: svobodno ot evreev*, directed by Yurii Kalugin (Dontelefil'm, 1992).
39 Yakov Krut, *Povest' o podarennoi zhizni* (Petakh-Tikva, 2009), 2–38; Yurii Burakovskii, "Pravokumka: poema," in Yurii Burakovskii, *Vremya dozhdyu razlivat'sya rekoi: stikhi* (Tel-Aviv: Dfus Shakhaf, 2004), 79–119.

5.1 "My Jews:" The Occupation and Holocaust History — 157

images of World War II and the Holocaust to life. As a rule, Holocaust survivors dedicate their works to their children and grandchildren, so that descendants do not forget the history of their family.

Non-Jewish citizens of Southern Russia also feel connected to the theme of the occupation in general, and the history of mass killings of Jews in their native locality, as part of the occupation. Many of them express a keen interest in the history of the lands where they were born and grew up, as does Svetlana Chechulina.[40] With no pretentions to elevated literary style, ordinary citizens take up their pens to record their knowledge, pain and suffering on paper. Notably, in the North Caucasian regions, where mostly Jewish evacuees were killed, it is local non-Jewish citizens who strive to restore the memory of Holocaust victims also by means of artistic responses. The occupation was a complex and diverse period in the history of World War II, and the Holocaust was an important, but not the only aspect of it. Accordingly, there are also stories about forced labour, mass killings of Soviet POWs, and the persecution of partisans and underground fighters in the literary works of Jewish and non-Jewish authors.

Jewish authors, on the contrary, mainly from the second generation of Holocaust survivors,[41] focus exclusively on Holocaust history in their imagery: it is important for them to understand the Holocaust as a historical phenomenon. From 2014 to 2018, the Israeli film director Boris Maftsir[42] worked on a series of documentaries for his project, Holocaust in the USSR.[43] This project appeared as a sequel to Yad Vashem's Shoah Victims' Names Recovery Project,

---

[40] Svetlana Chechulina, "Okkupatsiya: dokumental'naya povest' v stikhakh o sobytiyakh 1941–1943 gg. v khutore Bratskom Ust-Labinskogo raiona i ego okrestnostyakh," *Stihi.ru*, https://www.stihi.ru/2012/08/15/7764 (accessed March 23, 2020); Lidiya Stekhova, "U obeliska," *Vestnik Arzgirskogo raiona*, September 24, 2012.

[41] For the phenomenon of the second generation of Holocaust survivors and their works, see: Alan L. Berger and Naomi Berger, *Second Generation Voices: Reflections by Children of Holocaust Survivors and Perpetrators* (Syracuse: Syracuse University Press, 2001); Erin McGlothlin, *Second-Generation Holocaust Literature: Legacies of Survival and Perpetration* (Rochester, NY: Camden House, 2006); Marita Grimwood, *Holocaust Literature of the Second Generation* (New York: Palgrave Macmillan, 2007); Samuel Juni, "Identity Disorders of Second Generation Holocaust Survivors," *Journal of Loss and Trauma* 21, no. 3 (2016): 203–212.

[42] Boris Maftzir (born in 1947) is an Israeli documentary filmmaker. He was born in Soviet Riga. In 1970 he was arrested by the KGB and sentenced to one year in prison on charges of Zionist activity. In 1971, he emigrated to Israel. Between 2009 and 2017, he founded and chaired the documentary film department at the Haifa Women's International Zionist Organisation Academic Center. Throughout his career, he has produced over 200 documentaries and directed over 30 films.

[43] For more information about Maftsir's documentary project, see: *Holocaust in the USSR*, http://www.holocaustinussr.com/ (accessed March 21, 2020).

where Maftsir began his research. He attempted to reconstruct in detail and to film surviving evidence of the Holocaust in every city, town, or village on Soviet soil. He talked to eyewitnesses and professional historians and asked occasionally provocative questions in order to better understand the historical reality: What did Jews expect in 1942? How did perpetrators search for Jews? Who betrayed them? How did the mechanism of mass killing of Jews unfold? Why did the Germans single out Jews in 1942? Maftsir filmed each episode on the same streets and in the same houses where Jews had lived, at their gathering points, on the same road along which they were transported or driven, at the same time of year as the killing actions in a particular locality were carried out, and finally at the very killing site, which strengthens the presence effect. In the second film of the Holocaust in the USSR project, called *Holocaust. The Eastern Front*, Maftsir showed five episodes from different places in Russia, three of which took place in the North Caucasus: Rostov-on-Don, Kislovodsk/Mineralnye Vody, and Nalchik. Despite the fact that the main goal of Maftsir's project was to document killing sites in the Soviet Union, the Nalchik episode is the only case in his entire filmography that focuses on the reasons for the survival of Mountain Jews.

One of the most important outcomes of Maftsir's project is that in today's post-Soviet states there are a number of people willing to share their knowledge and stories about the Holocaust: "During the project, I have travelled through about 160 cities, towns, villages of the former USSR. This is a whole land of the 'hidden Holocaust,' as I call it. And in every city, there are activists, local historians, witnesses who have been interested in Holocaust history for years."[44] Furthermore, in each region there are artists, who evoke the tragedy of the Holocaust in their works. Igor' Khentov,[45] a poet from Rostov-on-Don, composed a number of poems gathered together in a "Judaic Cycle" [*Yudeiskii tsikl*], some of which are about the Holocaust in his native city.[46] In December 2014, his poem "Pain of the Earth" [*Bol' zemli*] – for symphony orchestra, mixed, and children's choirs – premiered at the Courage to Remember! [*Muzhestvo pomnit'!*] festival dedicated to the 70th anniversary of Victory in World War II and

---

44 Interview with Boris Maftzir, in *author's archive*, code 16-F-01, June 2, 2016, Rechovot, Israel.
45 Igor' Khentov (born in 1954) is a poet, prose writer, and playwright. He is a member of the Writers' Union of Russia and Israel and laureate of various musical and literary contests. The Holocaust in his native Rostov-on-Don is one of the main themes of his poetry.
46 Igor' Khentov, "Yudeiskii tsikl," in Igor' Khentov, *Vzglyad: stikhi raznykh let i novelly*, vol. 1 (Rostov-on-Don: Lunkina N.V., 2011), 21–48; Idem, *Stihi.ru*, https://www.stihi.ru/2009/09/06/1620 (accessed March 21, 2020).

the memory of its victims.⁴⁷ The culmination of the poem is "The Chorus of Children's Souls" [*Khor detskikh dush*], the chorus of those who were never fated to become adults:

And they were left lying	*I ostalis' lezhat' odni*
Our balls and skipping ropes,	*Nashi myachiki i skakalki,*
And the living thread broke	*I porvalas' zhivaya nit'*
in the frozen mud of Zmievskaya Balka.	*V styloy tine Zmievskoi Balki.*⁴⁸

The mass murder of children in this poem is represented as the breaking of a stretched thread, lives cut short, thus emphasising the Holocaust as a crime against humanity, childhood, and, therefore, the future.

The theme of Holocaust history, then, and the fates of thousands of Jewish victims in the North Caucasus are represented in artistic works of different genres. After the first attempts to capture the tragedy on paper and on screen, this theme was submerged throughout the Soviet post-war period in regional fiction and documentaries, and has re-emerged only in the last decades. The fates of Jews who perished in the North Caucasus have nowadays become an important theme for artistic works devoted both to everyday life under the occupation and to Holocaust history.

5.2 "Retribution Against Fascists and Collaborators" in Soviet Literature and Cinema

Even though the theme of the Holocaust did not obtain literary representation in Soviet times, there was a series of artistic responses to the so-called "Final Solution" on Soviet soil. The USSR became the first country in the world to bring perpetrators of Nazi crimes to trial. In 1943, 11 Soviet collaborators faced

---

47 Igor' Levin, *Bol' zemli: poem for symphonic orchestra, mixed and children's choir*, verse by Igor' Khentov, music by Igor' Levin, conductor Mikhail Kats, performed by Rostov Academic Symphony Orchestra, Academic Choir of Rakhmaninov Rostov State Conservatoire (artistic director Yurii Vasilev) and the children's choir "Solovushko" of Petr Tchaikovskii Children's Music School (director Yurii Vasilev). In 2015, the poem became a laureate of the Shostakovich Competition. For an analysis of this poem, see: Inessa Dvuzhil'naya, *Tema Kholokosta v akademicheskoi muzyke* (Grodno: Grodnenskii gosudarstvennyi universitet: 2016), 82–85.
48 Igor' Khentov, "Khor detskikh dush," in Igor' Khentov, *Rodoslovnaya* (Kyiv: Izdatel'stvo Dmitriya Yuferova, 2017), 94. Transl. by Tristam Barrett.

trial in Krasnodar, which became the first open tribunal[49] in a series of public and closed trials over the responsibility of Soviet citizens for complicity in the killing actions and other crimes against 'the Soviet people' during World War II. Almost all public trials were extensively covered by the Soviet press and attended by famous Soviet journalists and writers.[50] Indeed, newspaper articles, documentary short stories, and occasional newsreels are so far the only sources detailing the imposing history of Soviet trials against wartime Soviet perpetrators in today's Russia. The main theme of literary representations of the trials is the retribution against Soviet collaborators and traitors for their crimes against all "Soviet citizens." The responsibility of perpetrators for the Holocaust can be read only between the lines in these pieces. The Russian cultural historian Il'ya Kukulin identifies the attempts of Soviet poets and writers

---

[49] Since late-1942 special units of the internal affairs and state security apparatus could search for collaborators in the occupied Soviet regions. Thus, for example, by order of the regional branch of the NKVD of September 29, 1942 "On the organisation of intelligence work behind enemy lines in the areas of the 18th and 56th armies" special units were created. Their members found 827 traitors, accomplices, and spies in Krasnodar Krai from August 30, 1942 to January 20, 1943, of which 34 people were executed. In the Red Army the commander of a military unit was obliged to coordinate his activities with political bodies and special departments in order to identify German henchmen and their accomplices in the rear in order to eliminate them with or without due process. The legal basis of responsibility for collaboration activity was Decree no. 39 of the Presidium of the Supreme Soviet of the USSR of April 19, 1943 "On punitive measures for German-fascist villains guilty of killing and torture of the Soviet civilian population and captured Red Army soldiers, for spies, traitors among Soviet citizens and for their accomplices." See: Belyaev and Bondar', *Kuban' v gody*, 664–719; Sergei Stepanenko, "Sudebnye protsessy nad voennymi prestupnikami i ikh posobnikami kak akt realizatsii norm mezhdunarodnogo gumanitarnogo prava (na materialakh krasnodarskikh sudebnykh protsessov 1943–1974 godov)," *Istoricheskie, filosofskie, politicheskie i yuridicheskie nauki, kul'turologiya i iskusstvovedenie. Voprosy teorii i praktiki* 1, no. 5 (Tambov: Gramota, 2010): 161–165.

[50] Despite the press coverage of most public trials, we do not know the exact number of trials carried out on Soviet soil, nor do we know the names of all accused Holocaust perpetrators and collaborators. Novgorod researcher Dmitrii Astashkin is the curator of the 2016 exhibition "Soviet Nuremberg" ["*Sovetskii Nyurnberg*"], which was shown in several Russian cities. A group of historians from Russia and France conducted a project between 2016 and 2018, which aimed to study the most famous Soviet trials against Nazi perpetrators and collaborators. Meanwhile, researchers use the only available information in Russia, mainly articles in the Soviet media and Soviet newsreels. See: Dmitrii Astashkin, ed., *Protsessy nad natsistskimi prestupnikami na territorii SSSR v 1943–1949 gg.: katalog vystavki* (Moscow: Gosudarstvennyi tsentral'nyi muzei sovremennoi istorii Rossii, 2015); "Proekt 'Sovetskii Nyurnberg,'" *Portal istoriya Rossii*, http://histrf.ru/biblioteka/Soviet-Nuremberg (accessed March 21, 2020). For general information about public trials in Krasnodar, see: Avramenko et al., *Ekaterinodar-Krasnodar: dva veka*, 614.

to touch on censored themes,[51] especially that of the Holocaust, as a mechanism of "hidden amalgamation". Hidden amalgamation is a special artistic strategy by which an authorised memory of a collective traumatic experience is implicitly (metonymically, metaphorically, or through similar imagery) associated with the memory of other traumas and encoded through their description.[52] Developing and consolidating Kukulin's concept, it can be assumed that in literature on the trials of collaborators, crimes against the universalised victim category "Soviet citizens" became just such an authorised memory, through which the Holocaust trauma was encoded. The word "Jew" was not mentioned once – neither in the records of the 1943 Krasnodar Trial, nor in newspaper articles, nor in the 1943 documentary *People's Verdict* [*Prigovor naroda*][53] – although the main category of victims in Krasnodar and in other localities of the North Caucasus was Jews; including locals, but mostly evacuees. The poet and military correspondent Il'ya Selvinskii[54] attended the 1943 Krasnodar Trial. In the poem "Trial in Krasnodar" [*Sud v Krasnodare*][55] he depicted the personality of a gas van driver, the collaborator Rychkalov,[56] who was responsible along with other defendants for the mass killing of citizens in Krasnodar Krai, where "every seventh one is no

---

[51] For the history of censorial taboos related to World War II, see: Irina Scherbakowa, *Zerrissene Erinnerung: der Umgang mit Stalinismus und Zweitem Weltkrieg im heutigen Russland* (Göttingen: Wallstein, 2010), 7–62.
[52] Kukulin, "Sovetskaya poeziya," 342.
[53] "Sudebnyi protsess po delu o zverstvakh nemetsko-fashistskikh zakhvatchikov i ikh posobnikov na territorii goroda Krasnodara i Krasnodarskogo kraya v period ikh vremennoi okkupatsii," *Izvestiya*, July 15–20, 1943; Idem, *Pravda*, July 15–19, 1943; Elena Kononenko, "Pered sudom naroda," *Bol'shevik*, July 16, 1943; *Prigovor naroda*, shot by the film crew of the North Caucasus Front (Tsentral'naya studiya kinokhroniki, July 1943), 12 min. 46 sec.
[54] Il'ya Selvinskii (1899–1968) was a Russian and Soviet poet, dramatist, memoirist, and essayist. During World War II he served as a military journalist and combat political officer in his native Crimea and in the North Caucasus. For the poem "I Saw It!" and other pieces he was summoned to Moscow in 1943, punitively dismissed from the army, and subjected to repression. In April 1945 Selvinskii's status was restored, and he was allowed to return to the frontlines.
[55] The poem was first published in 1945. The 1947 publication was the second and last one: Il'ya Selvinskii, "Sud v Krasnodare," *Znamya* 11 (1945): 25–28; Idem, in Il'ya Selvinskii, *Krym: Kavkaz: Kuban': stikhi* (Moscow: Sovetskii pisatel', 1947), 147–155.
[56] Meant is Ivan Rechkalov, born in 1911. In August 1942, he avoided mobilisation into the Red Army, defected to the German side, enrolled in the police service, was transferred to SK 10a within a few days and served as a gas van driver in Krasnodar. On July 17, 1943, the Red Army's North Caucasus military tribunal sentenced Rechkalov to capital punishment by hanging. The execution of the sentence took place in the central square of Krasnodar in the presence of about 50,000 people on July 18, 1943. Nikolai Maiorov, "Krasnodarskii protsess," in Mikhail Karyshev, ed. *Neotvratimoe vozmezdie: po materialam sudebnykh protsessov nad izmennikami Rodiny, fashistskimi palachami i agentami imperialisticheskikh razvedok* (Moscow: Voenizdat, 1984), 160–170.

longer alive."⁵⁷ A native of the Crimea, Selvinskii visited an anti-tank ditch with 7,000 Jews buried in it in the village of Bagerovo west of Kerch (now in the outskirts of the city) in January 1942. He perceived the mass killing of Jews in his native peninsula as a personal tragedy. Then he wrote a poem called "I Saw It!" [*Ya eto videl!*], published in January 1942.⁵⁸ It became one of the first literary monuments to Holocaust victims in the USSR.⁵⁹ Selvinskii regarded the trial in Krasnodar first and foremost as retribution on Soviet perpetrators for the Holocaust. The auto-reference in the poem "Trial in Krasnodoar" to 7,000 corpses in Kerch . . .

There were children, women, invalids.	*Tam byli deti, zhenshchiny, kaleki.*
Perhaps THEY shot them too.	*Vozmozhno, ETI ikh i rasstrelyali.*⁶⁰

. . . was Selvinskii's way of indicating Jewish losses in the occupied territories while linking them directly to an early trial against Nazi collaborators and to a broader picture of Nazi genocidal crimes.⁶¹ The execution of the sentence is therefore especially important for Selvinskii:

And this feeling of the purity,	*I eto oshchushchenie chistoty,*
Innocence and holiness of all,	*Bezgreshnosti i svyatosti vsego,*
That was achieved in the name of the people,	*Chto sovershilos' imenem naroda,*
Burst within me a fervent wave	*Voshlo v menya goryacheyu volnoi,*
Revealed my whole life before me	*Vsyu zhizn' moyu raskrylo predo mnoi*
And lit my whole soul with a tear.	*I dushu vsyu slezoyu osvetilo.*⁶²

In the post-war works of Soviet writers, several open trials against Soviet collaborators are described, in which the key theme remains the one of retribution and fair judgment. The mass killing of Jews is depicted as one of the collaborators' crimes, however, the Holocaust as such was not mentioned. Soviet writers strove to portray "the monstrosity of the Nazi ideology" in contrast to the Soviet system. During the Cold War, trials of Soviet collaborators could serve as an instrument of ideological struggle. Soviet writers stressed in their works that if

---

57 Selvinskii, "Sud v Krasnodare," in Selvinskii, *Krym: Kavkaz*, 147.
58 Selvinskii, "Ya eto videl!".
59 For an analysis of this poem, see: Shrayer, *I Saw It*, 82– 116; Kukulin, "Sovetskaya poeziya," 329–338.
60 Transl. by Maxim Shrayer. Quoted in: Shrayer, *I Saw It*, 134.
61 For an analysis of the poem "Sud v Krasnodare," see: Shrayer, *I Saw It*, 133–135.
62 Selvinskii, "Sud v Krasnodare," in Selvinskii, *Krym: Kavkaz*, 154. Transl. by Tristam Barrett.

Soviet citizens made a mistake, betrayed their Motherland, and defected to the enemy's side, retribution would find them wherever they were. Therefore, the description of war crimes through the identity of the perpetrators themselves became an issue of great importance in fiction and non-fiction. While painting psychological portraits of Soviet collaborators, the authors of documentary fiction provided a deeper insight into the mechanisms of treason and the motivations of every perpetrator. The depiction of the subjective perspectives of perpetrators conveys to the reader a particular interpretation of their thoughts, motivations, memories and self-image.[63] An important feature of the Soviet literature on perpetrators is that it does not permit the readers make a free choice as to whose side they should take and with whom they should sympathise, and therefore to reach their own moral conclusions. The characters in these Soviet short stories are always sentenced to death for their crimes, that is, the evil in Soviet literary responses to crimes against humanity is inevitably punished, and the images of the defendants strengthen one's sense of the beastly nature of "Soviet collaborators."

The Soviet writer Lev Ginzburg[64] attended the 1963 Krasnodar trial against nine Soviet citizens, former members of SK 10a. He emphasised that the defendants "left their bloody mark in the Crimea, Mariupol, Taganrog, Rostov, Krasnodar, Yeysk, Novorossiysk, and later on, in Belarus and Poland." Ginzburg had the opportunity to familiarise himself with case records and attend the questioning of the defendants during the preliminary investigation.[65] In his short story "The Abyss" [*Bezdna*] he described in detail the work done by the investigative authorities in capturing perpetrators (the majority of them completely changed their identity, circle of contacts, and place of residence in the post-war period) along with the personal motives of perpetrators and the circumstances that caused their defection to the Nazis. The description of the crimes – including killings of "millions of our people" (Jews among them) under the occupation – reveals the universal image of

---

[63] Erin McGlothlin, "Narrative Perspective and the Holocaust Perpetrator: Edgar Hilsenrath's The Nazi and the Barber and Jonathan Littell's The Kindly Ones," in Adams, *The Bloomsbury Companion to Holocaust literature*, 161.
[64] Lev Ginzburg (1921–1980) was a Soviet translator and publicist. During World War II, he fought on the Far Eastern Front and printed poems in the front-line newspapers. He wrote several propagandic anti-fascist books, in which he described his numerous trips to the East and West Germany.
[65] Lev Ginzburg, *Bezdna. Povestvovanie, osnovannoe na dokumentakh* (Moscow: Sovetskii pisatel', 1967), 19–20.

Soviet collaborators in detail, thereby strengthening the rightfulness of the sentence or retribution.

Again, Ginzburg's short story is not about the Holocaust victims killed by the defendants of this and many other Soviet trials, and it is not even about perpetrators themselves. The protagonists are Soviet state security agents who have for decades hunted Nazi collaborators all over the Soviet Union and have managed finally to bring them to trial. Although little attention is paid to the figures of the Soviet Chekists in the tale, and we do not even know their full names, they were raised to the rank of heroes thanks to their successful activity in capturing the suspects. These Chekists embody the fair Soviet legal system and the socialist system in the wider sense. A paradox of "The Abyss" and other pieces about the trials of Soviet war-time perpetrators can be seen clearly: the main crime (mass killings of Jews and other Soviet citizens under the occupation) committed by the collaborators is overshadowed by the doubtful fairness of the Soviet criminal justice system; the charges against the perpetrators are secondary to the fact of collaboration, the defendants are judged not for committing homicide but for betraying the state. Literary responses to the Holocaust from this period thus shift the reader's attention from the tragedy itself to the inviolability of the Soviet system; a trial of crimes against morality is transformed into a political fable. In the post-Soviet period, several authors carry on this tradition. For example, a collection of short stories about the deeds of Chekists in Krasnodar Krai contains such tales as "The Retribution" [*Vozmezdie*] and "People as Beasts" [*Lyudi kak zveri*]. These two stories recapitulate the short story "The Abyss" with the addition of an episode about a trial of SS-Sturmbannführer SK 10a Kurt Christmann in West Germany.[66] The prison sentence of ten years trivialises his crimes, according to the authors of the collection: this "loyal" punishment characterises the capitalist legal system in contrast to the Soviet courts.[67]

The theme of good and evil in the opposition of heroes and perpetrators – who receive their just deserts in the form of a court judgement – is also apparent in the works of other Soviet writers. The non-fiction short story by

---

[66] On December 19, 1980 Munich Regional Court I sentenced Kurt Christmann to ten years imprisonment. *StA München*, Staatsanwaltschaft 35308.

[67] Aleksandr Belyaev et al., ed., *Bez grifa "sekretno": iz istorii organov gosbezopasnosti na Kubani: ocherki, stat'i, dokumental'nye povesti* (Krasnodar: Sovetskaya Kuban', 2005), 206–247.

Vladimir Gneushev[68] "Klavdiya Il'inichna" is devoted to the trial before an NKVD military tribunal in 1944 of Nikolai Engel, born in 1910.[69] During the occupation, the defendant was deputy commandant of the Gestapo prison in Stavropol. He himself participated in the arrest and shooting of "Soviet citizens." The short story shows the confrontation between two images of the Soviet people: Klavdiya Abramova, who stayed through the occupation and organised underground activity, and the collaborator Engel, who participated in the capture, interrogation, and killing of Abramova and her two young daughters. The retribution for the defendant's crimes, a death sentence, emphasises not only the triumph of good over evil but also the triumph of the Soviet system as a whole, because it raised patriots like Abramova. The murder of 660 patients of Stavropol's psychiatric hospital, and the registration and subsequent murder of over 3,500 Jews in Stavropol (including famous evacuated professors and medical workers) are among Engel's crimes. The short story aimed to depict the humanity of the Soviet Union, that had at its foundation the principles of friendship and equality between peoples, along with the superiority of socialism over fascism with its "philosophy of racism that did not tolerate the idea of the universal right to life."[70] It is remarkable that the author reports not only on Nazi crimes, personified in the defendant Engel, but also depicts the victims, including Jews, in detail. Thus, one of the story's plot lines addresses the fate of the Jewish Soviet professor and cancer specialist Brikker,[71] "whose knowledge could be useful to the Reich." Having had some privileges while in prison (walks, cigarettes, reading newspapers and books) he

---

[68] Vladimir Gneushev (1927–2011) was a Soviet and Russian poet, writer, and journalist. His literary works were mostly about his native Stavropol Krai.
[69] Vladimir Gneushev, "Klavdiya Il'inichna: dokumental'naya povest'," in *Vo imya zhizni: o podvigakh stavropol'tsev na fronte i v tylu Velikoi Otechestvennoi voiny* (Stavropol: Knizhnoe izdatel'stvo, 1987), 254–298.
[70] Gneushev, "Klavdiya Il'inichna," 267.
[71] Fedor Brikker (1895–1942) was a professor and head of the Department of Pathological Physiology. He graduated from the medical faculty of Kharkov University in 1920, defending his doctoral dissertation in 1925. In 1928, he became head of the Department of Pathological Physiology of Dnipropetrovsk Medical Institute, a position he retained in August 1941 when the Institute was evacuated to Stavropol Krai. During the occupation of Stavropol Krai, the Nazis did not kill Professor Brikker immediately alongside other Jews. He received permission to continue his research on cancer. However, he was later arrested and killed in a gas chamber on September 20, 1942 together with his wife and daughter. Andrei Kartashev, Svetlana Duginetz, and Anzhelika Bagdasarova, "Professor Brikker v shage ot gibeli: 'moya rabota polezna seichas kak nikogda . . . '," in Irina Rebrova, ed. *"Pomni o nas . . . ": katalog vystavki, posvyashchennoi pamyati patsientov psikhiatricheskikh klinik, detei-invalidov i vrachei-evreev, ubitykh v period natsistskoi okkupatsii Severnogo Kavkaza* (Krasnodar: Tipografiya "EdArt print," 2019), 185–190.

ultimately "was also included into the group for mass murder": "since Brikker is a Jew, the German authorities are not interested in his work."[72] Gneushev stressed once more the confrontation of two systems, the refined and educated Soviet system and the barbaric Nazi one. The reader's attention is thus shifted from the tragedy of the Holocaust to the elevation of science and education, which were among the pillars of the Soviet socialist system.

Lina Glebova[73] pursued similar aims in the short story "The Ploughed Grave" [*Raspakhannaya mogila*].[74] The inhumanity of the Nazi regime and the meanness of the Soviet collaborators were articulated in five chapters in the form of a confession of the defendant Aleksandr Kolesnikov, who served with EK 12 in Pyatigorsk. Glebova described in detail the generosity and culture of the Jewish intelligentsia who were killed by Nazis,[75] and thus provides a more emotional, or humanised, depiction of the Holocaust. At the same time, Glebova's main task was to show not the horror of the Holocaust but the terror of the Nazi regime in contrast to the Soviet system. Kolesnikov was not tried for the mass killing of Jews, but for killing his comrade and Red Army soldier Il'ya Gordienko, who had witnessed Kolesnikov's defection to the Nazis. The story's plot takes the form of an investigation into the patriot Gordienko's death, whose corpse was discovered by a villager, Garpina Glubokaya: "The victim was called Il'ya Gordienko, he was born the same year as my husband and he had a wife and children. And he had a baby born this year. And so, I cried so deeply over him."[76] She then buried his corpse in the steppe and tended his grave for many years, as if it were her husband's grave. In such a way, Gordienko is represented as a fallen hero, who died by "the perfidious hand of the collaborator Kolesnikov," who in turn received his just punishment thanks to the Soviet judicial system. That is, the collaborator had his retribution: Soviet society avenged a lonely hero, but not the thousands of Holocaust victims. The history of the mass killing of Jews is not denied, but it is not considered to be a unique or essential feature of the wartime crimes of the Nazis and their collaborators.

---

72 Gneushev, "Klavdiya Il'inichna," 269–272, 278–279.
73 Lina Glebova [Liana Goldman] (born in 1933) is a Soviet prosaist and novelist. She has lived in Israel since 1991.
74 Lina Glebova, "Raspakhannaya mogila ("Delo Kolesnikova" – povest' v pyati priznaniyakh)," in Lina Glebova, *Kogda ya snova budu* . . . (Moscow: Molodaya gvardiya, 1970), 79–204.
75 See the images of Polina Borisovna, a professor from the institute evacuated to Pyatigorsk, and Vladimir Izrailevich, an orchestral conductor from Pyatigorsk theatre: Glebova, "Raspakhannaya mogila," 106–108, 163–165.
76 Ibid, 200.

Only in literature published after the end of the USSR is the Holocaust, for the first time, treated as the main crime against humanity of the Nazi regime. The main theme of Vladimir Kolyabin's[77] documentary short story "The Betrayal" [*Predatel'stvo*][78] is the retribution by a Soviet court against the Soviet collaborator Nikolai Satin for his dealings as a police officer [*politsai*] in Blagodarnoe village of Stavropol Krai. This police officer participated in the shooting of more than 170 Jews. The defendant was sentenced to be shot in November 1959. The author of this short story quoted verbatim the transcript of Satin's interrogation and conducted his own investigation in order to find out the names of the killed Jewish evacuees in Blagodarnoe village. Kolyabin's aim was to draw the attention of local authorities to the killing site, which had become a rubbish dump instead of a commemoration site in the post-war decades. Kolyabin's investigation resulted in the official opening of a memorial obelisk at the killing site in 2010.[79]

The sovietisation of the Holocaust by the purging of its victims of their Jewishness was strengthened by the documentation and portrayal of Soviet criminals in fiction and documentaries on the Soviet trials. The perpetrators had attacked the ideals of the Soviet state. Soviet citizens attended the pronouncement of a verdict and its further execution with happy faces. They applauded long and hard and rallied spontaneously as a representation of the Soviet Motherland. All this depicted victory through the power of the united Soviet people.[80] In Soviet literature and cinema, accents were shifted from the perpetrators' crimes towards the honouring the Soviet socialist system with its inevitably correct judgments; this was the state with "a bright future." The Holocaust thus played an instrumental role in punishing erring Soviet citizens who (almost always) expressed repentance before they were sentenced. In "The Abyss," Lev Ginzburg depicted Mikhail Es'kov attempting to ingratiate himself with the Soviet court: explaining his betrayal by saying he "was young, stupid and could not find a way out,"[81] readily betraying his former accomplices, and assisting the investigation, if only in the forlorn hope of a small measure of mercy from Soviet justice. Ginzburg shared his feelings with the reader:

---

[77] Vladimir Kolyabin (born in 1955) is a lawyer and activist from Stavropol Krai. He has published dozens of short stories and tales on the website Proza.ru.
[78] Vladimir Kolyabin, "Predatel'stvo," *Proza.ru*, https://www.proza.ru/2014/11/30/182 (accessed March 23, 2020).
[79] Ibid.
[80] See: *Prigovor naroda*, min. 12:00–12:39.
[81] Ginzburg, *Bezdna*, 185.

"I am listening to his calm, dense voice, I am looking at the smile on his neat lips, and I understand that Es'kov is now absolutely sure of the opposite, that is, he is convinced that everything will be all right and that he has already won the saving 'sympathy' through his sad monologue and his repentance, a sympathy that can be sometimes stronger than the facts. But all this did not save him from his deserved punishment: the death sentence."[82]

Vladimir Kolyabin deliberately emphasised that Mikhail Satin, "who showed no mercy to his victims, pleaded before the court for mercy for himself after the verdict was announced." However, the Supreme Court affirmed its judgement, noting that "the residents of Blagodarnoe did not forgive the killer, they cursed him."[83]

Thus, Soviet literature on Soviet collaborators treated Holocaust history as a given, but placed the Jewishness of the victims in the background. The main goal of Soviet writers was to emphasise Nazi atrocities against the entire Soviet people, against the Soviet socialist system as a whole. Soviet collaborators were punished for their betrayal of the nation. Only post-Soviet writings based on Soviet court records highlight the Holocaust as a key wartime crime for the first time. In these pieces, Jews are seen as the main victims of Nazi perpetrators in the occupied regions of the North Caucasus.

5.3 "Accidental Survival:" Images of Holocaust Survivors and their Rescuers

The theme of Holocaust survival emerged in fiction after the acknowledgment of Holocaust history as a suitable theme in and of itself. The vast number of Soviet writers (especially those who described wartime events in the North Caucasus) introduced accounts of mass killings of Jews into their pieces to strengthen the perception of Nazi inhumanity. The lives of Jews who managed to survive, their suffering and fear of being caught under the occupation did not deserve specific attention in artistic responses to the Holocaust during the Soviet era. Perhaps, exceptionally is a story named "Alone in Krasnodar" [*Eine in Krasnodar*] by Rivka Rubina.[84] The story entered the collection of essays "*Jewish Women*" [*Idishe Froyen*] published in Yiddish in 1943.[85] This is a life

---

82 Ginzburg, Bezdna, 85–86.
83 Kolyabin, "Predatel'stvo".
84 Rivka Rubina [Rivke Rubin] (1906–1987) was a Soviet Jewish writer, literary critic, and translator. She was a member of the JAC. She wrote in Yiddish and translated into Russian.
85 Rivke Rubin, "Eine in Krasnodar," in Rivke Rubin, *Jidishe Froyen* [in Yiddish] (Moscow: Ogiz, 1943), 3–16. For the unpublished translation in Russian, see: Rivke Rubin, "A Page from the martyrdom in Krasnodar," in *GARF*, f. R-8114, op. 1, d. 200, l. 226–232.

story of Dvoria (Vera Isaevna) Edlin, "a woman with strong nerves and cool wits" in occupied Krasnodar. She had to endure torture by the Gestapo to save her family members: her mother-in-law, an 80-year-old woman, and her 8-year-old-granddaughter. The granddaughter looked Russian: "the child was a blonde little girl with blue eyes and fair skin", but the old woman looked typically Jewish.[86] Having done away with her documents, Dvoria had to make the Germans believe that she and at least her granddaughter were Russians. The woman and the child visited the Gestapo almost every day. The Nazi investigator arranged all kinds of tortures and tests for them. For example, he made them pronounce words with the letter "R," such as corn [*kukuruza*] or tractor [*traktor*], because Jews (he believed) could not pronounce it with its characteristic roll. He questioned the girl about their pre-war life: "in Poltava you lived on Sholem Aleichem Street, didn't you?"[87] He also checked their knowledge of Orthodox prayers. Local citizens – Dvoria's old acquaintances and even an Orthodox priest – attempted to help "the unfortunate woman" by testifying in her favour. As a result, Dvoria and her granddaughter survived and they managed to endure all the tortures during the occupation. Rubina gives an in-depth description of the life and suffering of the Jewish woman and child in occupied Krasnodar and emphasised that only they managed to survive in the city. This story was never translated into Russian, nor was it re-published in Yiddish.[88] Although, as one of the earliest literary responses to the Holocaust, Rubina's story is of great importance for the development of Holocaust memory in the region, it has not taken its rightful place in the regional dimension of Holocaust memory dispositif so far.

An autobiographical story, "Maruse. Tale of a Mother: Don't Die Before the Sun" [*Maruse. Povest' o materi: ne umirai ran'she solntsa*] by a Mountain Jew, Roman Bodalov, is about his and his mother's miraculous survival of the Holocaust.[89] Maria and her two-years-old son could easily have become Holocaust victims. They were caught by the Nazis and escorted to a killing site together with many other Mountain Jews from the Menzhinskii state farm. When the mass

---

86 Quoted in: Rubin, "A Page from the martyrdom," l. 226.
87 Ibid, l. 228 ob. Sholem Aleichem [Solomon Naumovich Rabinovich] (1859–1916) was a leading Yiddish author and playwright, who lived in today's Ukraine. Nearly every city in the former Pale of Settlement had a street named after him in Soviet times.
88 I learned about this story thanks to an interview with Sofiya Zambarnaya, conducted for the USCSF more than half a century after the events described by Rubina. Sofiya was that 8-year-old girl, who survived with her grandmother in occupied Krasnodar. Interview with Sofiya Zabramnaya, born in 1935, in *VHA USCSF*, code 33735, July 10, 1997, Poltava, Ukraine.
89 Roman Bodalov, "Maruse. Povest' o materi: ne umirai ran'she solntsa," in Roman Bodalov, *Andzhi-name: Rasskazy o gorskikh evreyakh* (Tel-Aviv, 2004), 6–122.

shooting started Maria (Maruse) jumped into the pit and covered her son. By good fortune, right after the massacre Soviet soldiers liberated this region and found Maria and her son alive. The plot of the short story is built around the personality of Maria, her deeds for the Mountain Jews (she was a prominent teacher), her courage and her decency. Her rescuers are not the main characters of the story, but without them she and her son might never have survived in the pit.

Writings by contemporary authors have enjoyed more success because they are published on the Internet. Published online, such pieces can be discovered by any readers, so the authors have acquired their own readership, followers, and, equally important, the attention of scholars and critics all over the world. Contemporary authors of literary works on the accidental survival of Jews (mostly children) in the occupied areas of the North Caucasus are, as a rule, local historians and writers, who heard the stories from locals and subsequently developed them artistically. Narrating the fates of Jewish survivors, such writings are above all a literary testament to the rescuers. A local historian from Arzgir and a member of the Russian Union of Journalists Anatolii Karnaukh entitled his story "The Incredible Rescue of the Boy Vanya" [*Neveroyatnoe spasenie mal'chika Vani*].[90] Evacuated from Kishinev together with his family, 8-year-old Ivan Bortnikov miraculously survived the mass killing in Zunda village of the Kalmyk ASSR. Wounded in the leg, the boy managed to get to Arzgir village, where he hid for a few days before being taken to the police again and imprisoned with other Jews. Before the Jews were sent to a killing site in Arzgir, Vanya was rescued by the local *politsai* Ivan Palaguta, "who specially came to work in the police a few days before [the killing action] to rescue the boy." Palaguta then went off to join the partisans while his wife hid Vanya in the cellar during the last months of the occupation. After the village's liberation, the Palaguta family adopted and raised the boy. "For the unprecedented deed they were awarded an honorary diploma and 'Righteous Among the Nations' medal in 2003."[91]

The notorious hostility of the Cossacks to the Jews lies behind the short story "Stolen from Hell" [*Ukradennaya iz ada*] [92] by Gennadii Guzenko-Vesnin.[93]

---

[90] Anatolii Karnaukh, "Neveroyatnoe spasenie mal'chika Vani," in *Pisatel' goda 2014*, vol. 16 (Moscow: Literaturnyi klub, 2015), 24–29; Idem, *Proza.ru*, http://www.proza.ru/2014/ 08/21/ 2040 (accessed March 23, 2020).
[91] Ibid, 27–28.
[92] Gennadii Guzenko-Vesnin, "Ukradennaya iz ada," *Proza.ru*, 2015, https://www.proza.ru/ 2015/06/23/1286 (accessed March 23, 2020).
[93] Gennadii Guzenko-Vesnin (born in 1938) is a poet and prosaist from Kislovodsk. Some of his poems are translated into German and Polish and others have become famous songs. The main theme of his literary works is the friendship of the Peoples of the Caucasus.

However, this piece is not about hatred and betrayal, but about the rescue of a Jewish baby girl by a Cossack family. The baby's mother managed to leave her outside the prison lying in a roll near a birch tree shortly before being killed in a mass shooting. The Cossack women took the roll and raised Aksin'ya as their own daughter and sister. Many years passed and the adopted mother decided to tell Aksin'ya the truth about her birth.

> Aksin'ya sat turned to stone. And then suddenly she fell on her adopted mother's neck crying: 'Mum! Mummy! I don't know anyone else, except you, my dear mum! I'm not an adopted child, I'm your real daughter and Olya's sister . . . ' So, they both sat silently for a while, crying and hugging each other. Each one seemed to be thinking her thoughts. Their faces revealed only tender emotion.[94]

This short story is about the bravery and courage of a Cossack woman, who had no fear of opposing Nazi rule, hiding and adopting the Jewish child at risk to her own life and that of her own daughter. Poorly written, the short story is deeply emotional in view of its truthfulness and, in many respects, uniqueness.

The story of the rescue by the Circassian villagers of Beslenei aul of 32 children who had been evacuated from Leningrad to the North Caucasus was widely known, even in Soviet times. Without paying attention to the ethnicity of the young survivors (the exact number of Jews among them remains unknown), nor to the reasons for rescuing children from the Nazis, the Soviet documentary *Children of our Village* [*Deti nashego aula*] highlighted "the friendship of peoples in the struggle against the enemy."[95] The documentary shows the fates of the Leningraders, who found their second home in this Circassian aul. The children's ethnicity is not mentioned in the film, but attention is drawn to the generosity of the "Peoples of the Caucasus who adopted the children, risking their own lives." The film centers on the happy post-war fates of the Leningrad children who stayed in the aul.

In the post-Soviet period, three more documentaries were filmed about this story, the performance "Beslenei" was staged, and the poem "Memory of the Heart" [*Pamyat' serdtsa*] was written. Each subsequent director brought new emphasis to the story in his or her own way. The main theme of the film *One Kin* [*Odna Rodnya*], filmed by Nalchik television studio in 2005,[96] is the friendship of peoples in a multi-ethnic Russian state. During the period of military

---

94 Guzenko-Vesnin, "Ukradennaya iz ada."
95 *Deti nashego aula*, directed by Yurii Gavrilov (Pyatigorskoe TV, 1970), min.18:02–18:28.
96 *Odna rodnya*, directed by Marina Sasikova (ORTK Nalchik, 2005), 21 min.

instability in the North Caucasus from the late 1990s to early 2000s,[97] the film highlights the best sides of the character and way of life of the Peoples of the Caucasus, their generosity, courage, and sacrifice for the sake of saving the children. Again, there is no mention of the children's ethnicity nor the reasons why they had to be saved from the Nazis in the film, because the interethnic relations within Russia during filmmaking were more important than the accuracy of the historical reconstruction. The film *Beslenei. The Right to Life [Beslenei. Pravo na zhizn']* directed by Vyacheslav Davydov in 2008[98] is the most objective and detailed retelling of the story. It mentioned for the first time that some children were Jewish and showed the fate of the remaining children, who left the village in the direction of nearby Teberda, where they were killed.[99] The themes of the Holocaust and Jewish survival also resound in the final soundtrack for the film:

| The calamity has neither kin, nor tribe | U bedy net ni roda, ni plemeni |
| The village of Beslenei knows it. | Eto znayet aul Beslenei.[100] |

This new knowledge increases the significance of the Circassians' deed: not only did they shelter children, hungry and exhausted after a long journey, but they also saved many of them from becoming victims of the Holocaust.

Director Nadezhda Popova shot the film *I Come from Childhood [Ya rodom iz detstva]* in 2013.[101] The film displays nostalgia for the Soviet past, for people of good morals and honour. In the story, modern-day schoolchildren at the secondary school in Beslenei write an essay about their relatives' deeds during World War II, seeking to answer the question of whether courage and compassion are necessary in life. Wartime shots morph into scenes of the children's drawings under the voice over of a story compiled from snippets of their essays.

---

97 Meant are the Chechen wars and a series of terrorist attacks in Moscow and other cities of Russia. The responsibility was placed on Chechen terrorists. See: Oleg Popov and Aleksandr Cherkasov, ed., *Rossiya – Chechnya: tsep' oshibok i prestuplenii* (Moscow: "Zven'ya," 1998).
98 *Beslenei. Pravo na zhizn'*, directed by Vyacheslav Davydov ("TONAP" film studio, 2008), 39 min.
99 Members of EK 12 directed the killing operation in the resort town of Teberda of Karachay AO in December 1942. As a result, the doctors and staff of the sanatorium and the children sick with bone tuberculosis became Nazi victims. Among them could have been the children evacuated from Leningrad. "Reports and lists of murdered medical workers (Jews) and children in Teberda of July 5, 1943," in *GARF*, f. R-7021, op. 17, d. 7, l. 3–12.
100 Soundtrack for the film *Beslenei. Pravo na zhizn'*, verse by Petr Sinyavskii, music by Aleksei Shalygin. Transl. by Anastasiya Toropko.
101 *Ya rodom iz detstva: dokumental'nyi fil'm*, DVD, directed by Nadezhda Popova (Studio "S.S.S.R.," 2013), 45 min.

The plot of the film also includes fragments from the play "Beslenei," staged by Artur Kudaev, the director of the Karachay-Cherkess Theatre in 2012. He believes that "there is a shortage of good deeds, charity, and simple human cooperation nowadays. Today, when inter-ethnic relations miss the point, this theme [the story of rescue of children by the villagers during World War II] is needed more than ever."[102] For this reason, the children, who remember their grandfathers' deeds, play the lead role in the film. The rescue of the Leningrad children is depicted as the rescue of Jews: "Between Germans and Allah, we have chosen our law [a Circassian cannot neglect a guest, especially a child], so it is not a heroic act for us, but a common occurrence,"[103] said Rima Patova, a Beslenei villager. Nikolai Chistyakov[104] also indicates the Jewishness of the surviving children and highlights the role of the traditions of Circassian society:

Those weakened children, knowingly	*Detei oslablennykh soznatel'no*
Were hidden in their homes.	*Vsekh razobrali po domam.*
For these Beslinites understood:	*Ved' Beslineitsy ponimali:*
Expect no mercy from the Germans:	*Ne zhdat' ot nemtsev poslableniya:*
Jewish youths, recognised	*Rebyat-evreev uznavali*
By their look and their talk.	*Po vneshnosti, proiznosheniyu.*
But compassion, warmth	*No miloserdie, serdechnost'*
In that moment knew no limit,	*Ne znali v tot moment predela:*
The Adygean code rose eternal	*Adyge Khabze*[105] *byl izvechno*
Above all this shooting.	*Prevyshe vsyakogo rasstrela.*[106]

No Beslenei villager who adopted Leningraders has hitherto received the honorary title 'Righteous Among the Nations,' since no documents can confirm the ethnicity of the survivors, many of whom could neither find their relatives after the war nor remember their real names. At the same time, this story is embodied

---

102 *Ya rodom iz detstva*, min. 20:00–23:03.
103 Ibid, min. 26:03–27:42.
104 Nikolai Chistyakov is a head of the Department of Theory and History of State and Law at the Financial University under the Government of the Russian Federation, doctor of historical sciences, and member of the Moscow Union of Journalists.
105 *Adyge Khabze* [Adyghe Habza] is the native belief system of Circassians, which includes a complex code of customs, social norms, and rules of comportment between people of different groups and ages in the Circassian society. The use of such norms of behaviour is one of the main indicators that a person belongs to the Circassian culture. See: Ruslan Khanakhu, ed., *Mir kul'tury adygov: problemy evolyutsii i tselostnosti* (Maykop: Adygea, 2004).
106 Nikolai Chistyakov, "Pamyat' serdtsa: istoricheskaya poema," in Nikolai Chistyakov, *Pamyat' serdtsa: leningradskie cherkesy Besleneya* (Saratov: OOO "Privolzhskoe izdatel'stvo," 2012), 131. Transl. by Tristam Barrett.

for good in the "monument not built with hands," and to paraphrase Pushkin, "common folk shall keep the path well-trodden."[107]

If these Circassian deeds are a more or less well-known story in Soviet and modern Russia, and above all in the Karachay-Cherkess Republic, the case of the rescue of Jews by the Karachays is a relatively new theme in contemporary works. Stalin's deportation of the Karachays in 1943 remains until the present day the main traumatic experience for this nation. Two documentaries, shot during the past years, seek to reflect on this sensitive historical past.[108] Cases of the courage of Karachay soldiers at the front line along with the stories of the rescue of children with disabilities from the sanatoriums of Teberda and the hiding of Jews aim to show the injustice of Stalin's directive against the entire population of Karachay. The main idea of these documentaries is that collaborators and traitors may be found among all ethnic groups, but all people should not be judged for the deeds of a few. In this context, the story of the rescue of the Geidman sisters by the Karachay families Khalamliev and Kubekov has a particular emotional resonance.[109] They survived the occupation, hiding in a house where a German officer had taken up residence in one room. After the liberation of Kyiv, they discovered that their family had become Holocaust victims in Babi Yar. The Geidman sisters tried their best to demonstrate the good deeds of the Karachay families, they visited them in exile, and in 1994 the Khalamlievs were honoured with the title 'Righteous Among the Nations.'[110]

Thus, the theme of Jewish survival in fiction and documentaries appeared relatively late, when most main characters (those who rescued Jews) were already dead. These pieces are about modern-day saints, who risked their own

---

[107] These lines come from Aleksandr Pushkin's celebrated poem about the role of literature: "A monument I've raised not built with hands, / And common folk shall keep the path well trodden." Aleksandr Pushkin, "Exegi Monumentum," transl. Avril Pyman, *RuVerses website* https://ruverses.com/alexander-pushkin/exegi-monumentum/1945/ (accessed February 11, 2020).

[108] *Voina i mir Karachaya*, directed by Vyacheslav Davydov, Episode 1 (Kinostudiya Tonap, 2011), https://www.youtube.com/watch?v=dy4VxSXRmUM (accessed February 11, 2020), 51 min. 48 sec.; *Prigovorennye k zabveniyu*, directed by Marat Duraev i Anton Stepanenko (2019), https://www.youtube.com/watch?v=3XuSN-ZqM-w (accessed February 11, 2020), 119 min.

[109] There are other stories of rescuing Jews in the films. See: *Voina i mir Karachaya*, min. 22: 34–28:21; 36:13–42:14.

[110] "Khalamliev Shamail and Khalamlieva Ferdaus," the Righteous Among the Nations Database, *Yad Vashem official website*, https://righteous.yadvashem.org/?searchType=righteous_only&language=ru&itemId=4045060&ind=6 (accessed February 15, 2020). See more about the rescue of Jews by Karachays: Rashid Hatuev, *Evrei v Karachae* (Chervessk: Karachaevskij NII, 2005), 48–50.

lives hiding and saving Jews in the occupation. The authors are mainly not Jews, and seek to memorialise the heroic deeds of those who did them, not for fame but for life itself.

5.4 A Look from Tomorrow into Yesterday: A Reflection on Artistic Representations of the Holocaust

Imagery is always an author's response to an historical fact, on the one hand, and a reflection of official memory, on the other hand. In Soviet times, most writers and poets consciously excluded and hid the theme of the Holocaust behind other military subjects or post-war stories, as, for example, behind the description of retribution against collaborators at the Soviet trials. In post-Soviet times, authors began to focus on the very fact of the Holocaust and sought to interpret this crime against humanity in contemporary light. Fiction and documentary became more sensitive, even reflecting upon problems in modern Russian society. Accordingly, the story of the removal of the commemorative plaque at the Zmievskaya Balka memorial in 2011 became a subject for interpretation by artists from Rostov-on-Don. In 2013, Yurii Kalugin made a new film *The Suffering of Memory* [*Stradanie pamyati*][111] 20 years after his first documentary *Judenfrei: svobodno ot evreev*. *Stradanie pamyati* tell us not only about the fact of the Holocaust in Rostov-on-Don. First and foremost, it is a declaration of remembrance of the Holocaust among the modern-day citizenry of Rostov-on-Don. It makes us to look at the figurative "Ivan, who does not remember kinship"[112] in a new light. The main idea of the film is to remind present generations that each person dies twice: once physically and a second time when he or she is forgotten. Indeed, for long decades after the mass killing of Jews in Rostov-on-Don, and even during the 20 years after Kalugin made his first film, a process of commemoration of Holocaust victims has been occurring in this southern Russian city. As of the time of filming, the Jewish community had managed to recover the names of more than 3,000 Jewish victims;

---

[111] *Stradanie pamyati*, directed by Yurii Kalugin (St. Petersburg, 2013), https://www.youtube.com/watch?feature=share&v=QRsMA1J65RQ&desktop_uri=%2Fwatch%3Fv%3DQRsMA1J65RQ%26feature%3Dshare&nomobile=1 (accessed March 23, 2020) 40 min, 14 sec.

[112] "Ivan, who does not remember kinship" is a Russian expression that in a figurative sense denotes people without convictions and traditions, who do not know and do not want to know their history, culture, and ancestors. Ol'ga Severskaya, "Ivan, rodstva ne pomnyashchii," *Russkii yazyk,* April 16–30, 2010, http://rus.1september.ru/view_article.php?ID=201000814 (accessed March 23, 2020).

a March of the Living has been held annually since 2012 on the anniversary of the mass killing of Jews in the city, in which hundreds of citizens and foreign guests participate; Zmievskaya Balka is gradually gaining a place as an unofficial symbol of Russia's Babi Yar. However, Kalugin shows the other side of the coin: everyday life in Rostov-on-Don has no place for memory of Holocaust victims. The "Case of Four Screws" is the title of the episode in the film in which Kalugin shows footage of the trial in Rostov-on-Don over the replacement of the commemorative plaque on the memorial in 2012.[113] *Stradanie pamyati* highlights the polarisation of memory and forgetting, how these two poles peacefully coexist in the modern Russian city and that the truth, as it turns out, can be different, just as its memory is multi-faceted. Although Kalugin's appeal to Holocaust eyewitnesses and relatives of Holocaust victims in Zmievskaya Balka to share their memories received responses from dozens of citizens of Rostov-on-Don, including those who live in Israel, the last scene of the film became its philosophical culmination. At the very end of the film, Kalugin filmed the memorial site at Zmievskaya Balka on International Holocaust Remembrance Day, January 27. The camera depicts the huge territory of the memorial complex, blanketed in a patchwork of snow and rain puddles (typical of winter in southern Russian cities). At the whole memorial site there is only one person, who places a pebble before the bowl with the Eternal Flame. But even this bowl is not ablaze. "There are 27,000 of us lying here and none of us can pay for the gas," says a voice-over.[114] One cannot avoid wondering: does the memory of Zmievskaya Balka remain? And if so, whose memory is it, and who needs it? Only the playing of final soundtrack, with the words "And you know, everything is yet to come" [*A znaesh', vse eshche budet*][115] inspires hope for change, and shows that someday Holocaust memory will "cease to suffer." In the film *Holocaust: The Eastern Front*, Boris Maftsir also showed footage from the lawsuit in the section about Rostov-on-Don.[116] The story he told is more life-affirming: filming an interview with the plaintiff Vladimir Livshits in the context of a new "compromise" plaque erected at the memorial site in 2014, which mentions Jews, but not the term Holocaust.

The replacement of the commemorative plaque in Zmievskaya Balka in 2011 also became a theme of several literary works. Igor' Khentov wrote the poem "Monument" [*Pamyatnik*] in August 2012, which raises the important question "What is more fatal than memory loss?" This rhetorical question is a

---

113 Kalugin, *Stradanie pamyati*, min. 29:40–34:25.
114 Ibid, min 36:00–38:00.
115 *A znaesh', vse eshche budet*, verse by Veronika Tushnova, performed by Fedor Pryanikov.
116 *Holocaust: The Eastern Front*, directed by Boris Maftsir, *Holocaust in the USSR official website*, http://www.holocaustinussr.com/films-en/ (accessed July 5, 2017), min. 45:00–67:00.

challenge to Rostov-on-Don's authorities and citizens, who have neglected history, and whom "Nazism rushed to visit again." Meanwhile, the poet seems optimistic about the future:

God gazes upon this earth,	*Bog vziraet na zemlyu pristal'no,*
He answers for victims and killers.	*On za zhertv i ubiyts v otvete.*
Everything passes, and only truth	*Vse prokhodit, i tol'ko istina*
Triumphs in this world.	*Torzhestvuyet na etom svete.*[117]

The celebrated Soviet poet Evgenii Evtushenko[118] also responded to the controversy over the memorial plaques in Rostov-on-Don and Holocaust remembrance in the city. In December 2014, he visited Rostov-on-Don, where he read his latest poem about the Holocaust in Zmievskaya Balka.[119] To him, the memorial site in Rostov-on-Don is "Babi Yar's sibling." In 1961, Evtushenko was one of the first poets who raised the curtain on the tens of thousands of Jews killed at Babi Yar. After that, the theme of the Holocaust entered literature and public debate for the first time following many years of post-war silence.[120] Back then, in the 1960s, Evtushenko's "Babi Yar" became a literary revelation for the Soviet public, making it possible to hear Holocaust victims' voices for a short time. His subsequent poem "Watchman of Zmievskaya Balka" [*Storozh Zmievskoi Balki*] did not cause a similar public response in modern Russia, but it became an event for Rostov-on-Don. Furthermore, it tells not only about Rostov-on-Don, but raises the problems of Holocaust memory in modern Russian society and the role of the official state ideology in remembering certain groups of victims of the war:

Why is it that at different times,	*Pochemu eto v raznoe vremya*
They made a fuss over nothing,	*Kolotilis', neznamo chego,*
Avoided the word "Jew,"	*Izbegayuchi slov "evrei,"*
And wiped him out?	*I vymaryvali ego?*
They didn't like his imperious looks,	*Tak ne shla k ikh nachal'nich'ei vneshnosti*
All this fuss over a word.	*Suetnya vokrug slova togo.*
And then they resurrected him in haste,	*A potom voskreshali v pospeshnosti.*

---

[117] Igor' Khentov, "Pamyatnik," in Khentov, *Rodoslovnaya*, 38. Transl. by Tristam Barrett.
[118] Published in English as Yevgeny Yevtushenko, Evgenii Evtushenko (1933–2017) was a well-known Soviet and Russian poet, and one of the authors politically active during the Khrushchev Thaw.
[119] Aleksandr Gamov, "Evgenii Evtushenko – 'Komsomol'skoi pravde': kupalsya v vannoi. Upal. Razbil golovu. Tol'ko ne podnimaite paniku!," *Komsomolskaya pravda*, December 14, 2014.
[120] Clowes, "Constructing the Memory of the Holocaust."

Resurrect one here at least.	*Voskresit' by zdes' khot' odnogo.*[121]

This poem is a call to think about the eternal, about the price of human life, about the idea that the memory of Holocaust victims, and more broadly, of all victim groups of World War II, the memory of every murdered person in general is more important than any "inscriptions" on a monument.

I would advise everyone spend their lives	*Ya by tratit' vsem zhizn' posovetoval*
On people, and not "inscriptions."	*Na lyudei, a ne na "nadpisya."*[122]

The theme of the responsibility of future generations towards all Holocaust victims and towards each person separately runs through writings devoted to the mass killings of Jews in the North Caucasus in 1942. The Russian poet of Mountain Jewish origin, Efrem Amiramov[123] dedicated his poem "At the Obelisk" [*U obeliska*] to the memory of 472 Mountain Jews buried in the village of Bogdanovka.[124] The psychological trauma from the loss of relatives during the Holocaust becomes a wound on the body of second-generation Holocaust survivors:

The rusty bullet wounds anew	*Pulya rzhavaya snova ranit*
Our souls.	*Nashi dushi.*
Piercing our memory,	*Vonzayas' v pamyat',*
Burning us with winds of sorrow.	*Vetrom skorbi nas obzhigaya.*[125]

A Rostov-on-Don writer and a Holocaust survivor Inna Kalabukhova[126] wrote a story called "A Memorial Prayer" [*Pominal'naya molitva*] about her relatives, who became Holocaust victims.[127] Literature was a way of putting her personal memory into words and at the same time express her disappointment over the quality of memorial work in her hometown. The story was written in 2002 after the author's participation in a commemorational gathering at Zmievskaya Balka.

---

121 Evgenii Evtushenko, "Storozh Zmievskoi Balki," in Gamov, "Evgenii Evtushenko." Transl. by Tristam Barrett.
122 Ibid.
123 Efrem Amiramov (born in 1956) is a Russian poet, author, singer-songwriter, and television producer. He is a Mountain Jew, born in Nalchik, and a well-known singer of Russian chanson.
124 Efrem Amiramov, "U obeliska," https://utices.com/page-text.php?id=20199 (accessed February 11, 2020).
125 Ibid. Transl. by Tristam Barrett.
126 Inna Kalabukhova (born in 1933) is a Jewish Soviet and Russian journalist and writer. She was born in Moscow, but most of her life is connected to Rostov Oblast, where she now lives. In 2008, she became a member of the Union of Russian Writers as a prose writer.
127 Inna Kalabukhova, "Pominal'naya molitva," in Inna Kalabukhova, *V proshchan'i i v proshchen'i* (Moscow: Dom evreiskoi knigi, 2008), 58–89.

> The empty, soulless museum and the state monument did not touch my soul; they had nothing in common with the cursed and murdered in Zmievskaya Balka. Well, what about this memorial prayer in Hebrew, these kippahs, these funerary pebbles? How do they mate with that special tribe of Soviet Jews, our Rostov Jews, natives of Ukraine and Belarus, whose forefathers' language was Yiddish, but they only knew a dozen or two different words and a couple of local jokes.[128]

Inna Kalabukhova raised the topical problem of the functioning of human memory. She stresses that it is not enough to install a monument, to establish a museum, to read a prayer on certain days, or to place memorial pebbles or flowers. More important is the memory of a soul, the memory of each person, of all humanity as a whole. Monuments continue to be installed; memorial gatherings are held on remembrance days, which are attended by Holocaust eyewitnesses, today's schoolchildren, representatives of the authorities, and ordinary citizens. But how deep does Holocaust memory exist in each of them, in each of us? The rhetorical question of this writer from Rostov-on-Don will remain topical in Russia for many years.

Artistic responses to Holocaust history are an important element in the functioning of a regional memory dispositif. The authors, natives of the USSR or of other post-Soviet countries, created their literary responses to the Holocaust at different historical times. This important fact suggests that the externalisation, universalisation, and sovietisation of Holocaust memory as the main feature of fiction and documentaries of Soviet and partially Russian (famous) artists could be here complemented by its localisation. As we can see, the theme of the Holocaust, though not always the main storyline, was present in the work of Soviet writers throughout post-war history. Depicting the occupation period, "the selfless struggle of Soviet patriots," or the necessity of retribution against Soviet collaborators for their crimes during World War II, Soviet imagery could not avoid one of the historical facts of the life of "Soviet citizens under the Germans," that is, the story of the Holocaust. For a long time, artists highlighted key points in their pieces "in a Soviet manner," which at the same time allowed readers to read between the lines. In modern Russian imagery, Holocaust history is represented as an integral part of the occupation. Professional writers and film directors are not always represented among today's authors. However, this fact does not degrade the quality and significance of the work. Artists frequently address their works to future generations using modern technology. Their writings are ever more published not on paper or videotape, but on the Internet, which significantly extends and increases their readership and audience. Special portals, where

---

128 Kalabukhova, "Pominal'naya molitva," 65.

literary works are published and documentaries are uploaded, provide an opportunity for many artists to be heard, avoiding the obstacles of publication expenses, promotion, and distribution. In the context of globalisation, works written or filmed in outlying areas can be viewed and downloaded in any corner of the world, which significantly expands the 'market' for artistic responses to the Holocaust in the North Caucasus. This localisation of Holocaust memory, then, after becoming a significant feature of the regional Holocaust memory dispositif, may turn into a globalised Holocaust memory. And artistic responses to the Holocaust in the North Caucasus will become an integral part of the world treasury of Holocaust imagery.

6 Reifying Memory? Representations of North Caucasus Holocaust History in Museum Collections

During my fieldwork in the North Caucasus region, I would contact the employees of particular state institutions and public organisations (schools, museums, Jewish communities) in advance and make appointments to meet with them. Our conversations were often personal in nature. During the interviews I was usually informed that I could ask any questions that were of interest to me, look through the paperwork, and make photocopies of any document. If I had a meeting with museum employees, the interview took place in the exhibition halls and was accompanied by stories about the creation of the particular exhibition. Despite my asking the museum employees to conduct their tour as if it were for an average group of visitors, we still had some sort of conversation concerning a specific topic. Being aware of my area of interest, my interview partners shaped their narrative according to my expectations. They consciously paid attention to those few showpieces dedicated to Holocaust history that are presented in the exhibition. For example, I was shown photocopies of the text of the "Appeal to the Jewish Population" in the majority of state museums of Stavropol Krai. More often, my interview partners usually stressed the reasons for the absence of Holocaust history in the permanent exhibitions about the "Great Patriotic War." They mentioned the lack of exhibition space, a different thematic focus of the exhibition, and sometimes even the absence of – and/or unwillingness of museum staff to seek out – physical evidence of the Holocaust. I thus received exhaustive information from the museum employees about how and according to what ideological background a museum exhibition on the war was created.

At the same time, my interviews with museum staff did not allow me to explore an even more important component of the work of the museum: the museum's visitors themselves. Visitors are the main reason for the creation of permanent or travelling exhibitions, for enriching the museum funds, and for the renewal of the exhibition halls. It is one thing to see the museum collection for myself and comprehend it according to my scholarly perspective or get answers to the questions I raise. It is quite a different experience to see the work of the exhibition in action, during a guided tour for visitors, when showpieces, arranged in a specific way by the curators, acquire a voice in the form of the guide's narrative. I was thus particularly encouraged when I visited the Stavropol State Historical-Cultural and Natural-Landscape Museum-Reserve (SGMZ) to find that, instead of giving me a personal interview, the official I was supposed to meet attached me to a group of schoolchildren from a Stavropol secondary school, which had come to visit the tour "Battle Glory of Stavropol" [*Boevaya slava Stavropolya*]. This exhibition was opened in 2015 and timed for the 70th anniversary of the Great Victory. Sergei Dvernik, the researcher at the museum's department of cultural and educational activities was the tour guide on the day of my visit. Some 25 teenagers in military uniform[1] listened to the guide for an hour (of course, only a few were paying attention while others were constantly distracted by chit-chat and flirting with each other, causing the group-leader and the guide to repeatedly call the adolescents to order). In his story, Dvernik dwelled at length on the military exploits of the residents of Stavropol Krai during the Great Patriotic War. The exhibits were organised according to the main historical stages of the war on Soviet soil with special attention to the history of the Battle for the Caucasus. A separate stand was dedicated to the occupation of Stavropol Krai, presented in the form of brief information and photos of well-known partisans and underground leaders, issues of newspapers *Russkaya pravda* [Russian Truth] and *Stavropol'skoe slovo* [Stavropol Word] printed under the occupation, as well as a photocopy of the "Appeal to the Jewish Population of Voroshilovsk" cited in Chapter 2. Standing near this particular exhibit, the tour guide moved on to the story of Klavdiya Abramova after telling us the dates of the occupation of Stavropol. This woman was an underground leader who was tortured in prison and then executed along with her young daughters.[2] Only after my question about the reasons for the arrest of Abramova – whether it was because of her Communist background or because of her ethnicity, since her last

---

[1] Schoolchildren from each Stavropol secondary school maintain a guard of honour at the Post No. 1 near the Memorial "Fire of Eternal Glory" [*Ogon' vechnoi slavy*] during one week. A visit to the museum is one of the forms of military-patriotic education of schoolchildren. The ritual of changing sentries at Post No. 1 was first held on January 27, 1924 at the Lenin Mausoleum. The first permanent Komsomol-Pioneer Post No. 1 was established in 1965 near the Eternal Flame in the center of the city-hero of Volgograd. In the 1970s and 1980s, Komsomol-Pioneer Posts were found throughout the country near the monuments and obelisks to the revolutionary, military or labouring glory of the Soviet people. In post-Soviet Russia, according to the state programme "Patriotic Education of Citizens of the Russian Federation for 2001–2005," the main purpose of which was the creation of a system of patriotic education, these Posts No. 1 were revived. In some cities, centers for patriotic education were founded on this basis, and they were given the status of an additional educational institution. "Istoriya Postov no. 1 strani," *website of Dom detskogo tvorchestva No. 1 goroda Irkutska*, http://38ddt1.ru/post-1-2/iz-istorii-postov-1-strany/ (accessed March 24, 2017).
[2] Klavdiya Abramova is the main character of the story by Vladimir Gneushev "Klavdiya Il'inichna," discussed in the previous chapter.

name sounds very Jewish – did the guide draw our attention to the text of the "Appeal" and said:

> Immediately after the occupation of the city on August 3, 1942, members of the Sonderkommando were the first to come to Stavropol. What did the people who were part of this unit do? Cleaning up [the city]. Whom did they clean the city of? The Jews. Here you can see [pointing to the text of the "Appeal"] 4,000 Stavropol and evacuated Jews were killed in total[3] [. . .] Gas vans drove up near Makarova Street with a few dozen people bundled inside, the door was closed and carbon monoxide came into the compartment. Sick people [i.e., patients of psychiatric clinics] could not work: they were of no value to the Hitlerites. The city was also cleansed of Communists; I already told you about Abramova, she was executed. Well, also [the city was cleansed] of all those who resisted the new [German] authorities. So, it was really difficult to survive.[4]

Thus, the tour guide mentioned the Holocaust and other Nazi crimes during the more than six-month occupation of Stavropol quite briefly and only after my intervention in his narrative. Again, the absence of details about the reasons for the mass killing of Jews in the guide's narration is astonishing, since at the same time he mentioned the reasons for the killing of mentally ill patients of the Stavropol psychiatric clinic and the killings of partisans. Dvernik gave out the anonymous general figure of Holocaust victims killed only in Stavropol, while the exploits and deeds of individual victims mainly party members and partisans were thoroughly described. Their portraits and full names are presented in the showcase (Figure 20). Hence, the objective of this and the majority of other similar exhibitions concerning the history of the Great Patriotic War in state local history museums in the North Caucasus – and in Russia more generally – is clear. They aim to show the heroism of compatriots and the significance of the Great Victory for the Russians. Therefore, the guide could devote a minute and a half of an hour-long narrative to name a couple facts about victims of the war, again using the same heroic-patriotic subtext: even the executed underground leaders and partisans were heroes; streets were named and monuments were installed in their honour in the post-war era. Holocaust victims, as we notice, remain outside the framework of the dominant heroic model through which war history is represented in Russian state museums.

---

3 He mentioned the exact number of Holocaust victims in Stavropol. According to the ChGK report, 3,500 local Jews were killed on 12 August 1942 at the airfield and about 500 Jews on 15 August 1942 near the psychiatric clinic. "ChGK report for Stavropol of February 23, 1943," in *GARF*, f. R-7021, op. 17, d. 1, l. 92.
4 Audio and video recording of the SGMZ guided tour "Battle Glory of Stavropol" and interview with the guide and researcher at the Department of Cultural and Educational Activities Sergei Dvernik, in *author's archive*, code 15-MEM-SK09, May 19, 2015, Stavropol.

Figure 20: The whole display case of the "Battle Glory of Stavropol" exhibition and a close-up view of the objects depicting the occupation of Stavropol at the SGMZ (author's photograph, 2015).

This chapter is devoted to the analysis of museum activities as one of the most important elements of the Holocaust memory dispositif. Unspoken in official discourse, historical facts are a significant feature of this memory dispositif in the North Caucasus. The few 'silent' Holocaust showpieces in museum exhibitions can be 'speaking objects,' provoking guides and visitors to pay attention to this non-heroic part of the history of the Great Patriotic War. A more detailed Holocaust history is presented in the Jewish museums (the Museum of Jewish Heritage and Holocaust in the Memorial Synagogue on the Poklonnaya Gora, and the Jewish Museum and Tolerance Center) that have opened during the last decade in the Russian capital, in some regional school museums and travelling exhibitions created by local historians, employees of the Holocaust Center, or independent researchers. Based on regional sources assembled as a result of several field research trips between 2013 and 2016, I discuss what kind of Holocaust history in the North Caucasus is represented in central and regional, public and private museums around Russia and in the permanent exhibitions of international Holocaust museums. It is important to understand how and through what means Holocaust history in the North Caucasus is (or is not) being written into the all-European Holocaust narrative as well as in the grand narrative of the "Great Patriotic War" in Russia. I am interested in how and in what circumstances the modernist museum[5] can become postmodern, or a space in which World War II history in Russia can be interpreted through a variety of stories, both heroic and non-heroic.

6.1 Holocaust History without its Story: (Mis-)representation of the North Caucasian Case in International Holocaust and Russian State Museums

The museum – in the form it took in the 19th century – is a special space, removed from everyday life. Michel Foucault viewed the museum as a heterotopia, a space of difference, a space that is absolutely central to a culture but in which the relations between elements of a culture are suspended, neutralised, or inverted. Heterotopias are real places, "designed into the very institution of society," where all the other real emplacements of a culture are represented, contested, and reversed at the same time. They exist outside everything, although they can be localised.[6] Representation is a crucial point in understanding the activity of any kind of museum, because visitors always perceive the historical past through a lens refracted by the museum's owners and staff, invited curators, and even the current political (state) ideology about how particular historical periods should be represented. At the same time, heterotopias are spatially isolated places that juxtapose incompatible objects and discontinuous time, having "the role of creating a space of illusion that denounces all real space, all real emplacements [. . .] as being even more illusory."[7] Thus, museums, like cinemas or meticulously organised gardens, present an illusory version of human life or nature, questioning and contesting the "real" order of things.[8]

---

[5] The museum in the form in which it was defined in the 19th century and existed until the last quarter of the 20th century is called classical or modernist. See: Eilean Hooper-Greenhill, *Museums and the Interpretation of Visual Culture* (London: Routledge, 2000), 126.
[6] Michel Foucault, "Different Spaces," in Michel Foucault, *Essential Works of Foucault 1954–1984*, transl. Robert Hurley, vol. 2 (London: Penguin, 1998), 178.
[7] Ibid., 184.
[8] For more about the concept of heterotopia in museum studies, see: Irit Rogoff, "From Ruins to Debris: the Feminisation of Fascism in German-history Museums," in *Museum – Culture: Histories, Discourses, Spectacles*, ed. Daniel J. Sherman (London: Routledge, 1994), 223–249; Tony Bennett, *The Birth of the Museum: History, Theory, Politics* (London: Routledge, 1995); Miriam Kahn, "Heterotopic Dissonance in the Museum Representation of Pacific Island Cultures," *American Anthropologist* 97, no. 2 (1995): 324–338; Hans Belting, "Place of Reflection or Place of Sensation?," in *The Discursive Museum*, ed. Peter Noever (Vienna: MK, 2001), 72–82.

A museum is a space where systems of rules enable words and things to hold together; it is systems of representation between words and things. Like discourse analysis, the museum can be read to reveal the corpus of rules that are used to bring words and things together.[9] Therefore, the important actors in the construction of stories in museums (the ordering of space, its filling with non-discursive showpiece-objects accompanied by stories written or told) are both curators (people creating exhibitions) and the historical time in which it is possible to create certain representations of the past. This characteristic of the museum as heterotopia is of great importance in the study of museums dedicated to Holocaust history.

The so-called "Holocaust museums boom" has emerged over the past thirty years, culminating in the establishment of dozens of museums and institutions to remember and tell the history of Nazi Germany's destruction of European Jews during World War II. Depending on where these museums are built, they remember this past through a variety of (trans)national myths, ideals and political demands.[10] Generally speaking, such museums can be placed in two categories. The first are located at the physical sites where collective violence took place. These include concentration camps, forced-labour camps, killing sites, and mass graves where Holocaust victims are buried. The second group consists of museums that re-create historical sites of violence within their walls, and are often removed from the locations where the events they describe originally happened. Often costly and based in large cities or capitals, they routinely attract hundreds of thousands of visitors per annum.[11] Regardless of the shape of particular national myths, international Holocaust museums of the second group (e.g., the USHMM in Washington, 1994; the Holocaust History Museum in Yad Vashem in Jerusalem, established in 1953, now housed in a new exhibition complex since 2005; the Memorial to the Murdered Jews of Europe in Berlin, 2005) aim to show the general history of the mass killing of the European Jews.[12] Holocaust victims

---

[9] Beth Lord, "Foucault's Museum: Difference, Representation, and Genealogy," *Museum and Society*, 4 no. 1 (2006): 11–24.

[10] The term "Holocaust museums boom" originated in the works of James Young. James Young, "Holocaust Museums in Germany, Poland, Israel, and the United States," in *Contemporary Responses to the Holocaust,* ed. Konrad Kwiet and Jürgen Matthäus (Wesport: Praeger Publishers, 2004), 249.

[11] Greig C. Crysler, "Violence and Empathy: National Museums and the Spectacle of the Society," *Traditional Dwellings and Settlements Review* 17, no. 2 (2006): 19.

[12] For more about Holocaust representation in international Holocaust museums see: Edward T. Linenthal, *Preserving Memory: The Struggle to Create America's Holocaust Museum* (New York; Chichester: Columbia University Press, 2001); Katrin Pieper, *Die Musealisierung des Holocaust: das Jüdische Museum Berlin und das U.S. Holocaust Memorial Museum in*

are remembered through their pre-war life, time in the concentration and death camps, through resistance and rescue, portraits of some of them in the Hall of Names, and occasionally through brief life stories of Holocaust victims. Using different visualisation models (showpieces or installations), Holocaust museums face the future with their universal idea of "never again:" each museum has educational centers, where seminars and workshops are held for teachers and scholars, dozens of groups of schoolchildren and students from all over the world have guided tours every day. However, if seeking to explore the Holocaust history of a particular region – for example the representation of the Holocaust in the North Caucasus in international Holocaust museums – it becomes apparent that the general picture has parts that remain completely hidden: the Holocaust in the North Caucasus remains overshadowed by the story of European ghettos and the gas chambers of Auschwitz.

Upon closer study, the universalisation of the Holocaust within museum exhibitions seems to dissolve and subordinate many "insignificant" historical facts that resulted in hundreds and thousands of victims. The Holocaust museum in its universal sense becomes the translator of the most common stereotypical symbols and images which society associates with the Holocaust tragedy. If historical events fall outside of these symbols and images, international remembering communities cannot recognise them. For example, the museum exhibition at the Memorial to the Murdered Jews of Europe in Berlin attempts to present Holocaust

---

Washington D.C. (Köln: Böhlau Verlag, 2006); Crysler, "Violence and Empathy," 19–38; Rick Crownshaw, "Photography and Memory in Holocaust Museums," *Morality* 12, no. 2 (2007): 176–191; Ana Carden-Coyne, "The Ethics of Representation in Holocaust Museums", in *Writing the Holocaust,* ed. Jean-Marc Dreyfus and Daniel Langton (London: Bloomsbury Academic, 2011), 167–184; Amy Sodaro, *Exhibiting Atrocity: Presentation of the Past in Memorial Museums* (New Brunswick: New School University, 2011); Young, "Holocaust Museums in Germany," 249–274; Idem, *Constructing Memory: Architectural Narratives of Holocaust Museums* (Bern: International Academic Publishers, 2013); Jennifer Hansen-Glucklich, *Holocaust Memory Reframed: Museums and the Challenges of Representation* (New Brunswick: Rutgers University Press, 2014); Michal Aharony and Gavriel D. Rosenfeld, "Holocaust Commemoration: New Trends in Museums and Memorials," *Dapim: Studies on the Holocaust* 30, no. 3 (2016): 162–165. See also the Catalogues of the museums' collections: Jeshajahu Weinberg and Rina Elieli, *The Holocaust Museum in Washington* (New York: Rizzoli, 1995); Dorit Harel, *Facts and Feelings: Dilemmas in Designing the Yad Vashem Holocaust History Museum* (Jerusalem: Yad Vashem Publications, 2010); Uwe Neumärker, ed., *Holocaust. Der Ort der Information des Denkmals für die ermordetet Juden Europas: Katalog* (Berlin: Stiftung Denkmal für die ermordeten Juden Europas, 2015); Ulrich Baumann, Paula Oppermann, and Christian Schmittwilken, *Mass Shootings. The Holocaust from the Baltic* to *the Black Sea 1941–1944: Exhibition Catalogue* (Berlin: Stiftung Topographie des Terrors, Stiftung Denkmal für die ermordeten Juden Europas, 2016).

history through its personal dimension. In the Room of Dimensions, 15 stories are told through the latest ego-documents (letters, diaries, notes) of Holocaust victims. The Room of Families shows various Jewish lifestyles, using the example of 15 families. Brief biographies of 700 killed and missing Jews from all over Europe can be heard in the Room of Names.[13] Despite the fact that the history of the Holocaust on Soviet soil is addressed in almost all exhibition halls – with closest attention devoted to events that occurred in Ukraine and Belarus – the history of the Holocaust in Russia is excluded from the general narrative. Of the 500 Holocaust sites presented on the map in the Room of Sites – small orange rectangles that indicate the locations of camps, ghettos, sites of mass shootings, and the starting points of deportations – little more than a dozen of these rectangles mark localities in the North Caucasus.[14] The absence of place-names on the map, and of descriptions of the places marked by rectangles, make it impossible for the average European visitor to 'read' the map and connect Holocaust history to events in the North Caucasus. The last room, an "Information Portal to European Sites of Remembrance" offers current and historical information concerning commemoration sites, museums and memorials and provides an insight into the world of European Holocaust remembrance. Visitors can access an electronic database of such sites here. Photographs of the most famous Holocaust memorial complexes throughout Europe hang on the walls of the room, but there are no photographs of Holocaust sites in the North Caucasus. Only the electronic portal contains information about Zmievskaya Balka in Rostov-on-Don, Petrushina Balka in Taganrog and the memorial complex in Krasnodar.[15] The Holocaust in the North Caucasus is forgotten three times: first, the sites in which hundreds of Jews were exterminated during the war; second, the oblivion into which these sites fell in the post-war period in the Soviet Union and later Russia (as noted in

---

[13] "Information Center under the Field of Stelae: permanent Exhibition," *Memorial to the Murdered Jews of Europe and Informational Center official website*, https://www.stiftung-denkmal.de/ausstellung/information-centre-under-the-field-of-stelae/?lang=en (accessed March 24, 2020).

[14] See the photograph of the map in the museum's catalogues: Neumärker, *Holocaust. Der Ort der Information*, 312–313.

[15] The Holocaust sites in Rostov-on-Don, Taganrog, and Krasnodar, *International portal to European sites of Remembrance*, https://www.memorialmuseums.org/denkmaeler/view/355/Denkmal-%C2%BBSmijowskaja-Balka%C2%AB; https://www.memorialmuseums.org/denkmaeler/view/812/Denkmal-%C2%BBSchlucht-des-Todes%C2%AB-f%C3%BCr-die-Juden-der-Stadt-Taganrog; https://www.memorialmuseums.org/denkmaeler/view/787/Denkmal-f%C3%BCr-die-Opfer-des-Faschismus-in-Krasnodar (accessed February 12, 2020). The museum's catalogue also features a photograph of the monument in Zmievskaya Balka. Neumärker, *Holocaust. Der Ort der Information*, 441.

the previous chapters); and third, as sites without a place in the global remembrance of the Holocaust in international museums and educational centers.

Representation always implies the perspective of one person or group of people on the events of the past. It also emphasises some facts at the expense of others to create a more or less holistic historical narrative. The creation of an exhibition story, which presents certain actors and specific events, is in many ways a subjective construction of the curators. Holocaust history in European countries is undoubtedly better documented and studied than the Holocaust on Soviet soil. The history of mass killings of Soviet Jews began to be actively researched only after the collapse of the USSR when scholars gained access to Soviet wartime sources. So far, the events in the North Caucasus are a grey zone: all the killing sites and the exact number of Holocaust victims are still unknown and the already known killing/grave sites are not always marked by a monument or a memorial stone. In many Holocaust museums in Israel, Europe, and in the United States, the Holocaust on Soviet soil tends to be overshadowed by what might be called the "Holocaust in Poland," characterised by large ghettos such as Warsaw and Łódź and killing centers like Treblinka and Auschwitz.[16] Hence, there is a complete absence of North Caucasian stories in the representation of Holocaust history in international Holocaust museums.

At the same time, the Holocaust museums boom is an integral part of Holocaust memorial work in most western countries and in Israel. There is still no special Holocaust museum in Russia, while museums or exhibitions dedicated to the history of the Great Patriotic War are in every city and district center. The most obvious Russian position in Holocaust representation can be traced on the work of the Central State Museum of the Great Patriotic War, or the Victory Museum. It was opened by the first Russian President Boris Yeltsin in 1995 and became the main object of the Victory Memorial Park on Poklonnaya Gora in Moscow.[17] There is little information about Holocaust history on Soviet soil in the museum's permanent exhibition "Exploits and Victory of the Great People" [*Podvig i pobeda velikogo naroda*]. A separate showcase covering the Nazi German occupation presents "items from the places of execution of Soviet citizens (Vitebsk region, Belarus)" alongside children's objects and toys from Majdanek

---

[16] Natan M. Meir, "The Place of Polish Jewry in a Russian Jewish Museum: The Case of the Jewish Museum and Tolerance Center in Moscow," *Studia Judaica* 32, no. 2 (2013): 112–113.

[17] "The national character of the Victory Museum was emphasised at the time of its opening by the presence of delegations from 55 countries, who left remarkable entries in the Book of Honoured Guests." "Istoriya muzeya", *Victory Museum official website*, https://victorymuseum.ru/about/history/ (accessed February 12, 2020).

concentration camp, and household items of Jews from the ghetto.[18] Annually since January 2015 – the 70th anniversary of the liberation of Auschwitz – temporary exhibitions related to Holocaust history in Western Europe and the liberation by Soviet soldiers are housed in the Victory Museum.[19] That is, the Soviet model of externalisation of Holocaust representation[20] continues to dominate in the main Russian state museum of the Great Patriotic War. The theme of mass killings of Jews is presented in the permanent and temporary exhibitions as one of the historical facts of World War II, but the displays speak about events that occur outside the borders of modern Russia.

In post-Soviet Russia, several museums of Jewish history were opened to represent the history and culture of the Russian/Soviet Jewry. In 1998, the Museum of Jewish Heritage and Holocaust was opened in the Memorial Synagogue on Poklonnaya Gora in Moscow. The museum was constructed and operated by the RJC.[21] The permanent exhibition is divided into two parts. The first is devoted to the history, religion and everyday lives of Jews in the Russian Empire and Soviet Union. There is an installation inspired by the images of Marc Chagall, paintings dedicated to the Jewish *shtetls*, which constituted an entire cultural domain in the life of Russian Jews. The world of the "shtetl with all its inhabitants was destroyed in the fire of the Holocaust, which is the subject of the second part of the exhibition."[22] Among the main showpieces there are official documents and photographs, lists of killed Jews and letters from the ghettos, designed to

---

18 *Victory Museum official website,* http://www.victorymuseum.ru/?part=57&id_single=3044 (accessed February 12, 2020).
19 There was information about past temporary exhibitions on the official website of the Victory Museum: A large-scale exhibition project dedicated to the 70th anniversary of Victory "Crimes of Nazism. Liberation Mission of the Red Army," June 22, 2014 – March 1, 2015; Exhibition "German Nazi camp – death camp 'Auschwitz'," January 15 – June 1, 2015; Art exhibition "In memory of the Holocaust," January 27 – February 9, 2016; Exhibition "Sheets of Sorrow," January 27 – March 19, 2017. See: *Victory Museum official website,* https://victorymuseum.ru/ (accessed February 12, 2020).
20 See more about externalisation as a characteristic of Soviet feature films in the works of Gershenson: Gershenson, *The Phantom Holocaust,* 2.
21 For more information on the history and activities of the museum, see: Ol'ga Sokolova, ed. *Muzei evreiskogo naslediya i Kholokosta: buklet* (Moscow: RJC, 1999); Natalia Anisina, "Muzei kak mesto pamyati Kholokosta v Rossii," in Al'tman, *Pamyat' o Kholokoste: problemy memorializatsii: materialy 6-i mezhdunarodnoi monferentsii,* 43–46.
22 "O memoriale," *RJC official website,* http://www.rjc.ru/rus/site.aspx?SECTIONID=415840&IID=416253 (accessed September 21, 2017). The official RJC website has been updated and historical information about the museum is not presented on it any more. See: "Memorial'naya Sinagoga," *RJC official website,* https://rjc.ru/ru/projects/project-35 (accessed February 12, 2020).

demonstrate the scale of the Holocaust. Like the exhibition in Yad Vashem,[23] Holocaust history presented here is the story not only of the victims, but also of heroes: Jewish soldiers of the Red Army, resistance fighters from the ghettos, and members of Jewish partisan groups. A special section of the exhibition is dedicated to the Righteous Among the Nations, who selflessly risked their own lives to save Jews during the war.

At the same time, the Museum in the Memorial Synagogue speaks of the fate of European Jews in general, consciously skipping not only the Soviet, but also the Russian context, which contributes to the sketchy and fragmentary representation of Holocaust history. The photo stands present photos of the exhumation of victims (most likely made by members of the ChGK) at the killing sites in different parts of the former USSR, mostly in the territory of modern-day Belarus and Ukraine. Only two photos are geographically related to Russia and the North Caucasus, and show corpses exhumed from mass graves in Armavir (Krasnodar Krai). Thus, a museum that purports to tell the story of Russian Jews continually expands the horizons of its task, extending its purview in somewhat colonialist style to represent the history of the Holocaust on the territory of former Soviet republics, which are now independent states. It sometimes deploys a universal European Holocaust narrative depicting death camps outside the former USSR and alternating this with the commonly accepted narrative of the glorification of the Soviet soldier (in this case, the Soviet Jew) in the fight against Nazism. Once again, regional Holocaust history and even its unfolding in Russia is almost entirely absent from this exhibition, in a museum that claims to be the main one representing Holocaust history in Russia.

A new Jewish Museum and Tolerance Center opened in Moscow in 2012. It was initiated by the FJCR, another Jewish organisation, which is the umbrella organisation for the Chabad Hassidic movement in Russia. The Museum was supported by the Kremlin and financed by a handful of Russian Jewish oligarchs.[24] The conceptual inspiration for the Museum was originally to commemorate Jewish

---

[23] For more about the museum in Yad Vashem see: Isabelle Engelhardt, *A Topography of Memory: Representations of the Holocaust at Dachau and Buchenwald in Comparison with Auschwitz, Yad Vashem and Washington, DC* (Bruxelles: Lang, 2001), 172–185; Amos Goldberg, "The 'Jewish Narrative' in the Yad Vashem Global Holocaust Museum," *Journal of Genocide Research* 14, no. 2 (2012): 187–213; Young, "Holocaust Museums," 263–267; Doron Bar, "Holocaust and Heroism in the Process of Establishing Yad Vashem (1942–1970)," *Dapim: Studies on the Holocaust* 30, no. 3 (2016): 166–190.

[24] For more about the history of the Jewish Museum and Tolerance Center see: Olga Gershenson, "The Jewish Museum and Tolerance Center in Moscow: Judaism for the masses," *East European Jewish Affairs* 45, no. 2–3 (2015): 158–173; Elena Rozhdestvenskaya, "Reprezentatsiya kul'turnoi travmy: muzeefikatsiya Kholokosta," *Logos* 27, no. 5 (2017): 118–123. See also the catalogue of the

victims, following the example of Yad Vashem, but the Russian state has supported a concept that combines it with a family activity center. The main principle of the exhibition design is that of "edutainment," so that it can address the widest possible audience in an accessible, fun, and educational way and "appeal not only to Russian Jews, a very small minority today, but also to non-Jewish Russians, tourists, and, most importantly, youth."[25] The museum represents the story of Jewish life in the Russian Empire, in the USSR, and in the Russian Federation in a very modern way using interactive technologies and multimedia. One of the central parts of the museum is the gallery "The Great Patriotic War and the Holocaust." The designers wanted to combine the Russian grand-narrative about the war with the Jewish presence in it, including mourning of Holocaust victims and recognising Jewish soldiers as war heroes. So, as Gershenson explained in her analysis of the museum's exhibition, the Jews are added to the great universalised narrative of the war, harking back to the idea of "international solidarity" of all Soviet citizens coming together to defeat the enemy.[26] Perhaps, such a representation of the heroic war and the inclusion of Holocaust history in it is one of the consequences of the fact that the museum was conceived as a Tolerance Center.

Importantly, the museum displays Holocaust sites on Soviet soil and partly on Russian soil. An important exhibition technique is the display of a chronological historical timeline, which consists of three thematic blocks: the history of Russian Jewry in the war, the important events of the Great Patriotic War, and World War II. Heroic wartime facts such as the first production of the Lavochkin La-5 fighter plane, designed by Jewish aircraft designer Semen Lavochkin, the activity of Jewish partisan groups, and the participation of Jews in the liberation of European countries alternate with the chronology of the Holocaust in the USSR: there are dates of mass killings and numbers of Holocaust victims at Babi Yar, Rivne, Rostov-on-Don, the dates of the establishment and liquidation of the Minsk Ghetto, and so forth. August 11, 1942 is marked as the date of "the killing of 16,000 (according to some other reports 27,000) Jews in Zmievskaya Balka (Rostov-on-Don). This became the largest killing site of Jews in Russia." The map "The Holocaust on Soviet soil" as part of the film *The Holocaust and the Great Patriotic War* complements the chronological timeline. It presents the geographical scale of the mass killing of Jews in the USSR alongside the total figures of Holocaust victims for each locality. More than a dozen localities belong to the

---

Museum: Irina Mak, *The Jewish Museum and Tolerance Center: Its History, Exhibition and Foundations*, transl. Ben McGarr (Moscow: August Borg, 2017).
25 Gershenson, "The Jewish Museum," 160.
26 Ibid, 166–167.

North Caucasus (Figure 21). Given the museum's avoidance of some of the difficult aspects of Holocaust history, such as the role of Soviet collaborators and perpetrators,[27] the exhibition is in fact the first more-or-less systematic representation of the Holocaust in the USSR in which the history of the Jews in the North Caucasus is represented by Zmievskaya Balka and some other Holocaust sites. Given the high visitor numbers at the Jewish Museum and Tolerance Center[28] and its edutainment approach, it is hoped that it may gradually incorporate Holocaust history into the general Russian narrative of the Great Patriotic War.

Figure 21: Map of the Holocaust on Soviet soil, displayed as part of the film *The Holocaust and the Great Patriotic War*, on show in the Jewish Museum and Tolerance Center (Mak, Irina. *The Jewish Museum and Tolerance Centre: its history, exposition and foundations*, translated by Ben McGarr. Moscow: August Borg, 2017. S. 246–247).

These two Jewish museums in Moscow evoke not only appreciation of collective pain and suffering among visitors, but primarily aim to disseminate knowledge of the active presence (and contribution) of Jewish communities in Russian and Soviet history, and their fate during the war. The geographical location of both Jewish museums in Moscow and their official and financial support by the Russian authorities[29] testify to the nominal acceptance of the Holocaust

---

[27] There is only one brief paragraph about local collaborators "in some Lithuanian and Ukrainian towns" in the entire exhibit. Quoted in: Gershenson, "The Jewish Museum," 168.
[28] According to the results of 2014 and 2015, the Museum holds 15th place in the ranking of the most visited museums in Moscow and St. Petersburg. "Samye poseshhaemye khudozhestvennye muzei Moskvy i Peterburga," *The Art newspaper Russia* 53 (2017), http://www.theartnewspaper.ru/posts/4529/ (accessed March 24, 2020).
[29] The first Russian President Boris Yeltsin supported the construction of the Memorial Synagogue at Poklonnaya Gora, and Vladimir Putin, in several meetings with Russia's Chief

narrative and to attempts to include the history of Jews as one of the representatives of multi-ethnic Russian society into an overarching all-Russian historical war narrative. At the same time, there is no such museum in the regions of Russia, where the history of Jews and their destinies during the war might be comprehensively explored. Exceptions do exist, for example, there are small museums or information stands in some regional synagogues. Their visitors are mostly Jews and invited honoured guests. That is, the history of the Jews and Holocaust history in the regions of Russia continues to be excluded from the social experience of the majority of residents.

At the same time, there is a local history museum in every major Russian city and district center telling the history, ethnography, and nature of each locality. The history of World War II is a mandatory part of permanent exhibitions in these museums. It presents the local dimension of the war and the story of the exploits of villagers in fighting against the enemy. In the last decade regional museums have updated their exhibitions. Despite the use of modern technologies (showing military newsreels, multimedia screens with maps of the conflict, dioramas and panoramas, recreation of the battlefield atmosphere), the ideological component of most regional state museums in representations of war history has remained unchanged since Soviet times. Stories about victims of the war are represented in a rather fragmentary manner in regional museums. The narratives of museum guides fail to provide a coherent and detailed story about the occupation, which is often omitted in favour of stories about war heroes, as already described for the tour of the SGMZ, above. The main characteristic of the vast majority of exhibitions in regional museums is the absence of Holocaust history itself, despite the fact that some of the showpieces are on display. For example, a new permanent exhibition "Kuban in the Years of the Great Patriotic War" [*Kuban' v gody Velikoi Otechestvennoi voiny*] opened in the Krasnodar State Historical and Archaeological Museum-Reserve named after Evgenii Felitsyn, hereafter Felitsyn Museum in 2015. The period of the occupation of Krasnodar Krai is presented on several stands with the following items:

– Orders and newspapers of the military administration authorities;

Rabbi Berel Lazar, repeatedly emphasised the importance of establishing a Jewish museum and Tolerance Center in Moscow, as it is "a project that seeks to ensure that all confessions lived in peace and harmony with one another. And this is the strength of our multi-ethnic and multi-confessional country." Putin donated one month of his salary towards the creation of the Jewish museum. "Nachalo vstrechi s glavnym ravvinom Rossii Berlom Lazarom," *Transcript of the meeting on the website kemlin.ru*, June 5, 2007, http://kremlin.ru/events/president/transcripts/24317 (accessed March 24, 2020).

- A mannequin of a German officer, personifying the Nazi military government;
- Photos and archival extracts from the ChGK reports testifying to the death of the Shishman family and Professor Vilik, who was a Chairman of the Jewish Committee appointed by the Germans, and on whose behalf the "Appeal to the Jewish population of Krasnodar" was printed;
- A picture and description of the gas van operated in Krasnodar (Figure 22).

Figure 22: Display cases of the exhibition "Kuban in the Years of the Great Patriotic War" about German rule in Krasnodar Krai at Felitsyn Museum in Krasnodar (author's photograph, 2015).

These showpieces are presented in the exhibition along with other photo-documentary evidence of Nazi crimes in the region, such as the burning of the working village [*Rabochii poselok*] Mikhizeeva Polyana with all its inhabitants,[30]

---

[30] The working village Mikhizeeva Polyana was razed along with its 207 inhabitants, including 29 evacuees from Leningrad on November 13, 1942. "ChGK report for Krasnodar Krai," in *GAKK*, f. R-897, op.1, d.1, l. 117. According to local historians, "the violence was provoked by the actions of neighbouring partisan groups, who used the help of villagers, distributed leaflets to them, and when the Germans appeared they just left." Georgii Pribylov, *Mikhizeeskaya tragediya ili "Kubanskaya Khatyn'"* (The village of Mostovskoi, 1993), 21–22; Krinko. *Zhizn' za liniei fronta*, 54.

which is considered to be the Kuban equivalent of Khatyn.³¹ However, none of these showpieces openly refers to the Holocaust, creating the false impression that the civilian population in general was victimised by the Nazis in Krasnodar Krai. According to the researcher and curator of the exhibition Nadezhda Suvorova, she "needs stronger evidence"³² in order to make explicit reference to the Holocaust. She never saw the text of the "Appeal to the Jewish population of Krasnodar;" and the ChGK final reports that she used in the exhibition mention the mass killing of "thousands of Soviet citizens," as stated in the reference relating to the death of members of the Shishman family.³³

Despite the fact that in recent decades regional Holocaust history has increasingly become the subject of research by activists, local historians, and many museum employees, more time is required for this theme to become a part of museum exhibitions. Konstantin Bandin, a researcher at the Kanevskaya regional local history museum is personally involved in the reconstruction of Holocaust history and the identification of its victims in Kanevskaya village (Krasnodar Krai). During my visit to the museum, he gave me a personal guided tour of the museum's exhibition "Kanevskaya during the Great Patriotic War" [*Kanevskaya v gody Velikoi Otechestvennoi voiny*].³⁴ It was put together in the early 2000s and, as usual, has a heroic orientation. The display "Memory and Pain" speaks of the heroic deeds of both partisans and underground leaders, who "sacrificed their lives in the fight against the enemy" in their native village. The theme of the occupation of the village is not heroic, but rather a tragic stage in the war's history, which delayed victory for more than half a year and, hence, is not presented in the exhibition at all. And yet, Bandin told the story of the killing of Jews in the village. Mikhail Esaulov, a researcher from the "Fortress" Local History Museum in Kislovodsk also focussed on the story of the mass killing of the Jews in the occupied city. It should be noted that this is probably the only museum in the North Caucasus where photos taken by ChGK during the exhumation of the corpses of Holocaust victims are openly displayed. Drawing attention to these photos, Esaulov mentioned that "among the people killed in the area of Mount Koltso the

---

31 Virtual tour "Kuban in the Years of the Great Patriotic War," *Felistyn Museum official website*, https://felicina.ru/virtualnaya-ekskursiya.html (accessed March 27, 2020).
32 Video interview with the senior research associate of the Felitsyn Museum Nadezhda Suvorova, in *author's archive*, code 15-MEM-KK06, June 5, 2016, Krasnodar.
33 The archival certificate presented in the exhibition refers to the ChGK report for Krasnodar. See: *GAKK*, f. R-897, op. 1, d. 2, l. 45.
34 A video interview with the research associate of the Kanevskaya Regional Museum of Local History Konstantin Bandin, in *author's archive*, code 15-MEM-KK01, June 4, 2015, Kanevskaya, Krasnodar Krai.

majority were Jews, but there were others who were suspected of disloyalty or of having connections with partisans. Policemen and local collaborators played the main role in the executions, according to eyewitness accounts."[35]

The mention of Jews as the main Nazi victim group and the involvement of local perpetrators in the Holocaust is an important step towards a fuller and more truthful representation of occupation history, including a variety of its unwanted and compromising stories. Incidentally, the exhibition in the Felitsyn Museum presents a 1943 documentary film *The People's Verdict* [*Prigovor naroda*] concerning the Krasnodar trial of Soviet collaborators.[36] It shows that eleven Soviet collaborators were punished immediately, already in the war years. Like every Soviet propaganda piece, the documentary highlights not the crime itself for which the perpetrators were judged, but revenge. So, displaying the film in this contemporary exhibition nevertheless underlines the Soviet model of remembering the war. In this way, the visual representation of the war in most local history museums contradicts the recent accumulation of knowledge about the Holocaust and other Nazi crimes in occupied Soviet regions. According to the majority of museum employees with whom I spoke, the main problem when creating a permanent exhibition and trying to reflect all the relevant stages of the war in the region is the shortage of exhibition space.[37] In my opinion, identifying the problem as a lack of space is a kind of window-dressing and actually belies an unwillingness to address the deeper problem of war representation in Russian state museums. The underlying concepts of most exhibitions rely upon the official state war narrative.

Among all state museums in the cities and villages of Russia's south, only one has a separate stand addressing the Holocaust in its permanent exhibition on the Great Patriotic War. This is the Municipal Local History Museum of Ust-Labinsk District. A display presents the story of Musya Pinkenson.[38] The visitor can find articles from the Soviet military press about the heroic death of Musya, along with photos of the boy, the killing sites of evacuated Jews, and the

---

35 Video interview with the research assistant of the "Fortress" Local History Museum in Kislovodsk Mikhail Esaulov, in *author's archive*, code 15-MEM-SK04, May 15, 2015, Kislovodsk.
36 Video interview with Suvorova.
37 Video interview with the head of the Department of Don History at Rostov Regional Museum of Local History Viktor Ushakov, in *author's archive*, code 16-MEM-RO05, September 8, 2016, Rostov-on-Don; Video interview with the employee of the Historical Museum of Taganrog Anton Kovalev, in *author's archive*, code 16-MEM-RO06, September 6, 2016, Taganrog; Video interview with Bandin.
38 Video recording of the guided tour in the Municipal Local History Museum of Ust-Labinsk District, in *author's archive*, code 15-MEM-KK03, May 29, 2015, Ust-Labinsk.

funeral of the "victims of fascism" in the liberated Ust-Labinsk. There is also a table of the numbers of victims in the region, in which a separate line indicates the number of Jewish victims (Figure 23). The story of the museum guide, which was engaging told, sought to combine the Soviet narrative about Musya as a heroic pioneer and as a hero of individual Jewish resistance to the Holocaust. Thanks to Ust-Labinsk's long tradition of studying Holocaust history, commemorative memorial events are regularly shown on the local television channel and museum staff present the story of Musya as part of Holocaust history. At the same time, objects presented in the display cases demonstrate the history of the occupation as a tragedy for all "peaceful citizens," and the portrait of the young Musya is designed to underline the inhumanity of the Nazi regime in general.

Figure 23: Display case about the Holocaust and the fate of Musya Pinkenson in the permanent exhibition "Kuban in the Great Patriotic War" at the Municipal Local History Museum of Ust-Labinsk District (author's photograph, 2015).

In 2015 a private museum dedicated to Sabina Spielrein was opened in Rostov-on-Don. It is located in the house, built in 1897, owned by the Spielrein family, where Sabina lived until the age of 19. The museum presents the perspective of its founder, regional artist Nikolai Polyushenko, on the life and fate of the first female psychoanalyst in Europe through interior objects, photographs and drawings of cities with which her life was linked.[39] This is the only museum in the North Caucasus and Russia more broadly, dedicated to the life of a Holocaust victim. However, there is no emphasis on the death of Sabina and her two daughters in Zmievskaya Balka in this particular exhibition.

There are still no Holocaust museums or educational centers near the killing sites of the North Caucasus. An exception is the memorial complex in Zmievskaya Balka, which has a museum room named the "Hall of Sorrow." The exhibition housed in it during Soviet times was dedicated to the exploits of Soviet people during the war and did not reflect Holocaust history. Currently, there is a permanent exhibition about the role of Rostov-on-Don in the fight against Nazi Germany and the "atrocities of fascism in the occupied city."[40] A separate stand called "Holocaust. Genocide" displays photographs of Holocaust victims with brief information about them, as well as photocopies of archival files on the mass killing of Jews in Zmievskaya Balka. Holocaust history is here exhibited through the personification of its victims. There are also showcases that tell the photo story of the creation and opening of the memorial complex in Zmievskaya Balka in 1975 and of the commemorative meeting in 2009.[41] The exhibition displays and banners are placed in the museum's main room. Several chairs are arranged in the center of the room for group tours. According to one of the museum employees, tours are held weekly throughout the museum's three opening days. On the day of my visit, there were no tours and we called in advance just to make sure that the museum employee was in (on his working day) and a visit to the museum was possible. Also, the absence of the head of the museum (as of 2013) and of a museum website casts doubt on the regularity of visits to the museum and memorial complex. The museum still does not have an educational center[42] unlike the vast majority of international Holocaust museums. It appears that the Russian public is not yet ready to address this issue. In 2007 the chairman

---

39 "V Rostove otkryt muzei uchenitsy Yunga i Freida Sabiny Shpil'rein," *Radio Svobody News*, November 7, 2015, https://www.svoboda.org/a/27351089.html (accessed March 23, 2020); Elena Potana, "Otkrytie muzeya Sabiny Shpil'rein," *REN TV Rostov*, November 11, 2015, https://www.youtube.com/watch?v=VutudOMYegQ, (accessed March 23, 2020).
40 "Muzei memoriala," *website "Holocaust. Rostov-on-Don. 1942,"* http://holocaust.su/memorial (accessed March 23, 2020).
41 A video recording of a visit to the Museum of the memorial complex in Zmievskaya Balka, in *author's archive*, code 13-MEM-RO02, February 27, 2013, Rostov-on-Don.
42 In 2011, Il'ya Al'tman, Yurii Dombrovskii, and Aleksandr Kozhin discussed plans to make the Zmievskaya Balka memorial one of the most important monuments to victims of the Holocaust, an object of international historical tourism and an information and educational center. Iya Shvets, "Memorial v Zmievskoi Balke mozhet stat' informatsionno-obrazovatel'nym tsentrom," *Gorod N*, November 1, 2011. However, these plans have not yet been implemented.

of the Jewish community of Krasnodar sought to establish a Jewish complex "consisting of a Holocaust Museum (Museum of Tolerance), an educational technology center of the Association of Skilled Trades [*Obshchestvo remeslenogo truda*], hereafter ORT,[43] and the Memorial near the old Jewish cemetery." Either the regional authorities or central Russian Jewish organisations did not support this initiative for the revival of Jewish history and culture in Krasnodar Krai.[44]

Holocaust history is thus fragmentally represented in region's local history museums. The inclusion of the history of the mass killings of Jews (mostly as "peaceful citizens") into the guide's story depends on the involvement of museum researchers in the study of this topic, as well as on the categorisation of the visitors. Pupils conducting the Memorial Watch at the Eternal Flame, according to Dvernik, should be told the heroic story of the war, and it is not considered worthwhile to tell primary school children about the tragic sides of the occupation.[45] Visitors of regional state local history museums, which have limited tour guides and no audio guides, would not be able to see or 'read' Holocaust history through the solitary objects on display. These museums present the official heroic concept of the history of the Great Patriotic War, focusing on regional stories of the exploits of fellow countrymen. This is designed to enhance the visitors' sense of pride for their forebears and compatriots, who struggled for the Great Victory. The stories of Soviet collaborators and Nazi accomplices are not always represented in regional museums, and if they are it is often in the context of the retribution of the Soviet state for their betrayal the Motherland. The idea of "national unity in the fight against the enemy" is still the leitmotif of exhibitions about the Great Patriotic War in local history museums. The emphasis is upon the importance of the victory: it is the outcome of the war, which is remembered in regional museums up until the present, not the historical process that led to the victory, which would necessarily include addressing the Holocaust among the main Nazi crimes, along with the other episodes of the war.

---

[43] ORT is a non-profit global Jewish organisation that promotes education and training in communities worldwide. It was founded in 1880 in St. Petersburg to provide professional and vocational training for young Jews. Today World ORT is a non-profit organisation with past and present activities in more than 100 countries. *World ORT organisation official website*, https://www.ort.org/en/ (accessed February 13, 2020).

[44] Interview with Teitelbaum; "Letter of the Executive Vice-President of the FJCR Valerii Engel to the Chairman of the Jewish community of Krasnodar Yurii Teitelbaum," May 21, 2007, in *personal archive of Yurii Teitelbaum*.

[45] Audio and video recording of the SGMZ guided tour "Battle Glory of Stavropol."

6.2 Regional Practices of Museification of Holocaust History in the North Caucasus

Representations of Holocaust history are more often seen in the temporary exhibitions of Russian state and regional museums and other educational and cultural centers than in their permanent collections. Depending on the organiser or curator of the exhibition, it is possible to single out the following areas of museification of regional Holocaust history.

Firstly, there is the activity of the Holocaust Center. In 2016, its staff prepared a historical and documentary exhibition "The Holocaust: Destruction, Release, Rescue" [*Kholokost: unichtozhenie, osvobozhdenie, spasenie*] with the support of the RJC and the Ministry of Foreign Affairs of the Russian Federation. The scale and features of the Holocaust on Soviet soil are presented in 15 mobile displays. Its main sources are "unique documents and photographs, including little-known evidence of the mass killing and rescue of Jews on the territory of the Russian Federation."[46] Holocaust history in the North Caucasus is revealed using a collection of primary sources about mass killings of Jews in Rostov-on-Don, Mineralnye Vody, Arzgir village, and in photographs of monuments installed in these locations. The geography of Holocaust history representation and its commemoration in Russia is the product of the activities of the Holocaust Center. The exhibition displays Holocaust history in those cities and villages where the Center has regional representatives, and where educational and other programmes are being implemented. It was no coincidence that the exhibition was first opened in the State Duma of the Russian Federation, instead of being presented to a wider Russian audience.[47] It was an opportunity to show the relevant authorities the fruits of their funding.[48] This was evidenced at further presentations with the text: "Opened on January 20, 2017 at the Cultural Center of the Ministry of Foreign Affairs of the Russian Federation by Minister of Foreign Affairs Sergei Lavrov."[49] During 2016–2019 the exhibition was shown in the buildings of the city administrations, archives, libraries, universities, and

---

46 Il'ya Al'tman, introduction to *Kholokost: unichtozhenie, osvobozhdenie, spasenie: buklet istoriko-dokumental'noy vystavki*, ed. Il'ya Al'tman et al. (Moscow: Tsentr "Kholokost," 2017), 1.
47 Vystavka "Kholokost: unichtozhenie, osvobozhdenie, spasenie," *Holocaust Center official website,* http://holocf.ru/электронный-вариант-выставки-холоко/ (accessed February 13, 2020).
48 State support funds in the form of a presidential grant were used to prepare the exhibition. Al'tman, introduction to *Kholokost: unichtozhenie, osvobozhdenie*, 1.
49 Ibid.

Jewish communities in more than 40 Russian cities.[50] In August 2017, as part of the memorial events in Rostov-on-Don, the exhibition was placed in the Don State Public Library and then in the library of the city of Taganrog.[51] In modern Russia, visitors of regional libraries are mainly elderly people or researchers. This is a consequence both of the capacity of regional libraries (they often have a good collection of Soviet literature, but little in the way of modern Russian literature or foreign scholarly holdings), and the fact that young Russians no longer use library resources to prepare for seminars (the majority of the requisite literature is normally provided by professors or it is in the public domain on the Internet). In contrast, regional local history museums often have links to schools and educational institutions, whose students are obliged to visit museums in some regions. The lack of agreements between the Holocaust Center and regional museums for the display of the exhibition reduces its audience among Russian citizens almost completely. According to the information on its website, the Holocaust Center has offered this exhibition for local museums and schools in digital form and/or stands for printing on the spot since early 2020.[52] The new version of the exhibition was designed for the Memory Week in January 2020, dedicated to International Holocaust Remembrance Day and consists of 11 stands. In this version, there is a new stand about Jewish children who became Holocaust victims on Soviet soil. Musya Pinkenson's story is presented here as well. It is too early to think about the results of such project in the Russian regions. In my opinion, such exhibitions must be accompanied by a guided tour to be effective. The historical literacy of average Russian citizens on the Holocaust has not yet reached the level of recognising historical events purely from a brief description and/ or photos of the killing and grave sites. We thus face the problem of Holocaust representation in the form of a travelling exhibition, which almost completely fails in its educational mission.

---

50 According to the information of Leonid Terushkin, the Head of the the Holocaust Center Archive.

51 "Mezhdunarodnaya konferentsiya," *Don State Public Library official website*, August 10, 2017, http://www.dspl.ru/news/7727.html (accessed March 23, 2020); "Istoriko-dokumental'naya vystavka. Kholokost: unichtozhenie, osvobozhdenie, spasenie," *Taganrog Central Public Library official website*, August 17, 2017, http://www.taglib.ru/news/Istorikodykymentalnaya_vis tavka__%C2%AB%20Holokost:_ynichtojenie,_osvobojdenie,_spasenie%C2%BB.html (accessed March 23, 2020).

52 Vystavka "Kholokost: unichtozhenie, osvobozhdenie, spasenie," *Holocaust Center official website,* http://holocf.ru/электронный-вариант-выставки-холоко/ (accessed February 13, 2020).

Secondly, there is the activity of local historians and independent researchers. For instance, Anatolii Karnaukh has coordinated the work of preserving the memory of Holocaust victims in Arzgir village for several years. He has donated his collection of materials to the village's local history museum. In 2015, the museum was presented with a small photo exhibition "Not to be Forgotten" [*Zabveniyu ne podlezhit*].[53] It consists of two stands in which photographs and names of Holocaust victims are placed alongside photographs of the killing site, the post-war monument and photographs of Holocaust survivors with their rescuers from Arzgir (Figure 24). These stands are displayed in the museum during commemorative events, usually in September. Since Arzgir is a small settlement, the work of Karnaukh is notable, and most importantly, it happens to be supported by the local authorities. The majority of villagers know about the mass killing of the Jews in Arzgirskaya Balka and actively participate in commemorative events. In this sense, a photo exhibition is designed to be presented to various guests. Delegations from the Holocaust Center and other institutions from abroad

Figure 24: Photo exhibition "Not to be Forgotten" at the Local History Museum named after Vasilii Ponomarenko in the village of Arzgir (author's photograph, 2015).

---

[53] Video recording in the Local History Museum named after Vasilii Ponomarenko in the village Arzgir, in *author's archive*, code 15-MEM-SK06a, May 15, 2015, Arzgir, Stavropol Krai.

have often visited Arzgir in recent years. A photo exhibition thus becomes a kind of 'report' for the honorary guests of the work done in the village.[54]

Such small mobile photography exhibitions – resulting from the research efforts of members of Jewish communities or local historians seeking to commemorate Holocaust victims in the region – are an important part of Holocaust remembrance in modern Russia. The Stavropol Jewish community headed by Efim Fainer has done a lot of work to commemorate victims of the Holocaust since the early-2000s. There is a small photo exhibition about the history of the Jews of Stavropol in the office of the Jewish community. A photo report of the opening of the Holocaust monument and of annual commemorative events is presented on a separate stand.[55] In 2009, the Krasnodar Jewish community displayed the exhibition "Jews in the Line of Fire" [*Evrei na linii ognya*] in the Center for National Cultures during the Remembrance Days.[56] This exhibition, unlike that in Stavropol, was of a universal character and local material was not represented in it.

The accumulated material on the history of mass killings of Jews in Zmievskaya Balka was incorporated into a travelling exhibition prepared by the German researcher Christina Winkler. The exhibition "The Forgotten: the victims of German occupation crimes in Rostov-on-Don 1941–1943" was opened in August 2017 simultaneously in Berlin's German-Russian Museum Berlin-Karlshorst and in Rostov-on-Don, in the Don State Public Library. The exhibition was sponsored by the Holocaust Center, the RJC, the Stanley Burton Center for Holocaust and Genocide Studies, and the German "Remembrance, Responsibility, and Future" Foundation (hereafter EVZ).[57] This is the first exhibition of its kind to show the fate of various groups of Nazi victims (handicapped people, Soviet POWs, forced labourers, and Holocaust victims) during the two occupations of Rostov-on-Don. Information about Holocaust victims is presented in the form of archival sources, photos and personal documents of the victims. Since a Western historian prepared the exhibition, it does not have the heroic and patriotic component that characterises all Russian museum exhibitions about World War II. On the contrary, the focus is not on

---

54 This also happened during my visit to Arzgir, when the museum's staff "was asked to put up stands, because a specialist [i.e., me] who is interested in the Holocaust is coming."
55 Interview with Fainer.
56 Interview with Teitelbaum.
57 "Untold Stories of the Holocaust: South Russia's Largest City to be Explored at Exhibition," *News and events of the University of Leicester on the university's official website*, 9 February 2016, https://le.ac.uk/news/2016/february/untold-stories-of-the-holocaust-south-russias-largest-city-to-be-explored-at-exhibition (accessed March 24, 2020).

heroes, but rather on forgotten victim groups of the war. This was possible because of the project's financial support from various European foundations. However, placing the Russian version of the exhibition in the library and not in the local history museum and the lack of both advertising and an information campaign in Rostov-on-Don has not led to large visitor numbers to the exhibition by the city's residents. The Anglo-German version was held in the Museum Berlin-Karlshorst for two months in 2017 and was then planned to be staged in the Rostov-on-Don's two twin cities in Germany (Dortmund and Gera) and in the UK (Glasgow). It was not clear if the Russian version would be displayed anywhere else in Rostov Oblast.[58] Thus, despite the existence of special thematic travelling exhibitions and exhibition displays on Holocaust history and Holocaust remembrance in the North Caucasus, the problem of their availability to the mass of visitors remains.

In late 2018 I opened a travelling exhibition in Rostov-on-Don, which I was able to design with support from EVZ, the Center for Research on Antisemitism at Technische Universität Berlin and the the Rostov section of the the All-Russian Society for the Protection of Monuments of History and Culture (VOOPIiK). It was devoted to the history and memory of the mass killing of patients of psychiatric clinics, people with disabilities, including children, and Jewish doctors who became Nazi victims in the occupied North Caucasus.[59] Holocaust history is only a small part of the exhibition. Three displays present the general story of the mass killing of local and evacuated Jews in the region, with a focus on the fates of Jewish doctors. I sought to exhibit photos of Jews who had been killed and to tell their stories as well-known and respected doctors, scientists, and persons. It was a difficult task, since we have only numbers, and sometimes lists with the names of victims in most localities. Even local Jewish communities are often unable to reconstruct the life stories of their former members. In this case, I was very surprised by the activity of colleagues from Stavropol Medical University, who have been reconstructing the history of their alma mater for the past decade. The fates of its staff and students during World War II was the central focus of their research. The university museum has unique documents and items that help to

---

58 Il'ya Al'tman said that the exhibition will "be definitely shown in Moscow and St. Petersburg" in an interview for the television company Don Pravoslavnyi. Anna Glebova and Konstantin Yakovlev, 'Zabytye' – vystavka pod takim nazvaniem otkrylas' v Rostove-na-Donu," *Don pravoslavnyi TV*, aired August 23, 2017, min. 21:21–32:37, https://www.youtube.com/watch?time_continue=7&v =F1LBSyPAJjQ (accessed March 26, 2020).
59 Rebrova, *"Pomni o nas . . . ": katalog vystavki*; Official website of the exhibition "Pomni o nas . . . ," https://nsvictims.ru/ (accessed February 13, 2020).

retell the history of this institution, including during the war.⁶⁰ Many of the findings of my Stavropol colleagues were incorporated into the travelling exhibition, which has been mounted in 20 local state, university and psychiatric clinic museums of 12 cities and villages in the North Caucasus since December 2018.⁶¹ Although it discusses more than Holocaust history, this exhibition could be a significant step towards the remembering of different groups of Nazi victims in modern Russia and shifting emphasis in the memorial culture from heroes to the victims. Positive and warm reviews by visitors, the obtaining of more than a dozen official partners of the exhibition, and willingness to host it not only in the North Caucasus, but also in Moscow, Central Russia, and, perhaps, in Israel and Germany, speak of the importance of the topic in Russia and internationally.

Thirdly, there are schoolteachers who are responsible for creating thematic school museums. In Soviet times, there was a Corner of Military Glory [*ugolok boevoi slavy*] in practically every secondary school, which could eventually be expanded into a school museum.⁶² A special programme for the development and support of school museums was created in post-Soviet Russia. Most school museums represent the history and ethnography of their native locality. The history of the Great Patriotic War holds the central place in most school museums. At the same time, Holocaust history is not officially present in any of the school museums registered in the North Caucasus.⁶³ As a result of my survey of

---

60 Andrei Kartashev, ed. *Stavropol'skii meditsinskii: dorogami voiny: biograficheskii spravochnik* (Stavropol: StGMU, 2015), 274–294; *Semestr, kotorogo ne bylo. Okkupatsiya*, directed by Andrei and Igor' Kartashevy (Stavropol: Tvorcheskaya Masterskaya "Brat'ya Kartashevy," 2018). 41 min. 47 sec., https://www.youtube.com/watch?v=B2QCN–IEPM (accessed February 13, 2020).
61 For reports about mounting the exhibition in different localities, see: "Geografiya demonstratsii vystavki," *Official website of the exhibition "Pomni o nas . . . ,"* https://nsvictims.ru/география-демонстрации-выставки/ (accessed February 13, 2020).
62 The "Regulations on the School Museum" were approved by the Secretariat of the Central Committee of the Komsomol, the Ministry of Education of the USSR and the Ministry of Culture of the USSR in 1974. According to it, a school museum can be opened with active schoolchildren and a fund of original materials corresponding to the museum's profile, and the equipment providing storage and display of the collections. "Polozhenie o shkol'nom muzee," *Prepodavanie istorii v shkole* 2 (1975): 12–14. For more details on the activities of Soviet and Russian school museums, see: Taisiya Kudrina, ed. *Muzei i shkola: posobie dlya uchitelya* (Moscow: Prosveshchenie, 1985); Valerii Tumanov, *Shkol'nyi muzei*, (Moscow: Tsentr detsko-yunosheskogo turizma i kraevedeniya, 2003); Anatolii Persin, *Kraevedenie i shkol'nye muzei: istoriya, teoriya, praktika* (Moscow: Federal'nyi tsentr detsko-yunosheskogo turizma i kraevedeniya, 2006).
63 The website of the Federal Center for Children and Youth Tourism and Local History provides a summary of all school museums registered in each subject of the Russian Federation. For example, almost 98 per cent of the 278 school museums in Rostov Oblast have a section on the Great Patriotic War. The history of the war is represented in about half of the school

9th and 11th-grade history teachers in the region,[64] I discovered that displays of Holocaust history are represented in a few school museums. As a rule, these are schools where teachers and local historians have researched Holocaust victims.[65] Thus, the Holocaust Museum of the "Or Avner" Jewish Gymnasium in Rostov-on-Don presents information about both groups of the city's Jews: the participants and heroes of the war and the victims of the Holocaust. "Methodological and research work and guided tours on preserving Holocaust memory in the city" have been carried out under the aegis of this museum, established in February 2000.[66] Holocaust history in the form of a clamshell folder displaying the results of Anatolii Karnaukh's research is presented in the museum of Arzgir Secondary School No.3. The museum's creator, history teacher Sara Konovalova organised an initiative group of schoolchildren who conduct a guided tour of the museum and in classrooms using such clamshell folders as visual material.[67]

The school museum of Beslenei aul tells the story of the rescue of children from a Leningrad orphanage. Several displays are filled with pictures of rescued children; most of them studied in this school and one of them, Musa Agarzhanokov, was its assistant principal from 1971 to 1989.[68] The presentation of the achievements of schoolchildren in the form of their competition projects, photo reports on past events, and a professional thematic portfolio of history teachers are important showpieces of any school museum. Holocaust

---

museums of the North Caucasian republics, giving way to the representation of the history of the village or aul. See: "Shkol'nye muzei Rossii," *Federal Center for Children and Youth Tourism and Local Lore official website*, http://turcentrrf.ru/shmrf (accessed December 13, 2019).
64 More about my survey of secondary schoolteachers in the North Caucasus and its results will be discussed in the chapter on teaching about the Holocaust at school (Chapter7).
65 There are thematic museums in the Jewish schools in Rostov-on-Don ("Or Avner" Gymnasium) and Pyatigorsk ("Geula" Secondary School) as well as in some secondary schools, for example, in Secondary School No.3, Arzgir village, in the school of aul Beslenei, and in Ust-Labinsk Secondary School No.1.
66 Lyudmila Miroshnikova, "Muzei v evreiskoi shkole kak metodicheskii regional'nyi tsentr sokhraneniya pamyati o Kholokoste," in Al'tman, *Pamyat' o Kholokoste: problemy memorializatsii: materialy 6-i mezhdunarodnoi konferentsii*, 156–160.
67 The 10th-grade schoolgirl Evgeniya Pisarenko was responsible for the preparation of stands on the story of the Holocaust in Arzgir in 2015. She also conducted a guided tour of the school museum for me on my arrival in Arzgir. Video recording in the museum "Origins" of Secondary School No.3 of Arzgir village, in *author's archive*, code 15-MEM-SK08, May 17, 2015, Arzgir, Stavropol Krai.
68 Video recording in the school museum of Beslenei aul, in *author's archive*, code 15-MEM-KCh02, May 15, 2015, Beslenei aul of the Karachay-Cherkess Republic.

history in school museums is represented through the comprehension of past events by modern schoolchildren with the help of their mentors/teachers. Since there is no Holocaust educational center in any state regional local history museum, school museums in which Holocaust history is represented fulfil the function of such centers. On the other hand, the presence of Holocaust-related showpieces in less than one percent of all school museums in the North Caucasus demonstrates the absence of a special educational programme for teaching about the Holocaust, as well as a lack of interest among the vast majority of schoolteachers in studying this topic together with schoolchildren.

To conclude, it can be seen that Holocaust representation in museums has appeared in Russia only in the last few years. If the world's leading Holocaust museums present their narratives through the eyes of its victims, Russian state regional and school museums aim to relay an image of a hero, a winner in the "Great Patriotic War." The few attempts to introduce Holocaust history into the narrative of the war in museum and thematic exhibitions are almost always accompanied by a representation of the struggle and resistance of Soviet Jews and their participation in the ranks of the Red Army. Separate exhibitions on Holocaust history in Russia, including in the North Caucasus are often not well known to the general public. Russian museums reflect the official state policy of memory of the war, designing and creating memories that can reinforce the acceptable past. On the other hand, museums represent the past through the prism of the present. This means that despite the first attempts to include Holocaust displays into museum exhibitions, a revision of the contemporary state heroic-patriotic representation of World War II history is still required. The charity organisation "Babi Yar Holocaust Memorial Charity Fund" was founded in Kyiv, Ukraine in September 2016. It is a non-governmental organisation with the ambition to build a center for documentation, commemoration, and education on the tragic events of September 1941 by the 80th Remembrance Day in 2021. One of the main tasks of the Babi Yar Holocaust Memorial Center (BYHMC) is to become a research center, specialising in Holocaust-related crimes in the former Soviet Union and Eastern Europe, to pay tribute to the victims, to tell the story of their lives, and to explain the relationship between Jews and non-Jews.[69] Perhaps, the creation of such a center will also be possible in Rostov-on-Don, the location of the largest killing site of Jews within the borders of modern-day Russia.

---

69 "About the project of BYHMC," *BYHMC official website,* http://babiyar.org/en/byhmc/about, (accessed March 23, 2020).

7 We Don't Need No . . . Holocaust Education?

In December 2011, a Russian entertainment TV channel, "Muz TV," released a new youth comedy show called "Insanely Beautiful" [*Bezumno krasivye*]. The concept behind the show was that three teams of young men would compete to comprehend the mysteries of female logic. To do this, the show's presenter asked two girls – in this case twin sisters, Moscow State Textile University students Evgeniya and Kseniya Karatygina – a set of questions, which they had to answer in a minute; and the men had to guess whether the girls would give the right answer. There are questions of varying complexity and thematic focus; most of them require the knowledge of the school curriculum, since generally participants and the target audience are recent school graduates, current students, and working youth. In this particular episode, the Karatygina sisters were asked what the Holocaust was. The male participants declined the presenter's invitation to explain to viewers the meaning of the term "Holocaust" before the girls started discussing it. "The audience is so smart, they probably know what Holocaust is," one of the participants said. Here is a fragment of the transcript of the girls' discussion:

> Evgeniya Karatygina: Glue . . . Maybe that's what it is.
>
> Kseniya Karatygina: To be frank this word says nothing to me. A piece of stationery . . . a tool.
>
> Evgeniya: No, it has to be glue. I don't know, maybe for wallpaper . . . a name for something . . . or probably an item of household equipment.
>
> Kseniya: Yeah, it might be that. So, something for the house, right?
>
> Evgeniya: Well it seems so. What do you think?
>
> Kseniya: I have no idea. Let's take your option . . . we think that Holohaust [sic] is something like wallpaper glue.[1]

Immediately after this episode was aired, the story made it to the Internet, where it was viewed by more than a million people and gained more than a thousand comments.[2] There followed a large-scale public discussion of the

---

[1] Fragment of the show on *Youtube.com*, https://www.youtube.com/watch?v=UGdmtDK9oYY (accessed March 23, 2020).
[2] Tetyana Kloubert, "Holocaust Education in postsozialistischen Ländern am Beispiel von Russland, Polen und Ukraine," in *Holocaust Education in the 21st Century*, ed. Eva Matthes, Elizabeth Meilhammer (Bad Heilbrunn: Verlag Julius Klinkhardt, 2015), 222. Currently, two short videos have been posted on Youtube, watched by more than 30,000 and 40,000 users respectively

sisters' statement that "the Holocaust is a wallpaper glue" ["*Kholokost – eto klei dlya oboev*"], in which the president of the Holocaust Foundation Alla Gerber participated as well.³ A Russian film director and journalist Mumin Shakirov made a documentary of the same title in 2013. He took the sisters to the Auschwitz-Birkenau State Museum, where they were shown documents about the history of the mass killing of the European Jewry during World War II.⁴ In this way, this one statement on an entertainment TV channel instantly made two sisters from a provincial Russian town famous. They came to conquer the capital, and wound up doing it in a manner they hadn't expected.

This story is thought provoking not because of the public reaction to the girls' interpretation, but rather because of the disregard of Russian youth for Holocaust history. In a later interview, the sisters confessed that they had, in fact, "heard of the Holocaust as a historical fact, but did not know what it was called." According to Kseniya, "the history lessons weren't good enough to make a serious impression and thus to remember the material, so we just forgot."⁵ Shakirov in his film, as if anticipating criticism of the system of Russian education, interviewed Svetlana Tikhonova, a history teacher from Boarding School No.1 in Vladimir, where the Karatygina sisters studied. The Holocaust, she said, was certainly taught at school. "Maybe they [the sisters] were absent, or perhaps they weren't listening at all. But as a term it was definitely mentioned." The teacher accepted no responsibility: she justified herself by the fact that she did not have appropriate handouts. "Maybe one day I'll be ready to give a separate lesson on the Holocaust . . . In order for me to come to this, and I certainly must come to this myself, I must find something that will shock me."⁶

---

and with more than 230 comments. See: "BuzUMno krasivye. Chto takoe Kholokost?", *Youtube.com*, https://www.youtube.com/watch?v=OVLx8YpGGyA (accessed March 23, 2020); "Kholokost – eto klei dlya oboev?!", *Youtube.com*, https://www.youtube.com/watch?v=UGdmtDK9oYY (accessed March 23, 2020).

3 See, for example: "Kholokost, dikhlofos, klei dlya oboev," *Radio Svoboda*, March 28, 2012, http://www.svoboda.org/a/24527650.html (accessed March 23, 2020); Mumin Shakirov's documentary "'Kholokost – klei dlya oboev?': diskussiya v mezhdunarodnom memoriale," January 9, 2014, *Website of the Society "Memorial" project "Uroki Istorii XX Vek,"* http://urokiistorii.ru/learning/edu/51970 (accessed March 23, 2020).

4 *Kholokost – klei dlya oboev?*, directed by Mumin Shakirov (Bacteria-film, 2013).

5 Ibid, min. 28:32–29:18.

6 Ibid, min. 31:58–32:45.

The story of the Karatygina sisters is quite didactic. It once again revealed the problems of history education in modern Russia, and revealed the attitude not merely of schoolchildren, but of teachers towards the past of the country and the fates of its constituent peoples. It is impossible to cover all the key topics of the history of the Second World War during the several lessons scheduled for the "Great Patriotic War" in school history courses. Nevertheless, the social memory of a historical event is also formed and developed in school history lessons.

Education thus becomes one of the central elements of the Holocaust memory dispositif. It reveals social development; the willingness to know and pass on to younger generations certain aspects of the historical past. The Karatygina sisters represent a kind of cut-off point for the average modern Russian youth, for whom "music, singing, and walking around Moscow"[7] seem to be more important than universal values. But are the Karatygina sisters the only ones to blame for their ignorance and historical illiteracy? Are there, maybe, other "perpetrators on the dock of historical memory?" This chapter studies teaching about the Holocaust in Russian history courses and local history courses in secondary schools of the South of Russia in an historical perspective.[8] I trace the place given to the study of the mass killing of Jews on Soviet soil in lessons on the history of World War II and the Great Patriotic War: in what context is this topic revealed, and what aims are pursued in teaching on the Holocaust in secondary schools? On the basis of textual materials in regional history textbooks I will establish what place is given to the history of the Holocaust on Soviet soil and whether it echoes in the dramatic wartime history of native Peoples of the Caucasus, such as the deportation of the Chechens, Ingush, Balkars, Karachays, and other peoples.

A textbook can be understood as a kind of core narrative, setting out the content and reflecting the trends of educational policy. Numerous everyday pedagogical practices contribute to the transformation of the textbook into an "objective reality" that is relevant to the life of a schoolchild.[9] This chapter is important not for just studying Holocaust memory in Russia in general and in

---

7 This is how the Karatygina sisters introduced themselves on the show.
8 Attention is paid only to school education, since school is the place where the "process of formation of a harmoniously developed personality takes place." 11 years of secondary education in Russia is compulsory in the Russian Federation in accordance with the state law "On Education." "Federalnyi zakon ot 29 dekabrya 2012 goda No. 273-FZ 'Ob obrazovanii v Rossiiskoi Federatsii,'" *Rossiiskaya gazeta*, December 31, 2012.
9 For example, the Russian historian Aleksandr Chashchukhin studies educational textbooks as a complex process of interaction between the normative instructions set by the authorities in Soviet times and the teachers' practices in the classroom. Aleksandr Chashchukhin, "Uchebnye teksty i professionalno-politicheskaya sotsializatsya shkolnykh uchitelei 1950-h gg.," in *Dorogoi*

the North Caucasus in particular, but also for understanding the historical amnesia in present-day Russia. Many professionals (especially teachers) neither know the subject nor are they willing to teach future generations about the Holocaust. This complacency, the passive acceptance in historical education of the official concept of the "Great Patriotic War" as well as of the Soviet People's "Great Victory," with almost no place allotted to the facts of the Holocaust, is an important feature of the Holocaust memory dispositif in Russia. On the other hand, existing pedagogical initiatives and the active participation of a small number of schoolchildren in national project competitions on Holocaust history are examples of the active inclusion of modern secondary schools into the formation and development of Holocaust memory in specific localities. Thus, the initiatives of regional history teachers, to teach (or not teach) the Holocaust demonstrate the practical application of the contents of history textbooks in the classroom.

The way in which the story of World War II or the Great Patriotic War is told in many ways depends on the personality of the teacher, his or her professional competence, and personal motivation. Under the authoritarian Soviet Union and then even in "democratic" Russia, where the Ministry of Education and Science develops national education standards with the required historical knowledge, a teacher becomes an intermediary between officialdom and the young generation. I developed a questionnaire to better understand the motivation of Russian teachers in educating schoolchildren about the war in general and the Holocaust in particular in the modern Russian secondary school. I conducted thematic interviews with some teachers who actively teach the Holocaust in history lessons. In this way, an analysis of daily teaching can serve as a litmus test for how the Holocaust is taught in Russian schools. Teachers become actors in shaping social memory of the Holocaust in the province; they are responsible for the memory of future generations. Holocaust history does not leave many schoolchildren indifferent, which later becomes noticeable in their classroom research projects. A thematic analysis of the materials of regional and all-Russian history project competitions for schoolchildren helps to trace Holocaust remembrance at an individual level.

---

drug. Sotsialnye modeli i normy v uchebnoi literature 1900–2000 godov: istoriko-pedagogicheskoe issledovanie, ed. Vitalii Bezrogov, Tatyana Markarova, and Anatolii Tsapenko (Moscow: Pamyatniki itoricheskoi mysli, 2016), 35–48.

7.1 Holocaust History in School Textbooks: From Depersonalisation to Recognition

The social memory of the past, and specifically Holocaust history in the USSR and Russia, is a temporal process. Moreover, memory is constantly being formed and reformed. There are many ways of forming memory, one of which is, quite simply, the school textbook. Textbooks are not the most effective way of forming memory. Mass media and television, in particular, have been demonstrated to be much more effective. However, "a textbook is more important than TV because of its normative nature."[10] In other words, a school history textbook is an official narrative about the past, designed to pass knowledge to a younger generation.

The Great Patriotic War in post-war Soviet and Russian history textbooks is commonly depicted as the main historical event of the last century. Despite the historical significance of the war for Soviet and Russian society, little time in the curriculum is allotted to its study.[11] The basic approach to teaching history in Soviet schools was linear,[12] involving a one-time study of world and

---

10 Nikita Sokolov, "Shkol'nyi uchebnik kak 'vmestilishche' prinyatoi v obshchestve istoricheskoi pozitsii", in *Materialy kruglogo stola "1937–2007: pamyat' i otvetstvennost'*," September 11, 2007, http://www.gorby.ru/activity/conference/show_685/view_27417/ (accessed March 23, 2020).
11 For example, in modern Russian secondary school about three academic hours are given to the study of World War II history in the 9th grade and five to seven academic hours to the study of the history of the Great Patriotic War. Three and five hours are devoted to these subjects respectively in the 11th grade. Whereas only 40 academic hours are given to the course "Russian History" in the 9th grade, in which the whole of Soviet and Russian history of 20th and early 21st centuries is studied. Tamara Gusenkova, ed. "*Rasskazhu vam o voine . . .* " *Vtoraya mirovaya i Velikaya Otechestvennaya voiny v uchebnikakh i soznanii shkolnikov slavyanskikh stran* (Moscow: Rossiiskii institut strategicheskikh issledovanii, 2012), 25. Nevertheless, if one compares the practice of teaching World War II in other European countries, it turns out that Russia has the greatest number of academic hours dedicated to it. Tetyana Kloubert, "Holocaust Education," 222.
12 In the years 1959–1965 an attempt was made to move to a concentric structure of school history education, when "in the 7th and 8th grades an elementary course of the history of the USSR was studied with the most important facts of the social and state structure of the Soviet state, as well as the main directions of the development of foreign countries in modern history. A systematic course of the history of the USSR and Modern and Contemporary history of foreign countries were taught in the 9th to 11th grades." Resolution of the Central Committee of the CPSU, Council of Ministers of the USSR of August 8, 1959, No. 1162 "O nekotorykh izmeneniyakh v prepodavanii istorii v shkolakh," in *KonsultantPlus official website*, https://www.consultant.ru/cons/cgi/online.cgi?req=doc&base=ESU&n=38088#0 (accessed March 23, 2020). However, the concentric structure of studying history in Soviet schools was rejected rather quickly and nobody returned to it until the collapse of the USSR.

national history in chronological order.[13] This means that the history of World War II was studied in the last (tenth) grade of secondary school. Soviet history textbooks were written by author collectives and underwent more than ten reprints. The "Great Patriotic War" section in Soviet textbooks is written uniformly with attention to the description of a certain number of military battles. The war is presented as a central and liberating event in human history.

In textbooks of the 1950s, the Great Patriotic War was covered in a single chapter, without division into sections.[14] From the 1970s onwards, the topic of the war was given a separate chapter consisting of thirteen sections.[15] The thematic grouping of military stories aimed to form schoolchildren's understanding of the historical significance of "the victory of the world's first socialist state over Fascism." The detailed depiction of battles was followed by information on the "selfless labour of the Soviet people in the rear" and "the heroic struggle of the Soviet partisans." Notable is the almost complete omission of killed soldiers and civilians in the concluding section, which discussed the "the price of victory," as well as the lack of any discussion of material damages caused on Soviet soil. Victory was for the living. Numbers of victims were replaced with the numbers of people decorated as Heroes of the Soviet Union, those who were honoured with state awards and obtained a place in the statistical data on Soviet achievements.[16] The numbers of losses were given only for Germany and its allies,[17]

---

13 Mikhail Studenikin, *Metodika prepodavaniya istorii v shkole: uchebnik dlya studentov vysshikh uchebnykh zavedenii* (Moscow: VLADOS, 2003), 29. For more information on the system of historical education in the Soviet school, see: Irina Ershova, "Evolyutsiya metodov obucheniya istorii v sovetskoi shkole, 1930-e – nachalo 1990's" (cand. of science diss. in pedagogy, Institute of Secondary School of Russian Academy of Education, 1994); Larisa Rudneva, "Evolyutsiya shkolnogo uchebnika istorii v 1940–80-e gody 20-go veka" (cand. of science diss. in history, Kursk State University, 2005).
14 Anna Pankratova, ed., *Istoriya SSSR: uchebnik dlya srednei shkoly*, 11th ed. (Moscow: Prosveshchenie, 1952), 366–416.
15 Maxim Kim, ed., *Istoriya SSSR (1938–1972): uchebnik dlya 10 klassa*, 4th ed. (Moscow: Prosveshchenie, 1975); Idem, ed., *Istoriya SSSR (1938–1976): uchebnik dlya 10 klassa*, 7th ed. (Moscow: Prosveshchenie, 1978).
16 Here are some quotations from Kim's history textbook: "Seven million soldiers, representing a hundred ethnicities of the Land of the Soviets, were awarded orders and medals for their combat feats. 11,603 brave soldiers were awarded the high title of Hero of the Soviet Union," "over 127,000 people were awarded a medal 'Partisan of the Patriotic War.'" Maksim Kim, ed., *Istoriya SSSR (1938–1978): uchebnik dlya 10 klassa*, 9th ed. (Moscow: Prosveshchenie, 1980), 78, 119.
17 In the ninth edition of Kim's history textbook, the following data are given: "In total, fascist Germany lost 13.6 million soldiers and officers, 10 million of them on the Soviet-German front, which was the main and decisive front of World War II." Kim, *Istoriya SSSR (1938–1978)*, 119.

which compared to the roll call of fallen Soviet heroes[18] provided ample testimony of the invincibility of both the Socialist system and the citizens of the Soviet state. Soviet history textbooks also did not provide information on reasons for the Red Army's setbacks at the beginning of the Great Patriotic War, Soviet collaboration – a feature of Nazi administration of the occupied zones, which had a bearing on the Holocaust – or Stalin's deportation of several Peoples of the Caucasus during the war. These wartime facts simply did not match the state-sanctioned heroic interpretation of the Victory of the Soviet People.

Nevertheless, the history textbook edited by Maksim Kim included a rather short discussion (just over one page of text) in a section on the partisans that covers the history of the occupation of Soviet regions. It recorded the plan of "Nazi Germany to destroy the Socialist system and annihilate millions of people," it mentioned a so-called "New Order" as "the German-fascist regime [committed] a bloody terror, looting, and violence in temporarily occupied territories."[19] The main victims of the "German fascist invaders" were communists, members of the Komsomol, activists, intellectuals, captured Red Army soldiers and commanders, old men, women and children according to the textbooks. The authors thus identified victims according to political, social, and gender classifications, but not ethnic ones. Moreover, Kim's textbook gave the number "of killed and tortured Soviet citizens" (about ten million people), of whom "about 200,000 people died in Kiev."[20] Yet the motives and the reasons for the mass killing of people by the Nazi authorities was not mentioned.

The non-ethnicity of Soviet citizens as Nazi victims was underlined by information that the Soviet Union had the "friendship of the peoples of our country as the source of its Great Victory [. . .] There are many names of Soviet patriots of different ethnicity among the Heroes of the Soviet Union," including "the Jew Gorelik."[21] It turns out that a decorated hero could be of any ethnicity,

---

18 Some names of the fallen heroes are given only in the history textbook of 1952. Among them are Zoya Kosmodemyanskaya, Aleksandr Matrosov, Panfilov's 28 Guardsmen, and others. Pankratova, *Istoriya SSSR*, 403.
19 Kim, *Istoriya SSSR (1938–1978)*, 73.
20 Ibid, 73–74. The number of victims is the most complicated question about the whole Holocaust history on Soviet soil. According to the latest existing data, about 100,000 Soviet people including Jews, Sinti and Roma, Ukrainian nationalists, patients of psychiatric hospitals, and Soviet POWs were killed and buried at Babi Yar. Norman M. Naimark, introduction to Hrynevych, *Babyn Yar*, vi.
21 At the same time, these Heroes of the Soviet Union, "the Jew Gorelik" and "Mary from Estonia" did not deserve the mention of their full names and the description of their feats in the history textbook, unlike the Russian pilot Aleksandr Pokryshkin, the Ukrainian partisan Sidor Kovpak, and the Kazakh Tudegen Tokhtarov. Pankratova, *Istoriya SSSR*, 402–403.

but a victim appeared only as a Communist Party member, a Red Army soldier, or "a defenceless woman, an old man, or a child." The shift from ethnic to gender- and age-related attributes of Nazi victims among the civilian population had an important ideological function. The authors of history textbooks, according to the official memory politics of the war, tried simultaneously to strengthen feelings of hatred towards the enemy (after all the killed civilians were mothers, brothers and sisters) and encourage awareness of the heroism of soldiers, who gifted life to future generations.

The narrative of World War II in school history textbooks begins to change after the collapse of the Soviet Union. All domains of public life in the Russian Federation, including education, were subject to reforms. In the academic year 1993–1994, the teaching of Russian history moved to a concentric structure,[22] which is still in use.[23] This is a combination of two linear courses taught in parallel: courses of general Russian and world history are taught from 5th to 9th grade followed by an in-depth study of the history of world civilisations and Russian history in the 10th and 11th grades.[24] In practice, this means that many topics, including military history, are studied twice. The legislative basis for this transition to a concentric structure in history education was the Law of the Russian Federation "On Education" [*Ob obrazovanii*] of 1992, where the mandatory basic nine-year school education was introduced. The history of the Second World War was divided conceptually in the history curriculum. World History textbooks discuss the main facts of World War II, and textbooks on Russian history focus on the Great Patriotic War.[25]

---

[22] Evgenii Vyazemskii, *Kak prepodavat' istoriyu v shkole: posobie dlya uchitelya* (Moscow: Prosveshchenie, 2000), 51.

[23] In February 2013, Russian president Vladimir Putin proposed "to think about unitary history textbooks on Russian history for secondary schools, designed for different grades, but built in the framework of a linear structure, within the framework of the unified logic of continuous Russian history, the interconnection of all its periods, respect for all events of our past." This initiative was implemented during the next two years and has been introduced into secondary schools with experimental status since the 2015–2016 academic year and from 6th grade only. Therefore, 20th-century Russian history will now be studied only in the 10th grade as of the 2019–2020 academic year and not in all Russian schools. For more details on the new Historical and Cultural standard for the Russian school education, see: "Kontseptsiya novogo uchebno-metodicheskogo kompleksa po otechestvennoi itorii," *Russian Historical Society official website*, https://historyrussia.org/images/documents/konsepsiyafinal.pdf (accessed March 23, 2020).

[24] Tatyana Tyulyaeva, ed., *Istoriya. 5–9 klassy: programmno-metodicheskie materialy* (Moscow: Drofa, 1999); Idem, ed., *Istoriya 10–11 klassy: programmno-metodicheskie materialy* (Moscow: Drofa, 1999).

[25] Gusenkova, *Rasskazhu vam o voine*, 25.

Yet, after the collapse of the Soviet Union, Soviet history textbooks continued to be used at schools for some time. New teaching handouts emerged in the mid-1990s. The content of Russian history textbooks of the transitional period (the first half of the 1990s) tried, with more or less success, to combine the incompatible; that is official Soviet history together with dissident readings of history that had for decades been considered a contra-history. Secondary schools had the right to choose their history programmes and history textbooks independently. Hence, a wide range of history teaching materials emerged in the absence of state educational standards in 1990s. It permitted the elaboration of new topics in courses on the Great Patriotic War. Among them were the Molotov-Ribbentrop Pact, the reasons of the Red Army's setbacks in 1941–1942, Stalin's wartime deportations, which profoundly affected the ethnic map of the Soviet Union, the problem of Soviet collaboration, and the role of the Russian Orthodox Church during the war.[26] Yet the core content of history textbooks of the 1990s, as well as of the Soviet period, still focused on military history, with the detailed description of the main campaigns and battles.[27] The authors of history textbooks during the transition period aimed to provide an objective history curriculum for schools, but in fact offered a new concept of the Great Patriotic War, still idealised, and sometimes including the contra-history.[28] In the history textbooks of the 1990s, the War remained a story of the heroism and fearlessness of the Soviet people, who "forged the Great Victory every day in the rear and at the frontline." Not even the history of collaboration and blunders of the Soviet command at the beginning of the war could bar the Soviets from Victory. The Great Patriotic War is shown as a still-glorious chronicle of the courage, bravery, and heroism of Soviet soldiers, with no mention of the sufferings of civilians nor of the ideologies that motivated mass killings (primarily of Jews) in occupied Soviet regions. Nevertheless, the emergence of the story of the "repressive politics of Stalinism," which affected the Peoples of the North Caucasus and the Crimea, is an important stage in the revision of the history

---

[26] Aleksandr Danilov and Lyudmila Kosulina, *Istoriya Rossii. XX vek: uchebnik dlya 9 klassa obshcheobrazovatel'nykh uchrezhdenii* (Moscow: Prosveshchenie, 1995); Vladimir Dmitrenko, Vladimir Esakov, and Vladimir Shestakov, *Istoriya Otechestva. XX vek: uchebnik dlya 11 klassa obshcheobrazovatel'nykh uchebnykh zavedenii* (Moscow: Drofa, 1995); Valerii Ostrovskii and Aleksei Utkin, *Istoriya Rossii. XX vek: 11 klass: uchebnik dlya obshcheobrazovatel'nykh uchebnykh zavedenii* (Moscow: Drofa, 1995); Andrei Levandovskii and Yurii Shchetinov, *Rossiya v XX veke: uchebnik dlya 10–11 klassa obshcheobrazovatel'nykh uchebnykh zavedenii* (Moscow: Prosveshchenie, 1997).
[27] Galina Klokova, "Rossiiskie uchebniki istorii o Velikoi Otechestvennoi voine," *Prepodavanie istorii i obshchestvoznaniya v shkole* 1 (2001): 42.
[28] Gusenkova, *Rasskazhu vam o voine*, 31.

of the War and of Stalinism in general.[29] The inclusion of this topic in school history textbooks resulted from the overall political situation in the 1990s, which intended not only to mark a clear break with the Soviet past, but to expose the blunders of the Soviet leadership to criticism in every possible way.

Holocaust history continued to be silenced in the textbooks of the transition period. Valerii Ostrovskii's textbook, for the first time, mentioned the "Master Plan for the East" [*Generalplan Ost*] for the colonisation of Central and Eastern Europe by Germans. Schoolchildren were informed that "the task of the Germans was to destroy 30 million Russians and 5–6 million Jews in the occupied Eastern territories."[30] The inhumanity of this plan, which "has no equals," was highlighted in the textbook, while nothing was written about the other Nazi plan, the "Final Solution to the Jewish Question." There was no information about the mass killing of Jews in Aleksandr Danilov and Lyudmila Kosulina's history textbooks either. Yet there was a note that "nearly 4 million partisans and underground fighters were killed in the enemy rear." According to Russian researcher Valerii Stolov, who examined the treatment of Jewish history in school textbooks of the 1990s, Danilov and Kosulina included in this figure the more than 2.5 million Jews killed on Soviet soil.[31] Nazi victims have acquired an even more universal character in the history textbooks of 1990s: from the depersonalised universal definition "Soviet citizens" they have become simply "people" or "local citizens."[32] This is another consequence of the collapse of the Soviet Union and the influence of the new government, which sought to draw a line under its Soviet past, but did not find the strength to tell schoolchildren about Nazi crimes against Jews, Sinti and Roma, and other victim groups.

The topic of Soviet losses, as well as Germany's losses during the war was not discussed at all in many textbooks of the mid-1990s. An important achievement of history textbooks of this period was the reflection of the liberal ideas of the Perestroika period. This could be seen in their main thesis that "it wasn't the authorities who won the Great Patriotic War, but the people full of heroism,

---

29 Dmitrenko et al., *Istoriya Otechestva. XX vek*, 336–337.
30 Ostrovskii, *Istoriya Rossii. XX vek*, 286.
31 Valerii Stolov, "Evreiskaya istoriya v rossiiskoi shkole," *Novaya evreiskaya shkola* 1 (1998): 179.
32 Here is a quote from Ostrovskii's textbook: "The fascists created death and concentration camps, thousands of prisons and ghettos in the occupied territories. Brutal massacres of the local population were carried out." Ostrovskii, *Istoriya Rossii, XX vek*, 287. See also the evolution of the definition of Nazi victim groups in the Soviet and Russian history textbooks in Table 5.

[who] in spite of unspeakable sufferings and losses defeated the enemy."[33] Thus, "Great Victory in the Great Patriotic War" excuses the crimes of the Soviet authorities during the 20th century, so that the memory of the Great Victory was completely colonised by the state.

With the adoption of amendments to the Law of the Russian Federation "On Education" in 2009, new federal state educational standards were developed.[34] Educational standards are learning objectives, detailing what schoolchildren should learn and be able to recall at each grade level. History curricula, educational literature, and tests are developed on the basis of these objectives.[35] According to the results of a regional survey of teachers teaching history to pupils in 9th and 11th grades, I selected the most frequently used history textbooks to analyse their treatment of Holocaust-related themes.[36] All the history textbooks studied had the stamp "Recommended by the Ministry of Education and Science of the Russian Federation." This means that they meet the requirements of federal and state educational standards for professional education and the compulsory minimum of secondary education, approved by the Ministry of Education.

The Second World War is treated in more detail than any other topic of twentieth-century Russian/Soviet history in contemporary Russian history textbooks.

---

33 Sokolov, "Shkol'nyi uchebnik."
34 Aleksei Maiorov, "Normativnye osnovy otbora soderzhaniya obrazovaniya v shkolakh Rossii," *Portal realizatsiya federalnogo zakona "Ob obrazovanii" v Rossiiskoi Federatsii*, http://273-фз.рф/publikatsii/normativnye-osnovy-otbora-soderzhaniya-obrazovaniya-v-shkolah-rossii (accessed March 23, 2020).
35 *Ministry of Education official website*, http://minobrnauky.rf/ (accessed March 23, 2020).
36 Aleksandr Danilov, Lyudmila Kosulina, and Maksim Brandt, *Istoriya Rossii, XX – nachalo XXI veka: 9 klass: uchebnik dlya obshcheobrazovatel'nykh uchrezhdenii* (Moscow: Prosveshchevie, 2013); Nikita Zagladin, ed. *Istoriya Rossii. XX – nachalo XXI veka: uchebnik dlya 9 klassa obshcheobrazovatel'nykh uchrezhdenii*, 11th ed. (Moscow: Russkoe slovo, 2013); Vladimir Shestakov, Mikhail Gorinov, and Evgenii Vyazemskii, *Istoriya Rossii XX – nachalo XXI veka: uchebnik dlya 9 klassa obshcheobrazovatel'nykh uchrezhdenii*, 7th ed. (Moscow: Prosveshchevie, 2011); Oleg Volobuev, Valerii Klokov, Mikhail Ponomarev, and Vasilii Rogozhkin, *Istoriya. Rossiya i mir: 11 klass: bazovyi uroven': uchebnik dlya obshcheobrazovatel'nykh uchrezhdenii*, 12th ed. (Moscow: Drofa, 2013); Nikita Zagladin, Sergei Kozlenko, Sergei Minakov, and Yurii Petrov, *Istoriya Rossii XX – nachalo XXI veka: uchebnik dlya 11 klassa obshcheobrazovatel'nykh uchrezhdenii*, 5th ed. (Moscow: Russkoe slovo, 2007); Andrei Levandovskii, Yurii Shchetinov, and Sergei Mironenko, *Istoriya Rossii XX – nachalo XXI veka: uchebnik dlya 11 klassa obshcheobrazovatel'nykh uchrezhdenii: bazovyi uroven'* (Moscow: Prosveshchenie, 2013); Vladimir Shestakov, *Istoriya Rossii XX – nachalo XXI veka: uchebnik dlya 11 klassa obshcheobrazovatel'nykh uchrezhdenii: profil'nyi uroven'*, 5th ed. (Moscow: Prosveshchenie, 2012).

The history of the war is presented in a separate chapter, consisting of seven sections in the textbooks for 9th grade and of four to five sections for the 11th grade. While preserving the periodisation and some clichés of the Soviet era, the content of history textbooks is enhanced with discussion of previously silenced themes. This was the result of the declassification of a variety of Soviet sources.[37] Among other wartime topics, the Molotov-Ribbentrop Pact, the reasons for the Red Army's setbacks in 1941–1942, the problem of collaboration, and Stalin's pre- and post-war plans for the administrative structure of Europe are discussed in the textbooks. There is data on human losses at the front and in the rear. Excerpts from historical documents are widely presented in the updated content of history textbooks, the analysis of which assists schoolchildren to understand historical facts more deeply. An attempt was made to present the war as a universal historical fact that affected all aspects of social life and, in turn, was influenced by the society. Compared with the textbooks of the Soviet period, much greater attention is given to the topic "people at war."

In all of the modern history textbooks studied, a separate paragraph is devoted to the topic of the occupation of Soviet territories.[38] However, Holocaust history is treated either superficially or not at all. Among Nazi victim groups the history textbooks mention "citizens of the Soviet Union, who became forced labourers in Germany,"[39] "Soviet POWs killed in Nazi concentration camps,"[40] old people and children, who "had no mercy from the Nazis,"[41] and "Communists and Soviet officials, whom the occupiers primarily identified and murdered."[42] There is little information about Jews, Roma and other "inferior peoples," who were subject to total annihilation. Only two of the history textbooks studied provide school students with statistics pertaining to the killing of Soviet Jews ("in the first six months of the war, the fascists killed up to 1.5 million Jews, practically

---

37 Guzenkova, *Rasskazhu vam o voine*, 34.
38 The only exception is Zagladin's textbook, in which the role of the USSR in World War II is discussed. With this approach, everyday life in the occupied territories turned out to be beyond the authors' view. Zagladin, *Istoriya Rossii. XX – nachalo XXI veka: uchebnik dlya 9 klassa*.
39 The textbook of Vladimir Shestakov for the 9th grade even introduces the term "Ostarbeiter," a term used to designate forced labourers from the occupied east of Europe, widely used in historical literature as a separate Nazi victim group. Shestakov et al., *Istoriya Rossii XX – nachalo XXI veka*, 184.
40 Zagladin et al., *Istoriya Rossii XX – nachalo XXI veka: uchebnik dlya 11 klassa*, 234.
41 Shestakov et al., *Istoriya Rossii XX – nachalo XXI veka: uchebnik dlya 11 klassa*, 235.
42 Volobuev et al., *Istoriya. Rossiya i mir*, 172.

one in every two of them on Soviet soil"[43]) and mention reasons for this Nazi policy against the Jews (Soviet Jews "ideologically supported the 'Judeo-Bolshevik' regime"[44]). The term "holocaust" (written in lowercase) is also mentioned in these history textbooks as "mass annihilation (genocide) of Jews."[45] In other history textbooks, where Jews are mentioned as a separate victim group,[46] the meaning of the Holocaust is debased by poorly written sentences,[47] which continue to highlight the suffering of all Soviet citizens, as in the history textbooks of the Soviet period. Only seven out of thirty history textbooks recommended for teaching about the Holocaust in Russia actually refer to the term Holocaust according to a report of the Organisation for Security and Cooperation in Europe (OSCE) on teaching about the Holocaust and antisemitism in European countries. The term itself is perhaps only precisely explained in a small minority of textbooks.[48] Despite episodic and not always correct representation of Holocaust history, the presence of the term Holocaust in modern history textbooks is an important step toward true depiction not only of the military theatres of operations, to which Russian textbooks still accord paramount importance, but also of life behind the front line and the fate of Soviet Jewry during the war, along with other Nazi victim groups on Soviet soil.

It is notable that the history of the occupation is written in general terms, and no documentary evidence or historical facts are provided. A fairly short, stiffly written paragraph on the occupation regime is replaced by an extensive story about the partisan movement behind the front line, containing tales of the heroic courage of Soviet partisans and underground fighters. The Holocaust is mentioned in modern history textbooks, as the required by the state standards, yet the author collectives continue to place the same emphases in the core narrative as during the Soviet period. It is much easier and emotionally more pleasant to write about exploits that glorified the Fatherland than to uncover disturbing aspects of the past. Nevertheless, if modern school history

---

43 Danilov et al., *Istoriya Rossii, XX – nachalo XXI veka: 9 klass*, 219.
44 Ibid, 219.
45 Shestakov et al., *Istoriya Rossii XX – nachalo XXI veka: uchebnik dlya 9 klassa*, 184–185. Incidentally, the terms Holocaust and/or genocide appear in the additional materials of several textbooks. These are textbooks by Danilov for the 9th grade and by Shestakov and Levandovskii for 11th grade. At the same time, the core narrative of Levandovskii's textbook does not mention Holocaust history at all.
46 See Table 5 "Definition of Nazi victim groups in Soviet and Russian history textbooks."
47 Literally the sentence is written like this: "Along with the Jews, who were victims of mass genocide, other peoples who lived in the USSR were also subjected to repression by the Hitlerite administration." Volobuev et al., *Istoriya. Rossiya i mir: 11 klass*, 172.
48 *Education on the Holocaust and Anti-Semitism: An Overview and Analysis of Educational Approaches* (Warsaw: OBCE Office for Democratic Institutions and Human Rights, 2006), 107.

education seeks to keep pace with the rest of the world, then it is the moral duty of every history textbook author to discuss the full extent of Nazi crimes.

Each paragraph of a school history textbook has a set of questions for revision or further in-depth study of primary sources, which are cited in each paragraph and after it. The questions draw pupils' attention to the main topics discussed in the paragraph. Because the history of the mass killing of Jews on Soviet soil is not represented in the majority of history textbooks, there are also no questions about this topic. The only exceptions are two textbooks. In the section "Themes for Research Projects" the first one suggests "to prepare a story about life under the occupation based on the recollections of your family members or using regional history materials,"[49] while the other textbook proposes the discussion topic "My Family during the Great Patriotic War."[50] However, these tasks do not necessarily include Holocaust history and the fulfilment of such tasks depends entirely on the teacher's interest in paying closer attention to the study of the occupation.

In 2007 on the initiative of the RJC, an expert group of researchers from the Institute of Oriental Studies and the Institute of World History of the Russian Academy of Sciences (hereafter, RAN) conducted a comprehensive analysis of Russian history textbooks for secondary schools. Their aim was to identify what elements of Jewish history were treated in world and Russian history courses.[51] They noticed some progress of the content of history textbooks in comparison with the Soviet period and the beginning of the 1990s,[52] insofar as Jewish history received some treatment. However, the researchers concluded that "none of the studied textbooks can be recommended to teachers or schoolchildren as a textbook that fully and correctly reveals the history of Russian Jewry on the basis of the latest research of Russian and international scholars."[53] They affirmed that there is no representation and understanding of Holocaust history in any textbook as "the only time in world history when a particular state attempted completely to annihilate a whole nation regardless of gender, age,

---

49 Shestakov et al., *Istoriya Rossii XX – nachalo XXI veka: uchebnik dlya 9 klassa*, 186.
50 Levandovskii et al, *Istoriya Rossii XX – nachalo XXI veka: uchebnik dlya 11 klassa*, 368.
51 "Evreiskii vopros v rossiiskikh uchebnikakh," *My zdes'*, February 22–28, 2008, http://newswe.com/index.php?go=Pages&in=view&id=128 (accessed March 19, 2020).
52 See similar studies of the Soviet period and the 1990s: Daniil Fish, "Evrei v shkol'nom kurse vsemirnoi i russkoi istorii (chto chitayut sovetskie shkol'niki o roli evreev v istorii)," *Evreiskii samizdat*, 15 (Ierusalim: Evreiskii Universitet v Ierusalime, 1978): 313–352; Moisei Chernyi, "Figura umolchaniya, ili 'Yudenfrai,'" *Evreiskaya shkola* 2–3 (1995): 59–66; Stolov, "Evreiskaya istoriya."
53 Aleksandr Lokshin, "Istoriya rossiiskikh evreev v sovremennykh shkol'nykh uchebnikakh Rossiiskoi Federatsii," *Evroaziatskii evreiskii ezhegodnik – 5768 (2007–2008)*, ed. Vyacheslav Likhachev et al. (Moscow: Pallada, 2008), 199.

place of residence, profession or religion of its representatives, and not for the sake of obvious material, territorial, or political benefits."[54]

My own analysis of the content of modern Russian history textbooks[55] shows that this problem has not yet been resolved. The topic of Holocaust history is either not addressed at all, or is given two or three general sentences; the term itself is sometimes given in the appendix section without any explanation in the core text. The reasons for the Holocaust and its impact on Soviet Jewry are often not clarified, and methods of killing Jews on Soviet soil are not mentioned at all. Hopefully, after the implementation of the new "Historical and Cultural Standard" and new history textbooks in the Russian school system,[56] Holocaust history will obtain a worthy place in the sections on covering World War II and the Great Patriotic War.

The Russian history school course includes a regional component that should represent not less than 10 per cent of the whole curriculum. The Law "On Education" specifies the unity of the educational space in the Russian Federation, as well as the protection and development of the ethno-cultural traditions of the peoples that live in the multi-ethnic state.[57] A regional history course [kraevedenie] has been a separate component of the secondary school programme since the mid-1990s. Specialists on local history in each Russian region

---

54 Aleksandr Lokshin, "Istoriya rossiiskikh evreev v shkol'nykh uchebnikakh RF," *Issledovaniya po prikladnoi i neotlozhnoi etnologii* 209 (Moscow: Institut Etnologii i Antropologii RAN, 2009): 18.

55 In addition to my analysis, see also modern studies of Russian history textbooks for secondary school: Irina Kamenchuk, and Evgeniya Listvina, "Kultura pamyati kak uslovie formirovaniya tolerantnosti: fenomen Kholokosta i problemy predodavaniya v rosiyskikh shkolakh," *Rossiisko-amerikanskii forum obrazovaniya: elektronnyi zhurnal* 4, no. 3 (2012), http://www.rus-ameeduforum.com/content/en/?task=art&article=1000933&iid=13 (accessed March 23, 2020).

56 According the new Historical and Cultural Standard the main thematic units for teaching the section "Nazi occupation regime" are the following: "Nazi occupation regime; Master Plan for the East; mass crimes of the Nazis against Soviet citizens; death camps; the Holocaust; ethnic cleansing in the occupied territory of the USSR; Soviet POWs; mass killing of Soviet POWs and medical experiments on prisoners; forced labour in Nazi Germany; looting and destruction of cultural property." "Historical and Cultural Standard of the Defence Ministry of the Russian Federation," *Russian Historical Society official website*, May 28, 2015, https://historyrussia.org/proekty/kontseptsiya-novogo-uchebno-metodicheskogo-kompleksa-po-otechestvennoj-istorii/istoriko-kulturnyj-standart.html (accessed March 23, 2020).

57 "Federalnyi zakon ot 29 dekabrya 2012 goda," art. 3, s. 4.4. Prior to this law's coming into force, the Federal Law of July 10, 1992, No. 3266–1 was in effect, which presupposed the existence of a Federal and Regional component of the State Standard of General Education. "Zakon ob obrazovanii Rossiiskoi Federatsii 1992 goda," *Rossiiskaya gazeta*, July 31, 1992.

have developed teaching handouts and textbooks for this course, aimed at inculcating a deep knowledge among school students of the nature and history of their native locality. In some constituent entities of the Russian Federation, for example, in Krasnodar Krai, there is a special obligatory course in the school programme called "History of Kuban Region or Krasnodar Krai" ["*Kubanovedenie*"]. It has become a priority area for regional history education, and author collectives have prepared a set of regional history textbooks for secondary schools. This course is taught from the 1st to the 11th grades in each elementary and secondary school of Krasnodar Krai.[58] Increased attention to regional history in Krasnodar Krai was associated with the Kuban Cossack revival movement in the early 1990s, which plays a significant role in the modern social structure of the region. In other regions, for example in Stavropol Krai and Rostov Oblast, regional history is included as an optional component of the curriculum and is introduced into teaching at the discretion of the individual school.[59] In the North Caucasian republics a new wave of attempts to develop local language, traditions and culture has also emerged since the early 1990s. In North Ossetia, for example, the study of regional history is conducted according to a concentric system: in the 5th grade and then in the 10th and 11th grades.[60] In Kabardino-Balkaria, this course is taught in 8th and 9th grades, followed by a special course "Traditional Culture of Kabardians and Balkars" at secondary school.[61] As a rule,

---

[58] See: "Metodicheskie rekomendatsii dlya obrazovatelnkyh organizatsii Krasnodarskogo kraya o prepodavanii predmeta 'Kubanovedenie' v 2015–2016 uchebnom godu," *Ministry of Education, Science, and Youth Policy of Krasnodar Krai official website,* http://www.minobrkuban.ru/obrazovanie/obsh-obrazov-shkoly/kubanoved-opk/ (accessed March 23, 2020).

[59] In Rostov Oblast, the course "History of Donskoi krai" (the river Don is the main river in Rostov Oblast) is taught from the 5th to 9th grades, and in Stavropol Krai, the regional history is a part of the 10th and 11th grade history course.

[60] Aleksandr Seravin and Igor' Sopov, *Ekspress analiz prepodavaniya istorii Rossii i regiona v sub"ektakh Severo-Kavkazskogo federalnogo okruga* (Vladikavkaz, 2014), 14–15.

[61] Relatively more attention is given to the study of native language, literature, and culture of the Peoples of the Caucasus in school education of the North Caucasian Republics, rather than the study of regional history, which is taught as part of Russian history. In this history course complex relations between the Peoples of the Caucasus and the Russian Empire or the USSR do not find critical reflection. Aleksandr Seravin and Igor' Sopov, "Ekspress analiz prepodavaniya istorii Rossii i regiona v sub"ektakh Severo-Kavkazskogo federalnogo okruga: doklad na vyezdnom zasedanii komiteta Soveta Federatsii po nauke, obrazovaniyu i kulture, Vladikavkaz, 15–17 maya 2014," *Tsentr sovremennoi kavkazskoi politiki,* http://politkavkaz.ru/pdf/istoriya-rossii.pdf (accessed March 23, 2020).

there is only one basic regional history textbook in each North Caucasian Republic.⁶² The most complicated situation was identified in the Karachay-Cherkess Republic, where a regional history textbook was not developed until 2019.⁶³

The history of the Great Patriotic War is again the central topic of all regional history textbooks. The history of military operations is studied on the basis of local sources with special attention paid to the countrymen "who have glorified their native locality for centuries." In the textbooks of the North Caucasian republics, the central topic is the role of the Peoples of the Caucasus in the war, "their military friendship and selfless service to the Motherland." Soviet ideas of unity, brotherhood, and friendship are extremely important for instilling a sense of tolerance between the Peoples of the Caucasus themselves and as a part of Russia, and the history of the war provides the best examples for this purpose. Even the history of the deportation of the Balkars, which the author of the regional history textbook of Kabardino-Balkaria called genocide, is described only as a result of "the unfair treatment of Soviet power" and is shown in terms of the heroic participation of Balkars in the Red Army.⁶⁴ Thus, the description of war history in school regional history textbooks also reflects the current political ideology at regional level.

The history of the war is represented in Cossack regions through the trope of the valour and courage of the Cossacks; cases of their collaboration are mentioned as isolated incidents of deviation that were not supported by most

---

62 Mukhamed Bekaldiev, *Istoriya Kabardino-Balkarii, 8–9 klassy: uchebnik dlya obshcheobrazovatel'nykh uchrezhdenii*, 4th ed. (Nalchik: "Elbrus," 2013); Kuchiev Vasilii, *Istoriya Osetii: XX vek: uchebnik dlya starshikh klassov obshcheobrazovatel'nykh shkol*, (Vladikavkaz: Ir, 2011); Aleksei Krugov, *Stavropol'skii krai v istorii Rossii (konets XVII – nachalo XXI v.): regionalnyi uchebnik dlya 10–11 klassov obshcheobrazovatel'nykh uchrezhdenii* (Moscow: Russkoe slovo, 2006); Andrei Zaitsev, ed. *Kubanovedenie: uchebnoe posobie dlya 9 klassa obshcheobrazovatel'nykh uchrezhdenii* (Krasnodar: Perspektivy obrazovaniya, 2012); Kazbek Achmiz, ed. *Istoriya Adygei: uchebnoe posobie dlya 9 klassa obscheobrazovatel'nykh uchrezhdenii* (Maykop: Adygeiskoe respublicanskoe knizhnoe izdatel'stvo, 2002); Sergei Kislitsyn and Irina Kislitsyna, *Istoriya Donskogo kraya: uchebnik dlya 9 klassa obscheobrazovatel'nykh uchrezhdenii* (Rostov-on-Don: Donskoi izdatel'skii dom, 2004).

63 Although the history of Karachay-Cherkessia is studied in the 10th and 11th grades, and a regional history textbook was published in 1998, today it is a bibliographic rarity. Because of its limited circulation and the absence of reprints in the 2000s, schools were not provided with it. Vladimir Nakhushev, ed. *Narody Karachaevo-Cherkessii: istoriya i kultura: uchebnoe posobie dlya 10–11 klassov obscheobrazovatel'nykh uchrezhdenii* (Cherkessk: Poligrafist, 1998).

64 Bekaldiev, *Istoriya Kabardino-Balkarii*, 279–281.

villagers.[65] The German occupation is treated most comprehensively in regional history textbooks; it is described in terms of the suffering it caused and, at the same time, as a "heroic struggle of local civilians against the enemy." The regional history textbooks of Krasnodar Krai continue to represent Nazi victims in the same terms as Soviet portrayals. The occupiers killed "unarmed civilians" in gas vans; the reasons for the mass killing of "more than 13,000 civilians by the Nazis in Krasnodar alone" remain unclear.[66] After the collapse of the Soviet Union, Krasnodar Krai has proven itself to be a hotbed of antisemitism, where decisions against Jews have been taken even at the legislative level.[67] Until now, Jewish issues are treated as undesirable in regional policy.

Only two of the regional textbooks studied pay attention to Holocaust history. There is information on the killing of "660 patients at the psychiatric hospital" and "of two groups of Jews (3,500 and 500 people) mainly professors of medical institutes" in the "Chronicle of Nazi Crimes under the Occupation" in the textbook for Stavropol Krai.[68] There is an excerpt from a wartime article by Aleksei Tolstoi "Brown Dope" [*Korichnevyi durman*] on the Nazi atrocities in Stavropol Krai among additional historical sources in the core text. The article was published in 1943 in the newspaper *Pravda* and reprinted in many regional newspapers.[69] There is also an illustration of the Nazi "Appeal to the Jewish population of Pyatigorsk" and a task for schoolchildren to prepare a story "about Nazi atrocities in your town or village." However, the author of this textbook did not use the term Holocaust and did not comment on the reasons for the mass killings of both Jews and people with disabilities. The way the history of the occupation in Stavropol Krai is presented in the regional history textbook again mirrors the local authorities' position towards the history of the war. The mass killing of Jews is recognised as a fact; at the same time the authorities of

---

[65] In the textbook *Kubanovedenie*, edited by Andrei Zaitsev, the history of Cossack collaboration is not represented at all, while Sergei Kislitsyn, although mentioning the Cossacks' support for the Nazi regime, notes that "there were relatively few Cossacks among Soviet collaborators." See: Kislitsyn, *Istoriya Donskogo kraya*, 88–92.

[66] Zaytsev, *Kubanovedenie*, 66–67.

[67] See, for example, an analysis of the activities of the former governor of Krasnodar Krai Nikolai Kondratenko (1996–2000), known for his antisemitic expressions: Roman Lukin, "Obyknovennyi natsional-kommunizm. Kubanskii, pravoslavnyi. Pri Nikolae Kondratenko v Krasnodarskom krae borolis' s 'sionizmom,' a pri Aleksandre Tkacheve vzyalis za iskorenenie 'sektantov,'" *Portal credo.ru*, https://www.portal-credo.ru/site/?act=comment&id=1019 (accessed March 23, 2020).

[68] Krugov, *Stavropol'skii krai*, 252–253.

[69] Aleksei Tolstoi, "Korichnevyi durman," *Pravda*, August 5, 1943; Idem, *Stavropol'skaya pravda*, August 11, 1943.

Stavropol Krai do not help, but also do not interfere with individual initiatives to preserve the memory of Holocaust victims in the region.

The mass killings of Jews because of their ethnicity is noted in the regional history textbook on Rostov Oblast and the memorial in Zmievskaya Balka is used to illustrate the core text. It also mentions that this memorial is "the only one in Russia dedicated to Holocaust victims."[70] This textbook reflects again the local authority's position: as already noted, in 2004 (the year of publication of this textbook) a memorial plaque proclaiming itself "the largest Holocaust Memorial in Russia" was installed on the Soviet memorial in Zmievskaya Balka by decree of the Mayor of Rostov-on-Don.

The French historian Mark Ferro believes that the main goal of Soviet history textbooks was "the preparation and interpretation of history that would adapt the past to the needs of the present."[71] This remark can be fully attributed to post-Soviet Russian history textbooks as well, both federal and regional. The analysis of regional history textbooks shows that history is represented according to the prevailing political views of state authorities and the set priorities in the regions. Table 5 "Definition of Nazi victim groups in Soviet and Russian history textbooks," filled out according to the discourse analysis of the war narrative in school history textbooks, clearly demonstrates the continued use of Soviet terminology in interpretations of Holocaust history. The politicised category "peaceful citizens" is still in use in the core narrative of these history textbooks. Jews are mentioned separately in history textbooks of the modern period but without explaining their particular significance to the Nazi regime. Holocaust history (if presented) is poorly developed and no attention is given to it in comparison with higher priority topics in regional history, such as the history of the Cossack movement and ethno-political stabilisation in the North Caucasus. Jews are presented in regional history textbooks as alien to the ethnic structure of the North Caucasus; their history during World War II little more than a footnote.

One of the important factors affecting the teaching of history in Russian schools is the attempt by institutions and social groups to break the traditional state monopoly on the interpretation of the past. Since the mid-1990s the Holocaust Center has sought to introduce teaching about the Holocaust into Russian schools. The staff of the Center have developed more than a dozen textbooks and teaching manuals for pupils and students, as well as for teachers and

---

70 Kislitsyn, *Istoriya Donskogo kraya*, 85.
71 Mark Ferro, *Kak rasskazyvayut istoriyu detyam v raznykh stranakh mira*, transl. Elena Lebedeva (Moscow: Vyshaya shkola, 1992), 158.

lecturers.[72] These manuals contain valuable and vivid materials on the history of the Holocaust on Soviet soil, based on variety of historical sources and wartime photographs. The Center's handouts give special attention to the history of the Jewish resistance in World War II, the life and daily struggle of Jews in ghettos, and the role of the Righteous among the Nations in rescuing Jews.[73] However, these materials represent Holocaust history with ghettos, concentration camps and death camps, and the history of the Jewish resistance movement is based mostly events in modern-day Ukraine, Belarus, and the Baltic states. The Holocaust in the North Caucasus is not represented in the Center's textbooks and manuals. Only Il'ya Alt'man's textbook for students contains information on the mass killing of Jews in the North Caucasus.[74]

7.2 The Spark that Starts the Fire: Regional History Teachers' Initiatives to Teach about the Holocaust

Developing an objective history textbook that would include the history of the Holocaust on Soviet soil is only half the solution to the problem of Russian history education. Another important component is the desire and ability of school history teachers to teach certain topics in the classroom. In practice, education planning is often quite different from actual teaching. The teacher acts as a mediator between the information given in the history textbook and discussions in the classroom, she or he is responsible for the lesson structure, developing narratives around certain topics, and setting priorities for their discussion.

Despite the presence of Holocaust-related topics in some modern history textbooks, this theme is excluded from the standardised questions on history

---

[72] Galina Klokova, *Istoriya Kholokosta na territorii SSSR v gody Velikoi Otechestvennoi voiny (1941–1945): posobie dlya uchitelya* (Moscow: Tsentr "Kholokost," 1995); Il'ya Al'tman, ed., *Prepodavanie temy Kholokosta v XXI veke.* (Moscow: Fond "Kholokost," 2000); Il'ya Al'tman, Alla Gerber, and David Poltorak, *Istoriya Kholokosta na territorii SSSR: uchebnoe posobie dlya 9 klassa* (Moscow: Fond "Kholokost," 2001); Il'ya Al'tman, *Kholokost i evreiskoe soprotivlenie na okkupirovannoi territorii SSSR: uchebnoe posobie dlya studentov vysshikh uchebnykh zavedenii* (Moscow: Fond "Kholokost," 2002); Il'ya Al'tman, ed., *Tema Kholokosta v shkolnykh uchebnikakh: posobie dlya uchitelya* (Moscow: Tsentr i Fond "Kholokost," 2010).
[73] More detailed analysis of the teaching materials published by the Holocaust Center see: David Poltorak, "Uchebniki i metodicheskie posobiya o Kholokoste," in Al'tman, *Tema Kholokosta v shkol'nykh uchebnikakh,* 17–21.
[74] Al'tman, *Kholokost i evreiskoe soprotivlenie,* 173–197.

in the Unified State Exam.⁷⁵ Russian schoolchildren do not learn about the Holocaust, though, mostly because of their history teachers. Teachers can help schoolchildren to develop their own opinion on different historical facts, including the Holocaust. In this case, it is the personality of the teacher that plays the decisive role.⁷⁶

I conducted a voluntary survey of regional history teachers for a better understanding the process of teaching about World War II and the Holocaust in history lessons at 9th and 11th grades. History teachers of secondary schools, gymnasiums, and lyceums of the North Caucasus participated in this survey during two academic years (2014–2015 and 2015–2016). Great assistance in conducting the survey was provided by the regional and republican Teacher Training Institutes, whose staff asked teachers to fill out the questionnaire. About 300 questionnaires were collected (43 per cent of respondents were teachers in Krasnodar Krai, 21 per cent in Stavropol Krai, 15 per cent in Rostov Oblast, 15 per cent in the Republic of Adygea and 6 per cent in the Karachay-Cherkess Republic).⁷⁷ The social characteristics of survey participants corresponds to the data for teachers in Russia as a whole.⁷⁸ They were predominantly women (86.7 per cent) of middle age (from 36 to 55 years old – 65.6 per cent), qualified in the "first" or "highest" professional category (38 per cent and 32 per cent of respondents respectively) with 16 or more years of general pedagogical experience (61 per cent). The survey was conducted among history teachers in urban (53 per cent) and rural (47 per

---

75 The president of the Holocaust Foundation Alla Gerber and co-chairman Il'ya Al'tman have repeatedly appealed to the Ministry of Education of the Russian Federation with the initiative to include Holocaust history into Russian history textbooks. In 2012, Valentin Shaulin, head of the Department for Assessing the Quality of Education, stated that 10 questions on Holocaust history featured in the Unified State Exam of 2011. However, teachers who cooperated with the Holocaust Center did not confirm this information as of 2016. "Otvet Valentina Shaulina," *Holocaust Center official website*, http://www.holocf.ru/pages/83 (accessed December 14, 2019).
76 Irina Kamenchuk, "Fenomen Kholokosta kak faktor vospitaniya tolerantnosti: rossiiskii i mirovoi opyt," *Proceedings of the 3rd International Academic Conference "Applied and Fundamental Studies"* (St. Louis, Missouri: Science and Innovation Center, 2013), 410.
77 The survey was conducted on a voluntary basis for history teachers of the 9th and 11th grades. Kabardino-Balkaria and North Ossetia were not included in the survey, since no contact with local education authorities was established. The result of the survey is part of the author's archive.
78 According to the latest international study of the Organisation for Economic Cooperation and Development, the average age of Russian teachers is 52 years. 85 per cent of teachers are women. The pedagogical experience of Russian teachers is more than 20 years, with more than 15 years in one workplace. "Portret rossiiskogo uchitelya," *RIA novosti*, May 26, 2015, https://ria.ru/sn_edu/20150526/1066505151.html (accessed March 23, 2020).

cent) secondary schools. The presence of young teachers (under 30 years old) was significant (about 10 per cent). This is explained by the fact that young teachers strive to obtain the "highest" qualification category as quickly as possible,[79] which can be achieved after attending special professional training sessions at regional Teacher Training Institutes. According to the Federal Law of the Russian Federation "On Education" all teachers are required to take such courses every three years to validate their professional category.[80]

The social characteristics of the history teachers who took part in the survey are useful for interpreting its findings. In general, both young and experienced teachers expressed personal ignorance of, and minimal interest in, the history of the Holocaust. Most questions in the survey were closed-ended, multiple-choice questions, and related to the teaching of the Great Patriotic War in modern schools. Teachers were asked to choose from a list (or add) those topics that are necessarily considered in their lessons, and those which they would make sure to address, even given constraints on time. Among such topics, teachers tended to choose the history of military operations and main battles, Soviet partisans, "the price of Victory," and "great Soviet generals and heroes." Such an approach to teaching about the war corresponds with the format of history textbooks at both federal and regional levels. The topic "life under the occupation" is mentioned only by half of the respondents, while only one third of history teachers list the Nazi policy of mass killings of civilians, primarily Jews. Half of the respondents declared that they use the term Holocaust in the classroom. This may well be the case, as the term appears in several modern history textbooks, so the teacher should at least raise the topic of the Holocaust with pupils.[81] It is another matter to dwell on Holocaust history more deeply. This requires the involvement of additional material and further preparation of the lesson by the teacher. In 2004, a schoolgirl of the Belorechensk Gymnasium in Krasnodar Krai, Ekaterina Kuznetsova revealed in her survey of 50 classmates that only six participants knew about the Holocaust, and two of them could

---

79 The qualification category affects the teacher's salary.
80 According to art. 47, s. 5.2 of the Federal Law of the Russian Federation of December 29, 2012, "pedagogical employees have the following labour rights and social guarantees: the right to additional professional pedagogical education no less than once every three years."
81 The OSCE report for 2006 says: "On average, one page of history textbooks is devoted to Holocaust history. According to the information provided, history teachers allocate 30 minutes to the topic." *Education on the Holocaust and on Anti-Semitism*, 107. Information for this report was provided by the Holocaust Center, which cooperates with a small number of teachers with high motivation of teaching about the Holocaust in the classroom. My own analysis of history textbooks and the results of survey do not support this conclusion.

explain it in detail. She then asked her teacher whether it was necessary to pay attention to the Holocaust in history courses: "She [the teacher] replied that there is a rather full programme in the 11th grade, three history textbooks must be studied, so she cannot give time to Holocaust history. For the Unified State Exam, students write down a definition of the Holocaust, and this, according to teachers, is enough."[82] Unfortunately, the situation over the past ten years has not changed. Teachers define the term, but they do not go into greater depth.

It is noteworthy that according to the survey results many young teachers demonstrated their personal and professional ignorance of Holocaust history. The answers "this topic does not attract me" and "I do not consider it necessary to pay attention to this topic in the classroom" were most common among teachers under 25 years old. Undoubtedly, such a reaction of young teachers is the result of their own historical education at the end of 1990s and in the early 2000s, when little attention was drawn to Holocaust history at schools and universities. An average young teacher is perhaps an exaggerated and more dangerous version of the Karatygina sisters, mentioned at the beginning of this chapter, one to whom society has entrusted the teaching of the next generation.

It is also thought provoking to consider the answers of teachers at rural schools. Historically, Jews settled in the large southern cities before World War II, so the Holocaust is not a relevant topic for the villagers: "If there were no local Jews in the rural locality, why do we need to teach about the Holocaust in the classroom?" This opinion was expressed by one third of rural teachers who participated in the survey. Teachers don't feel the need to teach topics that are distant (in geographical, political, and social senses of the term). Only in rural areas where evacuated Jews are commemorated as Holocaust victims, did teachers give proper attention to teaching the Holocaust. It is clear that for the topic to be presented in the classroom, there should already be some resonance of the topic locally, as well as some encouragement from the authorities.

A separate block of questions in my survey was devoted to special events and thematic lessons related to history of the Great Patriotic War. Since Soviet times, Russian schools have had a tradition of organising afterschool activities dedicated to Victory Day (May 9) and the Day of the Defender of Fatherland (February 23). Local activists, war veterans, and recently, children of the war are often invited to such events. The regional education authorities strongly recommend the inclusion of such activities in the calendar of school activities,

---

[82] Elizaveta Kuznetsova, "Ya pytalas' predstavit'," in *My ne mozhem molchat': shkolniki i studenty o Kholokoste*, ed. Alla Gerber and David Poltorak, vol. 1 (Moscow: Fond "Kholokost," 2005), 105.

offering pre-prepared plans for lessons on patriotic education and "remembrance evenings" [*vecher pamyati*].[83] According to the survey results, special meetings and thematic lessons are held annually in regional schools and devoted to various memorable dates. Only 25 respondents (9 per cent) organised special events on International Holocaust Remembrance Day. This demonstrates not only how little attention is paid to teaching the Holocaust in secondary schools, but above all, the lack of interest of teachers in increasing pupils' awareness of Holocaust history. On the other hand, it turns out that almost one in ten of the regional teachers who participated in the survey not only mention the Holocaust, but also introduce the study of Holocaust history in their teaching. Support for such activity is provided by the Holocaust Center in Russian regions, including the North Caucasus. Teaching about the Holocaust is undoubtedly the most important part of the Center's activities. After testing several ways of addressing its target audience (primarily schoolteachers), the Center concluded that this task would be best served by setting up thematic educational seminars.[84] The Center's staff annually organises up to ten seminars for teachers in different Russian cities. In the North Caucasus such activities have taken place in the cities of Krasnodar, Tuapse, Mineralnye Vody, Stavropol, Rostov-on-Don, Kislovodsk, Vladikavkaz, and Cherkessk in different years since early 2000s. These seminars consist of lectures and discussion sessions with regional history teachers, and propose a wide range of topics, including the history of the Holocaust on Soviet soil, the Jewish resistance, the participation of Jews in the partisan movement and in the Red Army, the problems of antisemitism in Russia since World War II, and Holocaust remembrance in Russia. The Center's staff uses engaging materials (thematic travelling exhibitions, audio-visual materials) and presents published textbooks and manuals at seminars.[85] Participants (usually between 20 and 30 regional teachers at each seminar) receive new knowledge, thematic literature and didactic materials, as well as the opportunity to further cooperate with the Holocaust Center. However, few teachers take up this

---

[83] Scenarios of such school events can be found on the websites of educational institutions and on specialised platforms, for example, on the online platform "Pervoe sentyabrya," http://festival.1september.ru.

[84] Kiril Feferman, "Preaching to the Converted? Teaching about the Holocaust in Modern-Day Russia," in *Holocaust Education in the 21st Century,* ed. Eva Mattes and Elisabeth Meilhammer (Bad Heilbrunn: Verlag Julius Klinkhardt, 2015), 235.

[85] Interview with Leonid Terushkin, in *author's archive*, code 16-MEM-MO04, December 15, 2016, Moscow, Russia.

offer: this depends entirely on their personal motivation to study and teach about the Holocaust.[86]

In 2011, the Holocaust Center collaborated with the Academy of Advanced Training and Professional Retraining of Teachers to develop a 72-hour professional training course for teachers, called "Teaching about the Holocaust as the Way to Tolerance."[87] For several years the Center has conducted distance courses for Russian teachers on Holocaust related topics. Attending distance courses is obligatory for further participation in the annual Summer School of Russian Teachers in Moscow. This summer school aims to enrich teachers' pedagogical skills and provide a forum in which to share best practice on teaching the Holocaust.[88] In order to promote this Summer School, and stimulate research interest among teachers, the most active participants of different educational programmes are selected by Holocaust Center staff for participation in seminars abroad at world-famous Holocaust educational centers,[89] including Yad Vashem in Jerusalem, Mémorial de la Shoah in Paris,[90] and the House of

---

[86] Interview with Terushkin.
[87] See the contents of the programme: Evgenii Vyazemskii, "Pedagogicheskie podkhody k izucheniyu kursa 'istoriya Kholokosta i sovremennost'," *Methodist* 10 (2011): 12–17. Unfortunately, data on the number of teachers, including those from the North Caucasus, who passed these courses, could not be found. The link to the electronic version of the programme on the Academy's website, to which Vyazemskii refers in his article, turned out to be non-working.
[88] "Letnyaya sessiya pedagogov po teme Kholokosta," *Holocaust Center official website*, August 20, 2018, http://www.holocf.ru/летняя-сессия-педагогов-по-теме-холок/?highlight (accessed March 23, 2020).
[89] It is important to note that the selection of participants for international seminars (if it exists at all) is conducted behind the scenes. The website of the Holocaust Center has information that participation in the Summer School for teachers in Moscow is a qualifying round for the formation of a participants' group in the seminar in Yad Vashem. It is not quite clear by what principle the group of participants in the seminars in Berlin and Paris is formed. As a rule, the group of participants in international teaching seminars from Russia includes not only schoolteachers, but also university lectures, employees of museums and archives, and even regional activists.
[90] This seminar is held annually for schoolchildren or students and their advisors who became winners of the annual International Schoolchildren Projects Competition "Holocaust Remembrance is a Way to Tolerance" [*Pamyat' o Kholokoste – put' k tolerantnosti*]. On the results of one of the trips, see: Evgenii Vyazemskii, "Seminar dlya rossiiskikh prepodavatelei po problemam Kholokosta," *Sovremennoe dopolnitelnoe professionalnoe pedagogicheskoe obrazovanie* 2 (2015): 76–81; Tatyana Bolotina and Elena Prigodich, "Istoriya Kholokosta i problemy prepodavaniya etoi temy v shkole," *Methodist* 5 (2013): 16–18.

the Wannsee Conference in Berlin.[91] The seminars abroad are organised on an annual basis and serve as a convenient platform for introducing international experience of Holocaust education to Russian history teachers.[92]

In spite of the educational activity of the Holocaust Center across the country, the North Caucasus cannot be considered a frontrunner in the promotion of knowledge about the Holocaust. A search for news in the section "Educational programmes – News from the Regions of Russia" on the Center's website lists only 19 posts (out of more than 200) relating to south Russia during the period 2007–2017.[93] These were mostly posts about methodological seminars for teachers and Holocaust regional conferences for schoolchildren, reports on hosting Holocaust thematic weeks at schools, and the outcome of the regional stage of the annual International Schoolchildren's Projects Competition "Holocaust Remembrance is a Way to Tolerance."[94] These posts came from Krasnodar and Stavropol Krais and Rostov Oblast.

In order to assess the results of cooperation between teachers and the Holocaust Center, I conducted ten thematic interviews with educators: employees of the regional teacher training centers, and history teachers of secondary schools, gymnasiums, and specialised private schools mainly for Jewish children. I also had an opportunity to study didactic literature for teaching World War II history and to visit thematic school museums. Geographically, the results of my

---

91 Seminars in Germany for Russian teachers have been held since 2011. "Seminar v Berline: itogi," *Holocaust Center official website,* May 10, 2017, http://www.holocf.ru/семинар-в-берлине-итоги/?highlight (accessed January 28, 2020).

92 This task is complicated by the fact that seminars for Russian teachers are conducted in Russian. The lack of foreign language skills of most participants is an obstacle for accessing the accumulated methodological literature on the teaching about the Holocaust. For more methodological literature on the teaching about the Holocaust in the Western world, see: Philip Rubenstein and Warren Taylor, "Teaching about the Holocaust in the National Curriculum," *British Journal of Holocaust Education* 1, no. 1 (1992): 47–54; Ido Abram and Matthias Heyl, *Thema Holocaust: Ein Buch für die Schule* (Reinbek bei Hamburg: Rowohlt, 1996); Rachel Feldhay Brenner, "Teaching the Holocaust in Academia: Educational Mission(s) and Pedagogical Approaches," *The Journal of Holocaust Education,* 8 no. 2 (1999): 1–26; Thomas D. Fallace, "The Origins of Holocaust Education in American Public Schools," *Holocaust and Genocide Studies* 20, no. 1 (2006): 80–102; Matthes, *Holocaust Education in the 21st Century.*

93 The website of the Holocaust Center provides the official and most complete information on the Center's activity.

94 The articles and reports were posted in the section "Educational Programmes – News from the Regions of Russia" on the old version of the Holocaust Center official website: http://www.holocf.ru/pages/26 (now in the section "News" and "Education Programs": http://holocf.ru/). See also the section "News" on the official website of the SKIRO PK and PRO, http://staviropk.ru/index.php.

research matched the data of the Holocaust Center. I affirmed that teaching about the Holocaust is present to some degree in secondary schools in Krasnodar and Stavropol Krais and Rostov Oblast. Krasnodar Krai was the leading region for Holocaust education in the 2000s, when the Holocaust Center had strong cooperation with Andrei Eremenko, the former head of the regional Department of Social Science Disciplines at the Institute of Further Professional Education.[95] One of his responsibilities was the development of history courses using new methodological approaches. During that period, regional methodological seminars for history teachers were regularly held and many regional teachers engaged in research on Holocaust history including its local dimension. A group of enthusiastic schoolteachers taught the Holocaust in the classroom. Currently, all Holocaust-related programmes in Krasnodar Krai are closed. Eremenko's retirement[96] deprived him both of a platform from which to implement new ideas and the power to influence history teaching at regional schools. In recent years, Stavropol Krai has become the most active regional provider of Holocaust education in the North Caucasus. Natalia Masyukova, the head of the Department of Humanities at the Institute for Education Development, Professional Development, and Retraining of Teachers in Stavropol Krai (SKIRO PK and PRO) became a regional representative of the Holocaust Center in 2012 and since that time she has overseen the teaching of the Holocaust at regional schools.[97]

An educative mission to teach about the Holocaust is carried out by individual activists in Rostov Oblast, who represent the Holocaust Center in the region. Sergei Shpagin, a former senior history lecturer at Southern Federal University, now retired, continues to teach about the Holocaust as a museum director at the "Or Avner" Jewish gymnasium in Rostov-on-Don.[98] Igor' Polugorodnik who received

---

[95] Interview with Andrei Eremenko, in *author's archive*, code 15-MEM-KK07, June 15, 2015, Krasnodar.

[96] For some time, Eremenko has continued introducing Holocaust history to the regional history teachers as director of the Felitsyn Museum between 2009 and 2013 and using his network of contacts. From 2014, when he left the post of museum director and moved to work as a college teacher, he stopped this activity.

[97] The Department of Humanities of the SKIRO PK and PRO together with the Holocaust Center organised a regional seminar for history teachers in the city of Mineralnye Vody in 2014. Its topic was "Teaching about the Holocaust on the Stavropol Krai: Problems and Prospects." Another teacher's seminar and a regional scientific and practical conference "Stavropol Krai in the Great Patriotic War: A View from 21st Century" was held in Kislovodsk in 2015. Interview with Natalia Masyukova, in *author's archive*, code 15-MEM-SK11, May 21, 2015, Stavropol.

[98] Interview with Sergei Shpagin, in *author's archive*, code 16-MEM-RO-06, June 9, 2016, Rostov-on-Don.

the honourable title Righteous Among the Nations often gives talks about his family's rescuing of a Jew and about Holocaust history in general in the classrooms of Taganrog secondary schools.[99] During his talks he uses visual materials organised in clamshell folders. Among them are photos of commemorational gatherings and other Holocaust related events held in Taganrog.

With a few exceptions, I was not able to identify any attempts to introduce Holocaust history into the classroom in the North Caucasian republics. When teaching the history of the Great Patriotic War, teachers of Beslenei aul always tell the story of how Leningrad orphans were rescued and adopted by villagers in 1942.[100] In the Republic of Adygea, a 9th-grade schoolgirl, Anzhela Kade, successfully defended her project "The Leningrad Cherkess of Beslenei" at the "Fatherland" Republican Competition of Projects on Regional History in 2013.[101] Unfortunately, these are individual initiatives and are not further developed by republican authorities in the North Caucasian republics. Teachers who participate in regional education management and are interested in teaching about the Holocaust help disseminate the Holocaust Center's educational materials in school libraries and organise teachers' initiative groups. The latter are willing to introduce Holocaust history in the classroom. The results of my survey confirm this finding. 60 per cent of regional teachers in Stavropol Krai and almost 25 per cent of teachers in Krasnodar Krai indicated that they know, participate and inform schoolchildren about the educational programmes of the Holocaust Center, while respondents in the Republic of Adygea, the Karachay-Cherkess Republic, and Rostov Oblast mainly answered that they do not know, but would like to get involved with the Holocaust Center's teaching activities.

On the basis of sources collected during my field research it is possible to define the main approaches of history schoolteachers to teaching about the Holocaust in the North Caucasus. Among them are the following:

– Giving thematic lessons in (regional) history courses and in the special course "History of the Great Patriotic War."[102] Methodological recommendations and pre-prepared lesson plans about Holocaust history are of great

---

99 Interview with Igor' Polugorodnik, born in 1927, in *author's archive*, code 13-X-RO-08r, March 1, 2013, Taganrog.
100 Interview with Aishat Meidshakhova, in *author's archive*, code 15-MEM-KCh-02, May 15, 2015, Beslenei aul of the Karachay-Cherkess Republic.
101 Kade Anzhela, "Leningradskie cherkessy Besleniya" (Maykop, 2013). Anzhela's mother Irina Kade kindly provided me with this work in 2015.
102 This special course is taught only in the schools of Stavropol Krai.

importance for an aspiring teacher.[103] According to Elena Okhmat, the history teacher at Arzgir village school, "a lot depends on the teacher in teaching about the Holocaust, the way she or he perceives the information, whether she or he can properly introduce this topic in the classroom, whether she or he actually wants to explain it at all."[104] Arzgir teachers discuss Holocaust history in their village in a special course on the history of the Great Patriotic War and give schoolchildren an idea of the concept of genocide. The prevailing opinion, as stated by Irina Svetlichnaya from Ust-Labinsk, is that "the use of the term Holocaust with an unprepared audience should not be allowed, because the teacher is unable to predict the consequences that might result from a superficial introduction to Holocaust history, especially when teaching teenagers."[105] But, she concludes, "gradual introduction to Holocaust history gives notable results in the formation of a harmoniously developed and tolerant personality."

- Organisation of thematic school events. On average, an experienced teacher can prepare up to ten Holocaust-related events during one academic year. Demonstrative is the teaching experience of Svetlichnaya, who gives two or three thematic lessons in autumn semester: "I give these lessons during the Week of Tolerance and on Kristallnacht Remembrance Day [November 9]. Then begins a Month of Patriotic Work (January–February), where we organise three or four events, including open lessons, a commemorative evening in memory of Musya Pinkenson, and meetings with Holocaust survivors. And then, of course, there are a number of events during the week before the celebration of Victory Day."[106] Thematic events are less formal if compared to regular lessons, so it is important to gain pupils' interest in Holocaust history.[107] The Ministry of Education and Science of the Russian

---

103 See, for example: Ol'ga Luchina, "Tolerantnost' – usvoenie urokov Kholokosta: urok pamyati, posvyashchennyi mezhdunarodnomu dnyu pamyati zhertv Kholokosta (9–11 klassy)," *Obrazovanie v Kirovskoi oblasti* 2 (2012): 54–58; Irit Abramski and Noa Sigal, "Tri kukly: plan Uroka," *Yad Vashem official website*, https://www.yadvashem.org/ru/education/educational-materials/lesson-plans/three-dolls.html (accessed March 22, 2020). A selection of materials for developing lessons about the Holocaust can be found on the official website of the Holocaust Center, section "Teaching Materials": http://holocf.ru/образовательная-программа/методические-материалы/ (accessed Merch 22, 2020)
104 Interview with Okhmat.
105 Interview with Svetlichnaya.
106 Ibid.
107 Methodological recommendations for teaching about the Holocaust at school are developed by many history teachers who have accumulated sufficient personal experience in this field. See some examples of such recommendations: Ekaterina Petrova, "Osobennosti prepodavaniya

Federation has recommended organising special educational events and giving thematic lessons around the International Day of Tolerance (November 16) and International Holocaust Remembrance Day (January 27) since the academic year 2016–2017.[108] Perhaps these recommendations will help to increase interest in the regions in Holocaust education.

- Development of elective courses and authorised programmes on Holocaust history. In the process of teaching about the Holocaust, teachers gradually collect historical sources. In the course of time one might want to present all the collected materials as a separate course, giving schoolchildren detailed and systematic knowledge. The school curriculum has a so-called variable component, which provides for teaching hours that may be spent at the school administration's discretion. For instance, Tatyana Belyaeva, a teacher at Krasnodar Gymnasium No.23 has developed a course entitled "The Holocaust," which runs for 216 academic hours (3 hours per week during the academic year). She taught this course in Krasnodar Municipal Institution of Additional Education for Children "Small Academy" [*Malaya Akademiya*] (hereafter "Small Academy") in the mid-2000s. Belyaeva considers the main aim of this course as providing "the possibility for schoolchildren to expand their knowledge of World War II and to take a fresh look at the history of the Great Patriotic War with a view to understanding the functioning of repressive consciousness in the society."[109] In 2010, Svetlichnaya developed an elective course, "Holocaust History," for schoolchildren in the 10th and 11th grades to give them "systematic knowledge about the Holocaust."[110] These and other teaching initiatives of regional

---

temy 'Kholokost' v starshikh klassakh," *Prepodavanie istorii v shkole* 6 (2011): 51–54; Irina Averyanova, "Prepodavanie temy 'Kholokost' na urokakh istorii," in *Rossiya v perelomnye periody istorii: nauchnye problemy i voprosy grazhdansko-patrioticheskogo vospitaniya molodezhi*, ed. Marina Dmitrieva (Ivanovo: Institut razvitiya obrazovaniya Ivanovskoi oblasti, 2012), 341–347; Il'ya Al'tman, ed., *Kholokost: vzglyad uchitelya: metodicheskoe posobie – sbornik rabot pedagogov Rossii* (Moscow: Tsentr i fond "Kholokost," 2006); Natalia Anisina, ed. *Uroki Kholokosta i narushenie prav cheloveka v sovremennoi Rossii: pedagogicheskii aspekt* (Moscow: Tsentr i Fond "Kholokost," 2014).
108 "Pismo Ministerstva obrazovaniya i nauki RF, No. NT-943/08 of July 5, 2017 'O kalendare obrazovatelnykh sobytii na 2016/17 uchebnyi god," *Elektronnyi fond pravovoi i normativno-tekhnicheskoi dokumentatsii*, http://docs.cntd.ru/document/420373119 (accessed March 22, 2020).
109 Tatyana Belyaeva, "Programma 'Istoriya Kholokosta,'" in Al'tman, *Kholokost: vzglyad uchitelya*, 162–180.
110 Irina Svetlichnaya, "Ne stat' ravnodushnymi nablyudatelyami: elektivnyi kurs 'Istoriya Kholokosta,'" *Uchitel'skaya gazeta*, October 26, 2010.

teachers are universal in character, since they cover the whole Holocaust history with almost no attention to regional events.
- Supporting schoolchildren to conduct research projects and present their findings at regional, republican, all-Russian, and international research project competitions. Research work and local historical study, with close attention to Holocaust events in one's native locality, is an important part of a history teacher's work. Further on in this chapter, I will pay special attention to the results of projects in which schoolchildren research Holocaust history together with their teachers.
- Participation in scientific and practical conferences and seminars on Holocaust teaching. Annual conferences held by the Holocaust Center usually discuss different aspects of the history of the Holocaust on Soviet soil. They have become a platform for teachers to exchange their experience. There are also opportunities for schoolteachers to learn about the latest Holocaust research by Russian and European scholars. Proceedings of a series of international annual conferences, "Holocaust Lessons and Contemporary Russia" [*Uroki Kholokosta i sovremennaya Rossiya*], have already been published in nine collections, and include papers by regional history teachers.[111]
- Thematic design of school museums, which I have already discussed in the previous chapter.

To conclude, it should be noted that curricula for teaching about the Holocaust in history lessons have been successfully developed in some regions of the North Caucasus. This activity would have been impossible without the awareness-raising mission of the Holocaust Center. And yet, according to a former employee, Kiril Feferman, "given the country's enormous size and the paucity of resources available to the Center, the number of teachers, the educational materials available to them, and their overall knowledge of the Holocaust continue to be largely inadequate."[112] As the results of my research indicate, teaching about the Holocaust in Russian regions is possible thanks to the involvement of individual representatives of education authorities and the personal initiative of history teachers, who are themselves willing to learn and pass their knowledge on to younger generations.

---

111 See, for example the articles by regional teachers: Sergei Shpagin, "Problemy memorializatsii zhertv Kholokosta: sootnoshenie istorii i postistorii," in Al'tman, *Pamyat' o Kholokoste: problemy memorializatsii: materialy 6-i mezhdunarodnoi konferentsii*, 22–25; Miroshnikova, "Muzei v evreiskoi shkole," *Idem*, 156–160.
112 Feferman, "Preaching to the Converted?," 235.

The main disadvantage of the Holocaust Center in the regions is its limited reach. The Center has only a few personal contacts in the regions, mainly with the individual employees of the teacher training institutes, who play the role of mediators in spreading educational materials and the Center's publications. A kind of "reward" for their activity is their participation in educational seminars in Moscow and abroad, organised by the Holocaust Center. However, if the Center's regional representative changes his or her workplace, as in the case of Andrei Eremenko, the Center's activity in the region is immediately ceased.[113] From the perspective of the Holocaust Center, the successful extension of Holocaust education in the regions requires an extensive network of influential regional representatives capable of effective cooperation with the regional authorities. In this sense, the experience of the International Historical and Educational, Human Rights and Charitable Society "Memorial" (hereafter, Society "Memorial") is useful. It has a network of regional staff and offices, websites, and sometimes even printing houses, all over Russia in addition to its head office in Moscow.[114] Among the tasks of regional offices are the development of educational programmes and organisation of commemorative activities remembering victims of political repressions in the USSR and in present-day Russia. The centralisation of the Holocaust Center in Moscow is the main obstacle to their attempts to disseminate knowledge about the Holocaust in the regions. Inclusion of topics related to the Holocaust into the mandatory curriculum of regional teacher training seminars would also make it possible to bring teaching about the Holocaust in schools to a new level.

7.3 Initiatives "from below:" Schoolchildren's Research on Holocaust History in the North Caucasus

A child's personality is largely formed during her or his school years. During this time, children form their own identity and learn various social roles. Their worldview is influenced by how they learn to treat themselves, people around

---

[113] A similar situation occurred in Rostov-on-Don. I was not able to contact Ol'ga Vityuk, the former coordinator of the Holocaust Center, nor to discover the results of her activity on teaching about the Holocaust in the region as a former employee of the Rostov Institute of Advanced Studies and Professional Retraining of Educators.

[114] Memorial has 66 offices throughout the Russian Federation. See the list of Society "Memorial" organisations in Russia: *Society "Memorial" official website,* http://old.memo.ru/s/309.html (accessed March 23, 2020).

them, and the world in general. Teaching tolerance at school leads above all to the formation of personality. The teacher can help develop a schoolchild's interest in history by organising various forms of school activity, such as arranging meetings with Holocaust survivors and Holocaust eyewitnesses, and supervising historical research on Holocaust history. Gradually, schoolchildren become interested in the detailed study of the past, including Holocaust-related themes.

World War II history has been one of the most popular research fields for history teachers and their pupils since the Soviet period, when almost every Soviet school had a research team or a military-patriotic club. The trend of youth and schoolchildren to study war history and plan out routes for visiting memorial sites began in the 1960s. At that time schoolchildren under the teachers' supervision searched for the remains of Red Army soldiers and took care of the Grave of the Unknown Soldier in their local school district. Patriotic Museums and Rooms of Glory were founded at schools, memorial plaques and obelisks were installed as a result of youth research activities across the whole country.[115] Schoolchildren and students worked to identify unknown names and episodes of the war. Some teachers continue to organise such afterschool activities in modern Russia, paying particular attention to the regional history of the Holocaust. For instance, older pupils of Kislovodsk Secondary School No.2, under the supervision of their history teacher Irina Mazanova, have been working on a research project named "Kislovodsk Valley 1942–1943: memory, pain and mercy" [*Kislovodskaya dolina 1942–1943 godov: pamyat', bol' i miloserdie*] for several years.[116] The aim of the project is "to draw people's attention to the problem of genocide during World War II and to illustrate the horrors of the Nazi regime through Holocaust history in order to raise awareness of the underestimation of neo-Nazi organisations in the modern world."[117] Pupils in the "initiative group" (the website gives information on 18 participants of the project) study the history of the mass killing of Jews in Kislovodsk, learn to work with sources, and conduct oral history interviews with Holocaust survivors and eyewitnesses. Based on the results of

---

[115] Yuliya Melikhova, "Opyt poiskovo-kraevedcheskogo dvizheniya sovetskogo perioda i perspektivy razvitiya voenno-istoricheskogo turizma v Kurskoi oblasti," *Uchenye zapiski Rossiiskogo gosudarstvennogo sotsialnogo universiteta* 4 (2011): 72.
[116] *Website of the project "Kislovodskaya valley 1942–1943: memory, pain and mercy,"* https://sites.google.com/site/school2historyproject/ (accessed March 23, 2020).
[117] Ibid.

their activities, some participants of the project have presented research papers at regional and all-Russian historical competitions.[118]

Another important form of afterschool activity is tending to Holocaust memorial sites, whether there is a pre-existing monument or not. For example, schoolchildren and college students are responsible for taking care of a small Soviet monument near the glass factory in Mineralnye Vody, site of the largest mass killing of Jews in Stavropol Krai. Each September they clean the surrounding area of weeds[119] and paint the monument and fence before commemorative events are held.

Schoolchildren are frequent guests and participants of commemorative events held near Holocaust monuments or at memorial sites. Over the past several years, groups of schoolchildren led by teacher-activists have visited Holocaust sites mainly on International Holocaust Remembrance Day and the Remembrance Days of the mass killing of Jews in particular localities.[120] For instance, the opening ceremony of the renovated memorial in Arzgirskaya Balka was held in autumn of 2016 in Arzgir village.

> Here schoolchildren of village secondary schools were honoured guests along with representatives of local authorities, members of Jewish communities, and representatives of the Holocaust Center. During the ceremony, schoolchildren of Secondary School No.2 performed the song "Buchenwald Nabat" in memory of innocent victims of fascism.[121]

After the event "schoolchildren release white balloons in the sky as symbols of peace in our land."[122] Schoolchildren also formed the honour guard at the

---

[118] 10th-grade schoolgirl Svetlana Tyutyunnikova became a laureate of the Competition "Holocaust Remembrance is a Way to Tolerance" in 2015. Svetlana Tyutyunnikova, "Emil Genrikhovich Zigel'," in *My ne mozhem molchat': shkolniki i studenty o Kholokoste*, ed. Maria Gileva and Tatyana Pasman, vol. 12 (Moscow: Tsentr "Kholokost," 2015), 39–46.
[119] "Nikto ne zabyt," *Vremya: gazeta goroda Mineralnye Vody*, April 11, 2015. This monument is located in a wasteground. To get there one must walk almost two kilometres off-road. Local public figures have tried unsuccessfully to obtain municipal funds to pave at least a small road to the monument. Interview with Mikhail Akopyan, in *author's archive*, code 15-MEM-SK01, May 13, 2015, Mineralnye Vody, Stavropol Krai. To open a new memorial complex at the killing site near the glass factory in 2019, a gravel road was laid to facilitate the approach to the old and new memorials.
[120] Inna Kalinicheva, "27 yanvarya – mezhdunarodnyi den' pamyati zhertv Kholokosta," *Pul's litseya: gazeta litseya No. 3 goroda Mineralnye Vody*, 5 (January 2016).
[121] Nadezhda Klimchenko, "Budem pomnit'! Pomnit' vsegda!," *Zarya: obshchestvenno-politicheskaya gazeta Arzgirskogo raiona Stavropol'skogo kraya*, September 13, 2016.
[122] "Otkrytie memorialnykh plit," *Official website of the Arzgir municipal district*, http://arzgir admin.ru/news.htm (accessed December 24, 2019).

opening of six more Holocaust monuments in Stavropol Krai in 2016.[123] Kislovodsk schoolchildren hosted and participated in the unveiling of a memorial plaque for Holocaust victims at the monument near Mount Koltso in May 2015. The presence of youth at commemorative events at Holocaust sites indicates the transmitting through the generations and the "vitality" of the memory of Holocaust victims.

The most effective form of schoolchildren's activity in studying Holocaust history is conducting research. Under the supervision of their history teachers, schoolchildren present the results of their research at project competitions on regional, Russian, or world history and at special competitions on World War II and Holocaust history. As a rule, participants who present on Holocaust-related topics do so because their history teachers teach about the Holocaust in the classroom and have established connections with the Holocaust Center. But there are also independent projects initiated by schoolchildren themselves. For example, the aforementioned project "Leningrad Cherkess of Beslenei" by Anzhela Kade emerged as a result of her personal interest when she visited her relatives in Beslenei aul and first learned the details of the story. She then conducted research on this topic under the supervision of her teacher from a Center of Additional Education in Maykop.

The supervisors of schoolchildren's Holocaust projects are often teachers from Centers for Additional Education. For example, a special course "The School of the Young Researcher: Young Historian, Social Scientist, and Lawyer" is taught in Krasnodar "Small Academy." A 10th-grade schoolboy, Mikhail Khristoforov, won the second all-Russian Youth Competition of Research Projects "The Young Archivist" in 2014. He presented a paper "The history of mass killing of Jews in the village of Novoderevyankovskaya, Krasnodar Krai during the Great Patriotic War."[124] In his work Mikhail analysed archival sources (mainly ChGK records) and data collected during his fieldwork in the village including materials collected by a local historian Aleksandr Deinevich, who has reconstructed Holocaust history in his village over the past several years.[125]

---

123 Nadezhda Klimchenko, "Otkrytie pamyatnika 'zhertvam Kholokosta' v sele Serafimovskom," *Zarya: obshchestvenno-politicheskaya gazeta Arzgirskogo raiona Stavropol'skogo kraya*, December 2, 2016.
124 "Vospitanniki tsentra 'Malaya Akademiya' stali pobeditelyami i prizerami II vserossiiskogo konkursa yunosheskikh uchebno-issledovatel'skikh rabot 'Yunyi Arkhivist,'" *Official website of "The Small Academy,"* October 17, 2014, http://m-academ.centerstart.ru/index.php?q=node/497 (accessed March 23, 2020).
125 Aleksandr Deinevich, *Tam za povorotom* (Maykop: Poligraf yug, 2012).

An all-Russian Internet competition with international participation named "A Forgotten Memorial of an Unforgettable War" [*Zabytyi pamyatnik nezabytoi voiny*] was held on the portal "Innovative Teachers Network" during the 2009–2010 academic year and was dedicated to the 65th anniversary of Victory Day.[126] One of the nomination categories of this competition was "The Tragedy of the Holocaust."[127] Out of eight projects submitted in this category, one was presented by schoolchildren of Ust-Labinsk's Gymnasium No.5 and supervised by Irina Svetlichnaya. The project's aim was to study a monument to Holocaust victims near Ladozhskaya village, Krasnodar Krai. Through their participation in this competition, the schoolchildren hoped to draw public attention to the need to restore this monument.[128]

Since the late 1990s, the Society "Memorial" has organised an all-Russian School Competition of Historical Research Projects, "Man in History: Russia in the 20th Century" [*Chelovek v istorii. Rossiya – XX vek*]. Its main goal is "to awaken the interest of young people to the destinies of ordinary people, their everyday life, to those who make up the 'grand history' of the country."[129] Hundreds of research projects from all over Russia are annually sent to the competition committee. As a result, all participants and their supervisors receive certificates of participation and small prizes, historical books as a rule. The winners are invited to Moscow to participate in the annual awards ceremony. The awarding of participants in any research competition is an important and memorable event in the life of each young researcher. Even if the project was not given an award, the encouragement of all participants of the Competition is intended to mark the student's research achievements at any level, to encourage him or her to pursue further scientific activity. The winners have an opportunity to travel to Moscow and to publish their papers in a separate volume and on the competition's website. Holocaust history is a constant theme of participants' research. There have been more than a hundred works on Holocaust history sent from various Russian regions to the competition since

---

[126] This portal was created with the support of Microsoft Corporation to enable teachers within Russia and abroad to communicate and exchange information and didactic materials on the use of information and communication technologies in education. Thus, the Innovative Teachers Network has been created. See: *Innovative Teachers Network official website*, http://стуархив.рф/wa-itn/20170624075416/https://it-n.ru/default.aspx (accessed March 23, 2020).

[127] *Regulation on the Internet project "A forgotten memorial of an unforgettable war,"* http://it-n.ru/communities.aspx?cat_no=163670&tmpl=com (accessed March 23, 2020).

[128] The project "Zabyt' nel'zya . . . ," *Internet project "A forgotten memorial."*

[129] "Polozhenie o konkurse," *Project website of the Society "Memorial" – "Uroki istorii XX vek,"* http://urokiistorii.ru/konkurs (accessed March 23, 2020).

it was established.¹³⁰ In 2006, the special collection of research projects in the section "Friends and Strangers" [*svoi – chyuzhie*] was devoted to the study of Jewish history, which, according to the organisers of the Competition, remains with few exceptions "a big blank spot in Russian history."¹³¹ A schoolboy of Krasnodar Secondary School No.10, Vadim Ivanov, was the prize-winner for the 2000–2001 academic year. He conducted research on the history and memory politics of Holocaust victims in Krasnodar under the supervision of his history teacher and father Leonid Ivanov.¹³²

The Holocaust Center is undoubtedly the leader in attracting schoolchildren and students to the study of Holocaust history. It has been organising an annual International Competition of Research Projects on Holocaust history since the early 2000s. In 2018, the fifteenth issue of the collection of school and student projects "We cannot be silent" [*My ne mozhem molchat'*] was published. The themes of research projects are wide ranging. In the form of essays, schoolchildren present their philosophical thoughts and reflections on the tragedy of Jews during World War II after studying Holocaust history in the classroom or watching and discussing Holocaust feature films and documentaries. Historical projects of schoolchildren are often based on the study of local Holocaust history. Various literary works mainly represent the author's artistic response to the Holocaust. Schoolchildren from North Caucasian secondary schools have been among the prize – winners of this Competition in different years.¹³³ For example, the project of Anna Arsenova discussed the mass killing of Jews in Zmievskaya Balka and the preservation of Holocaust memory in today's Rostov-on-Don. Alena Mitenko studied the tragedy of Jews in Arzgirskaya Balka. The history of the occupation of Kabardino-Balkaria became the subject of Estela Kasimova's project.¹³⁴ Varying in the content and depth of research, these

---

130 Archive of the school competition "Chelovek v istorii. Rossiya – XX vek" of the Society "Memorial."
131 *Polustertye sledy: Rossiiskie shkol'niki o evreiskikh sud'bakh: iz rabot uchastnikov vserossiiskogo konkursa starsheklassnikov "Chelovek v istorii. Rossiya – XX vek"* (Moscow: Mezhdunarodnoe istoriko-prosvetitel'skoe i pravozashchitnoe obshchestvo "Memorial," 2006).
132 Ivanov "Eto nuzhno ne mertvym."
133 It is not possible to identify the total number of schoolchildren's projects received for the Competition from each Russian region and their thematic diversity based on the data of the official website of the Holocaust Center. I was not allowed to work with the archive of this Competition since it is not systematised.
134 Kuznetsova, "Ya pytalas' predstavit'"; Anna Arsenova, "Tragediya Zmievskoi Balki i sokhranenie istoricheskoi pamyati o nei," in *My ne mozhem molchat': shkolniki i studenty o Kholokoste*, ed. Il'ya Al'tman and Dmitrii Prokudin, vol. 9 (Moscow: Fond "Kholokost," 2012), 110–115; Alena Mitenko, "Tragediya v Arzgirskoi Balke," in *Ibid*, ed. Dmitrii Prokudin and Maria Gileva, vol. 10

projects are important in drawing public attention to the history of their native locality and to Holocaust history in general.

Active forms of participation in Holocaust commemorative events such as conducting research, writing a paper, tending the grounds of the killing sites, and passive forms such as attending commemorative meetings as spectators are among the main forms of youth activity in Holocaust remembrance. Education in this sense can be continued outside the classroom, in museums, at thematic meetings with Holocaust survivors, or during commemorative events at the killing sites. Sometimes direct interaction with history on the spot is more important and memorable for youths than a standard classroom lesson. Schoolchildren's research and commemorative activity goes beyond the bounds set by history lessons, brings to life the abstract formulations of history textbooks, and enlivens the study of their native locality and the lives of their fellow countrymen. What cannot be contained in the usual 45-minute interaction between a schoolchild and a teacher in the classroom is successfully implemented in afterschool activities with enthusiastic youths.

***

Irina Kalashnikova, a member of a protestant church, conducted awareness-raising work in schools in Novorossiysk for several years, where she lived in the 2000s. According to my knowledge, her initiative was the only one that was conducted with the support of the "Operation Jabotinsky" foundation[135] in the North Caucasus. For several years Irina gave lectures on Holocaust history in the city's schools and colleges. Thematically, the lectures were about the history of the mass killing of European Jewry and did not cover regional cases. For a provincial city, her altruistic initiative was a deed of great importance, as her lectures about the Holocaust were recognised not only by schoolchildren and students, but also by history teachers themselves. Irina filled out notebooks,

---

(Moscow: Fond "Kholokost," 2013), 35–42; Estela Kasimova, "Kholokost na territorii Kabardino-Balkarii," in *Ibid*, ed. Dmitrii Prokudin and Maria Gileva, vol. 11 (Moscow: Fond "Kholokost," 2014), 20–23; Tyutyunnikova, "Emil Genrikhovich Zigel'."

[135] Ze'ev Jabotinsky, born Vladimir Zhabotinskii, (1880–1940) was a Russian Jewish revisionist Zionist leader, author, poet, orator, soldier, and founder of several Jewish organisations in the Russian Empire and Palestine. In 1993, the Word of Life church established the "Operation Jabotinsky" foundation to help Jews from the former USSR move to Israel via Sweden. The followers of this mission believe in spreading knowledge of Holocaust history, because "if we forget this story, it will happen again." Yurii Gorbanyuk, *I will send lots of fisherman* (Moscow, 2004), 3.

where she recorded the dates of the lectures and pasted photos of schoolchildren who were present at the meetings. On these photos we can see children of different ages, from 5th-grade schoolchildren to college students. The main content of Irina's notebooks is the answers of schoolchildren and teachers to the question "Do we need to know about the Holocaust today?" Answers were written by students of different ages and at different years, but they all share important thoughts, such as "knowledge about the tragedy of the Jews during the war will not allow us to repeat such horrors in the future," "this topic should concern all people of any ethnicity," and "the task of adults is to pass the truth about those times to children."[136] This task could become a priority in the implementation of history education in modern Russia, but as this chapter has indicated, it remains dependent on the personal initiative and motivation of the teachers in question. As a field of interaction between state-sanctioned historical narrative, local historical memory, and local and civil society initiatives, historical education is one of the key elements in the Holocaust memory dispositif in Soviet and post-Soviet Russia. Yet, despite a few separate initiatives, the Holocaust still plays a marginal role in teaching about the Second World War in modern Russian schools, where the hegemonic military-patriotic narrative of the Great Patriotic War remains dominant.

---

136 Quotations of schoolchildren and teachers of Novorossiysk schools in the notebook of Irina Kalashnikova, 2004–2005 academic year. I thank Irina, who sent me two notebooks through the Jewish community of Novorossiysk in 2016.

8 Revisiting Experience: Personal Narratives of the Holocaust in the North Caucasus

In 2017, I was a Margee and Douglas Greenberg research fellow at the Center for Advanced Genocide Research at the University of Southern California (USC). I had sought this unique opportunity for a couple of years, as it would permit me to continue studying video testimonies with Holocaust survivors, conducted for the USC Shoah Foundation.[1] I had spent much of the previous year watching video interviews, selecting and transcribing those relevant to my study. Out of more than 300 interviews which mention the localities of the North Caucasus during World War II, I chose the 80 that address Holocaust history in this region. The more I transcribed the testimonies of Holocaust survivors, the more I thought about the details: the time the interview was conducted, the influence of the interviewer upon the narrator, the role of the questionnaire in the core conversation between two people in front of a camera, and the meaning of both events in the lives of the narrators – the Holocaust itself and the survivor's opportunity to pass his or her life experience to future generations.

I analysed only Russian language interviews, the majority of which were conducted in the late 1990s in the post-Soviet countries and in the countries of expatriation of Soviet Jewry including Israel, Canada, and the USA.[2] The USC Shoah Foundation's Oral History project was among the first large-scale efforts to seek the voices of Soviet Holocaust survivors at a time when the individual experience of surviving the Holocaust on Soviet soil became a subject of interest not only to Holocaust survivors and their family members but also larger social groups. Most Soviet Jews had lived their whole lives (or a significant part their adult lives before expatriation was possible) in the Soviet Union, where Holocaust history had been silenced for decades. The Shoah Foundation's project is a unique and largely successful effort to make private memories of the most traumatic experience of the Soviet Jewry open to the public. At the same

---

[1] During the filming of *Schindler's List* in Krakow (Poland) Holocaust survivors expressed their desire to talk about their memories in front of a camera. Inspired by this, the film director Steven Spielberg founded the non-profit Survivors of the Shoah Visual History Foundation (Shoah Foundation) in 1994. It aimed to document, catalogue, and index as many personal accounts of witnesses of the Holocaust as possible. 52,000 testimonies were videotaped in 56 countries and 32 languages between 1994 and 1999. "USC Shoah Foundation. The Institute for Visual History and Education," *Website of the VHA at Freie Universität Berlin*, http://www.vha.fu-berlin.de/en/archiv/index.html (accessed March 24, 2020).
[2] I found no Russian-language interviews conducted in Germany relevant to my study, although Germany has been a destination for emigrating Soviet Jewry since the late 1980s.

https://doi.org/10.1515/9783110688993-008

time, there was no developed tradition of doing oral history with Holocaust survivors in the post-Soviet academy in the 1990s.[3] So the interviewers conducted interviews and narrators[4] gave testimony mostly for the first time in their lives. They knew that the final product of their activity, the video testimony itself would become a part of the research archive or would be even included into Spielberg's future documentary or feature film.[5]

Considering the prevalence of antisemitism in the USSR, the little scholarly attention given to Soviet Jewish history, and the weak oral history tradition in historical research in the USSR, I wanted to find out whether there were special regulations or prescriptions for how to conduct interviews with Soviet Holocaust survivors in the late 1990s. Although a lot of time was devoted to the war period in the retelling of the narrator's life, the interviews also give us information about the time when they were conducted. It is obvious that the narrator always

---

3 The first oral history association in the Soviet Union was founded in 1989. This organised three all-Russian conferences: two in Kirov and one in Kaliningrad in 1989, 1991, and 1992. The participants were mainly historians from central Russian universities and high schools. Russian historian Dariya Khubova analysed the first results of doing oral history in the USSR/Russia in her PhD thesis in 1992. The first oral history projects did not touch upon the experience of Holocaust survivors. Dmitrii Ursu "Metodologicheskie problemy ustnoi istorii," in *Istochnikovedenie otechestvennoi istorii: sbornik statei*, ed. Vladimir Kuchkin (Moscow: Nauka, 1989), 3–32; Dariya Khubova, "Ustnaya istoriya i arkhivy: zarubezhnye kontseptsii i opyt" (cand. of science diss. in history, Russian State University for the Humanities, 1992).

4 I will use the term "narrator" instead of "interviewee." An oral interview is a process of interactive communication between a researcher and a narrator. As oral historian Barbara Allen states, the terms "interviewer" and "interviewee" – implying a do*er* and a do*ee* – clearly define the active and passive roles in the interview: the interviewer is the actor, in the sense of initiating and directing the interaction; the interviewee is the reactor. The terms "researcher" and "narrator" avoid these connotations. Barbara Allen, "Re-creating the Past: The Narrator's Perspective in Oral History," *Oral History Review* 12 (1984): 5.

5 I thank USCSF staff for providing me with access to the recorded training sessions with Russian-speaking interviewers in Moscow and in Kyiv in 1996. This permitted me better understanding the motivations of the interviewers, their training in history, and their expectations of the project. Future interviewers received information about the history of the Shoah Foundation during the trainings, which they most likely told their future narrators during the preliminary interview session. Many Holocaust survivors and the interviewers were often convinced that the interview would be included in one of Spielberg's future films. That's why some interviewers behaved artistically during the interview, turning the recording of Holocaust survivors' memories into a kind of performance; their attitude resembled the host of a television programme rather than a researcher interested in what the narrator has to say. "Interviewer training session in Moscow," August 1996, 37 videocassettes, in *VHA USCSF*, Digital and remote storage, ALT03014; "Interviewer training session in Kiev," 1996, 24 videocassettes, in *VHA USCSF*, Digital and remote storage, ALT02567.

speaks from today's perspective about his or her past.⁶ Going beyond oral history, this insight is also relevant for any other type of ego-documents or personal narrative. Diaries, military correspondence, written memoirs and oral interviews always carry imprints of the time during which their creation was possible.

All these types of war-related narratives, including Holocaust narratives, are acts of commemoration.⁷ As such, they are an important element of the Holocaust memory dispositif. Its regional dimension is no exception, although there are few personal narratives of Holocaust history in the North Caucasus. Despite the fact that the North Caucasus was one of the priority evacuation destinations of high schools and medical institutions with a high percentage of Jewish intelligentsia, the majority did not survive the Holocaust. Holocaust survivors in the North Caucasus are mostly Jews from the Ukrainian and Belorussian shtetls who either did not have writing skills or never felt the need to keep a diary. The constant fear for their lives and the lives of their loved ones under the occupation also prevented any possibility of possessing Jewish identification papers and other personal items such as photos, letters, and Yiddish books. A diary written in Russian or Yiddish could play against its owner if found. In the post-war years, a few Holocaust survivors wrote memoirs; trying to preserve their military experience, the bitterness of the loss of their relatives and loved ones, and the circumstances of their survival on paper. They wrote memoirs either following their hearts or in response to reading a magazine or newspaper article, or someone's memories that somehow echoed their fate. Often the memories of Holocaust survivors became part of thematic collections published in different regions of Russia and abroad. Usually they were published in Russian, Hebrew, or English as first-person narratives or in edited collections. These testimonies are important for the general history of the Holocaust, where each name is valuable and each fate should be remembered.

---

6 For more about the oral history method and its usefulness not only in reconstructing historical reality but especially in investigating the work of individual memory about past events, see, for example: Alessandro Portelli, "What makes oral history different," in Alessandro Portelli, *The Death of Luigi Trastulli and other stories: Form and Meaning in Oral History* (Albany: State University of New York Press, 1991), 45–58; Elizabeth Tonkin, *Narrating our Past: The Social Construction of Oral History* (Cambridge: Cambridge University Press, 1994); Paula Hamilton and Linda Shopes, ed., *Oral History and Public Memories* (Philadelphia: Temple University Press, 2008); Robert Perks and Alistair Thomson, ed. *The Oral History Reader* (London and New York: Routledge, 1998); Donald A. Ritchie, ed. *The Oxford Handbook of Oral History* (Oxford: Oxford University Press, 2011).
7 See more about each type of war narrative and their specification: Samuel Hynes, "Personal Narratives and Commemoration," in *War and Remembrance in the 20th Century*, ed. Jay Winter and Emmanuel Siran (London: Cambridge University press, 1999), 209.

The testimonies of Holocaust eyewitnesses[8] are another important group of ego-documents. Demanded during the war, these testimonies were subsequently removed from the official memory of World War II in the USSR. Nevertheless, decades later Holocaust eyewitnesses describe mass killings of Jews in their native village or town in detail. Their memoirs provide unique information helpful for the reconstruction of a particular mass killing. Yet another important group of personal narratives is the testimony of Righteous Among the Nations or rescuers, people who saved Jews during the Holocaust. However, I did not find any personal testimony of those who saved Jews in the North Caucasus.[9]

The main paradox of personal narratives is their ambivalence regarding historical time. On the one hand, the time at which diaries or memoirs are written, or an interview conducted, affects the narrative. The authors and narrators speak on different topics using different phrases and terminology. They may or may not follow the state ideological concept of the war, highlighting particular aspects of their past in different life periods. On the other hand, the language of the narrative about the war and the specific events experienced or observed by the narrators remains unaffected by the influence of time and official ideology. Thus, every ego-document has two levels of information: direct personal wartime story, which remains practically unchanged throughout the life of the narrator, and reflection about the personal wartime experience, which can be re-thought under the influence of time, political atmosphere, the educational level of the narrator, and his or her involvement in studying Holocaust history. Ego-documents as an element of the regional dimension of the Holocaust memory dispositif allow

---

[8] There is a debate among scholars about the correct term for those who witnessed the Holocaust but did nothing to help or hinder it. Historians turned to "onlookers," "eyewitnesses" and even "neighbours" after using the term "bystanders." Raul Hilberg, *Perpetrators, Victims, Bystanders: The Jewish Catastrophe* 1933–1945 (London: Secker and Warburg, 1995); Henrik Edgre, ed. *Looking at the Onlookers and Bystanders. Interdisciplinary Approaches to the Causes and Consequences of Passivity* (Stockholm: The Living History Forum, 2012); Jan Grabowski, "The Role of 'Bystanders' in the implementation of the 'Final Solution' in occupied Poland," *Yad Vashem Studies* 42, no. 1 (2015), 113–117; Gunnar S. Paulsson, "Bystanders to the Holocaust: A Re-Evaluation," *Holocaust and Genocide Studies* 18, no. 1 (2004): 110–112. I use the term "eyewitnesses" to refer to local citizens who had somehow observed the gathering and killing of Jews in a particular locality and then gave their testimonies about what they had witnessed.

[9] The achievements of these people were of little or no interest to Soviet and later Russian society. Yad Vashem awarded the title of Righteous Among the Nations to many of them when they were already elderly people and their deeds were written about by historians, journalists or local historians. I was lucky to interview two of them. They saved Jews on the territory of today's Ukraine but now live in Rostov Oblast. There are about 100 interviews in Russian with Holocaust rescuers in the USCSF archive, which describe the rescue of Jews mainly on the territory of modern Ukraine and Belarus.

the influence of wartime events on the fate of Holocaust survivors and Holocaust eyewitnesses to be traced, offering the possibility and means of passing their experience on to future generations. In this chapter, I will analyse personal narratives of the Holocaust in the North Caucasus, as well as the circumstances of their emergence. I will particularly focus on a thematic analysis of oral interviews with Holocaust survivors, as this is the most representative group of personal narratives. It provides information both on the fate of Jews in the occupation, and the time when these sources were created. Another equally important objective of this chapter is to understand how personal narratives of the Holocaust in the North Caucasus are (or are not) used in remembering Holocaust victims and in teaching about the Holocaust in the region.

8.1 "We used to have a big family; now I am the only survivor:" Historical Overview and Thematic Analysis of the Personal Narratives of Holocaust Survivors

Personal narratives are an important element of the Holocaust memory dispositif: they help one to understand the meaning of Holocaust history in the lives of survivors and the importance of the survivors' experience for future generations. Their role in the regional Holocaust memory dispositif is ambiguous, since they are still rarely in demand at the regional level. Yet, at the same time they give the international remembering community the opportunity to better understand the experience of Holocaust survivors in Southern Russia.

One can rarely find diaries of Holocaust survivors in the region. A diary is a daily (sometimes intermittent) record of information about the diverse events that occurred in the life of its author with an "occasional link" to the calendar date. The actual absence of a time distance between the event and its documentation on paper allows the author to reflect "here and now." The most important characteristic of a diary is that it is self-addressing; the writer and the reader share the same knowledge and the same feelings.[10]

---

[10] For more about the Holocaust diaries see: James E. Young, *Writing and Rewriting the Holocaust: Narrative and the Consequences of Interpretation* (Bloomington: Indiana University Press, 1990); Victoria Stewart, "Holocaust Diaries: Writing from the Abyss," *Forum for Modern Language Studies* 41, no. 4 (2005): 418–426; Alexandra Garbarini, *Numbered Days: Diaries and the Holocaust* (New Haven: Yale University Press, 2006); Amos Goldberg, *Holocaust Diaries as "Life Stories"* (Göttingen: Wallstein, 2008); Frank Bajohr und Sybille Steinbacher, ed. *". . . Zeugnis ablegen bis zum letzten": Tagebücher und persönliche Zeugnisse aus der Zeit des Nationalsozialismus und des Holocaust* (Göttingen: Wallstein Verlag, 2015).

8.1 "We used to have a big family; now I am the only survivor" — 253

There are very few diaries by Holocaust survivors on Soviet soil. Researchers and a wide range of readers have come to know of many of them only recently.[11] Most Holocaust diaries belong to authors who survived the Holocaust in the ghettos on the territory of modern Lithuania and Latvia. Diaries of those who survived the Holocaust in Russia were almost not preserved. This is partly because Jews of the occupied Soviet Union kept diaries only on their personal initiative. Such an initiative did not stem from a "cultural life" and an organised collection of witnesses as it was in the ghettos of Poland and in the territories annexed to the USSR in 1939–1940.[12] Instead, Holocaust survivors in Southern Russia could not record the thoughts and details of their everyday struggle for life because there were no ghettos: they had constantly to change hiding place, and they lived in perpetual fear of being found out. The only known diary mentioning the Holocaust in the North Caucasus is one by the famous Karaite scholar Boris Kokenai. His diary is a kind of ethnographic observation on the history and culture of the Karaite communities in the South of Russia in 1930s and 1940s.[13] At the same time, there were no records in the diary from 1937 to 1946. Kokenai wrote down his notes about the war period and the fates of members of the Karaite community of

---

11 The first diary published in the USSR was that of Masha Rolnikaite and is called *I Must Tell You* [*Ya dolzhna rasskazat'*]. It contains diary entries written by the author between the ages of 14 and 18 in the Vilnius ghetto and two Nazi concentration camps. As she wrote, she simultaneously memorised her entries. Her diary was literarily edited in the post war years. Masha Rolnikaite, *Ya dolzhna rasskazat'* (Leningrad: Soviet writer, 1976). The diary was first published in Yiddish: Mashe Rolnik, *Ikh muz dertseyln* (Warsaw-Moscow: Yidish Buch, 1965). See information about this diary in English: Laurel Holliday, ed. *Children in the Holocaust and World War II: Their Secret Diaries* (New York: Pocket Books, 1995), 185–200. See other diaries of the Holocaust survivors on Soviet soil: Tamara Lazerson-Rostovskaya, *Zapiski iz Kaunasskogo getto (Katastrofa skvoz' prizmu detskikh dnevnikov)* (Moscow: Vremya, 2011); Filipp Fridman, *Gibel' l'vovskikh evreiev*, http://samlib.ru/h/hohulin_aleksandr_wasilxewich/gibelxlxwowskihewreew.shtml (accessed March 17, 2020); Dnevnik Sheyny Gram (22 iyunya – 8 avgusta 1941 goda), in *Evrei v menyayushchemsya mire: materialy 8-i mezhdunarodnoi konferentsii, Riga, 30 iyunya – 4 iyulya 2011*, ed. German Branover and Ruvin Ferber (Riga: Latviiskii universitet, 2015), 90–96.; Edward Anders, ed., *Nineteen Months in a Cellar: How 11 Jews Eluded Hitler's Henchmen: The Diary of Kalman Linkimer (1913–1988)*, transl. [from Yiddish] Rebecca Margolis (Riga: Jewish Community of Riga: Museum "Jews in Latvia," 2008).
12 Il'ya Al'tman, *Dnevnik Lyusi Kaliki "Odessa. 820 Dnei v Podzemel'e" kak istoricheskii istochnik*, https://docplayer.ru/38571483-Dnevnik-lyusi-kaliki-odessa-820-dney-v-podzemele-kak-istoricheskii-istochnik.html (accessed March 17, 2020).
13 Boris Kokenai, "Dnevnik," in *personal archive of Sergei Shaitanov*.

Feodosia (Crimea) and Krasnodar only after the war. His written entries were mainly based on knowledge obtained from other people. Thus, the record in Kokenai's diary on the destruction of the Krasnodar Karaite community under the occupation should be considered as written memories containing information from other sources.

The memoirs of Holocaust survivors are a somewhat more representative type of personal narrative in the North Caucasus than diaries. Not all memoirs waited long to get written, but the time separating events from their retelling is clearly an important, distinguishing feature of this type of ego-document. The author of a memoir selects and organises the information that his or her memory has preserved, through whatever processes. Memoirs usually have an addressee or reader. These can be readers of a narrow, predetermined circle, such as family members, but memoirs are more often addressed to a large readership: contemporaries, descendants, or historians.[14] The primary post-occupation memories of Holocaust survivors collected by the ChGK were analysed in Chapter 3. I found no memoirs about surviving the Holocaust in the North Caucasus written before the mid-1960s. This reticence of Holocaust survivors during the first post-war decades could be due to several reasons. First, NKVD-MGB[15] officers investigated citizens of the liberated Soviet regions to assess if the local population had collaborated with occupation authorities. The purpose of such investigations was to bring Soviet traitors and collaborators to trial. Many Holocaust survivors were under suspicion by the Soviet security agencies. They were often asked how they had been able to survive

---

14 For more details on Holocaust memoirs, see: Pyotr Kukhiwtsak, "Mediating Trauma: How do we read the Holocaust Memoirs?," in *Tradition, Translation, Trauma: The Classic and the Modern*, ed. Jan Parker and Timothy Mathews (Oxford: Oxford University Press, 2011); Alyson Miller, "Stylised Configurations of Trauma: Faking Identity in Holocaust Memoirs," *Arcadia: International Journal of Literary Culture* 49, no. 2 (2014): 229–253; Jeremy D. Popkin, "Holocaust Memories, Historians' Memoirs: First-person Narrative and the Memory of the Holocaust," *History and Memory* 15, no. 1 (2003): 49–84; Jared Stark, "Broken Records: Holocaust Diaries, Memoirs, and Memorial book," in *Teaching the Representation of the Holocaust*, ed. Marianne Hirsch and Irene Kacandes (New York: Modern Language Association of America, 2004), 191–204; Robert Rozett, "Published Memoirs of Holocaust Survivors," in *Remembering for the Future: the Holocaust in an Age of Genocide*, ed. John K. Roth and Elisabeth Maxwell, vol. 3: *Remembering for the Future* (Basingstoke, Hampshire: Palgrave, 2001), 167–171; Laurence Kutler, "Holocaust Diaries and Memoirs," in *Holocaust Literature: a Handbook of Critical, Historical, and Literary Writings*, ed. Saul S. Friedman (Westport: Greenwood Press, 1993), 521–532.
15 The NKVD was renamed the Ministry of State Security [MGB] in 1946. In 1954 it was reorganised into the Committee for State Security [KGB]. It was the main security agency of the Soviet Union until its dissolution in 1991. The FSB preserves its functions in modern Russia.

the occupation when most Jews had been killed. Jews were suspected of having had interactions with the Nazis. Boris Kamenko, the Holocaust survivor in Stavropol whom I discussed in Chapter 3, had to undergo special inspections by state security agencies for almost 20 years after the war. He was called both as a witness and as a suspect in interactions with the Nazis during the occupation. According to the Holocaust survivor Inna Veisburd, who preferred to hide her military past and the circumstances of her survival in occupied Taganrog, "the KGB wanted to finish what the Nazis could not."[16] Several Holocaust survivors stated that surviving the Holocaust was in some ways easier than living in the USSR under the constant suspicion of the Soviet security agencies. Secondly, since the late 1940s when state antisemitism peaked with open campaigns against Jews – the defeat of the JAC, the campaign against rootless cosmopolitans, the Doctors' Plot[17] – most Holocaust survivors tried not only to refrain from recalling their military experience but if possible, to not highlight their Jewish identity in public at all. Thus, the ideology of the Soviet state stood in the way of comprehending the wartime experience of Holocaust survivors.

The authors of the first post-war written memoirs were expatriates from the USSR to Israel in the early 1970s. The Yad Vashem Archive (YVA) has been constantly updated with letters from Holocaust survivors since the 1980s. They become a part of the Testimonies, Diaries and Memoirs Collection of the YVA.[18] These letters often contain the recollections of Holocaust survivors.[19] The memoirs of Tankha Otershtein, for instance, seek to attract the readers' attention to Holocaust history of the North Caucasus, as it is quite different from "the well-known spots of the Jewish tragedy, such as Treblinka, Auschwitz, the Warsaw Ghetto, and Babi Yar."[20] A number of memoirs of Soviet Holocaust survivors were published in the

---

[16] Interview with Inna Vaisburd, born in 1922, in *VHA USCSF*, code 46991–13, June 30, 1998, Haifa, Israel.

[17] Gennadii Kostyrchenko, ed. *Gosudarstvennyi antisemitizm v SSSR: ot nachala do kul'minatsii, 1938–1953* (Moscow: Materik, 2005); Idem, *Tainaya politika Stalina. Vlast' i antisemitizm* (Moscow: Mezhdunarodnie otnosheniya, 2015); Gerd Koenen and Karla Hielscher, *Die schwarze Front: der neue Antisemitismus in der Sowjetunion* (Reinbek: Rowohlt, 1991); Leonid Luks, ed. *Der Spätstalinismus und die "jüdische Frage": zur antisemitischen Wendung des Kommunismus* (Köln: Böhlau, 1998).

[18] *YVA*, O.33 "Various Testimonies, Diaries and Memoirs Collection."

[19] Sarra Kreymer, "Po kazhdomu moyemu sledu ezhednevno shagala smert'," 1991, in *YVA*, O.33, no. 5029; "Pis'mo-vospominanie Zel'dy Al'tgauz v Yad-Vashem," 1999, in *YVA*, O.33, no. 5761; Yurii Prisov, "The Reminiscences of a Young Holocaust Survivor," in *USHMMA*, access no. 2011.337.1.

[20] Tankha Otershtein, "Vospominaniya o perezhitom evreiskoi sem'ei na okkupirovannoi territorii i v fashistkoi tyur'me," in *YVA*, O.33, no. 4026.

early 2000s.[21] They are sometimes published as collected short stories of Holocaust survivors, accompanied by brief biographical information about their authors. More often, they are thematic collections of memories with a detailed story of surviving the Holocaust in different parts of the former Soviet Union, including the North Caucasus.[22] Wartime and Holocaust history can be a central thematic thread of memoirs, around which the author weaves the narrative. More often, the memoirs represent the life story of the author, so that the Holocaust becomes just an episode in a broader tale. But mostly, the regional complex of written memoirs of Holocaust survivors consists of short stories. They are interesting for researchers because of the historical details they give. Holocaust survivors wrote down such brief recollections to record their experience for future generations. These memories are important for reconstructing Holocaust history in the North Caucasus. At the same time, they play only a minor role in the regional Holocaust memory dispositif. Published often abroad and in small print runs, they circulate mostly among Jewish communities and family members of Holocaust survivors, seldom reaching a wider reading audience. Unpublished memoirs become part of the archival collections of international Holocaust museums and research centers awaiting their researcher. The memories of Holocaust survivors have never been collected in the regional archives of the North Caucasus; unlike a large-scale campaign to collect the memories of partisans in the 1960s and 1970s.[23]

---

21 *Vyzhit' i pomnit': vospominaniya byvshikh uznikov getto i kontslagerei – zhitelei g. Karmielya, Izrail'* (Carmiel, 2014); Svetlana Danilova, ed., *Deti voiny: vospominaniya* (New-York: Design-ER, 2015); Aleksei Khveister, ed., *"Kto vyzhivet – rasskazhet": vospominaniya poslednikh svidetelei Kholokosta* (Wismar: Verlag Koch und Raum Wismar, 2015); Lyubov Rozenfeld, ed., *Voennoe detstvo: sbornik rasskazov detei, nynche zhitelei Ashkelona (Izrail'), perezhivshikh Velikuyu Otechestvennuyu voinu* (Ashkelon – Tel-Aviv, 2015).

22 For instance, the memoirs of the Odessa historian Saul Borovoi, published in the early 1990s, contain a small fragment of his life during the evacuation to Krasnodar Krai in the autumn of 1941. The author describes relationships with local residents, the fates of other Jews evacuated to the North Caucasus, and emphasises the comfortable life conditions in Krasnodar Krai. At the end of 1941 the author was evacuated further to the Central Asia. Saul Borovoi, *Vospominaniya* (Moscow-Jerusalem: Hebrew University in Moscow, 1993), 248–258.

23 Aleksei Kurnosov, "O memuarakh uchastnikov partizanskogo dvizheniya (1941–1945 gg.)," in *Istochnikovedenie istorii sovetskogo obshchestva*, ed. Nikolai Ivnitskii, vol. 1 (Moscow: Nauka, 1964), 289–319; Idem, "Priemy vnutrennei kritiki memuarov: vospominaniya uchastnikov partizanskogo dvizheniya v period Velikoi Otechestvennoi voiny kak istoricheskii istochnik," in *Istochnikovedenie: teoreticheskie i metodologicheskie problemy: sbornik statei*, ed. Sigurd Shmidt (Moscow: Nauka, 1969), 478–505; Irina Rebrova, "Velikaya Otechestvennaya voina v memuarakh: iIstoriko-psikhologicheskii aspekt" *(*cand. of science diss. in history, North-West Academy of Public Administration, 2005).

The most representative group of Holocaust survivors' narratives is oral testimonies, which have been conducted since the 1970s.[24] The first narrators were, again, former Soviet Jews who had expatriated to Israel. The YVA has a collection of first oral interviews with Holocaust survivors, including those who survived the occupation in the North Caucasus. These interviews are preserved in the form of transcripts of a question-and-answer thematic conversation about the war period in the narrator's life. The transcripts are accompanied by a list of topics addressed and an index of persons and places mentioned. Most oral testimonies in the YVA, dealing with the history of the Holocaust in the North Caucasus described events in Stavropol Krai.[25] An important characteristic of the first oral interviews with Holocaust survivors from the North Caucasus is that the narrators paid attention to the ethnic history of the region and the attitude of the Peoples of the Caucasus to the Jews during the war. The narrators emphasised the fact that many representatives of the Peoples of the Caucasus had resisted Soviet power, which subsequently led to their collaboration with the Nazi authorities. Speaking of life during the occupation of Kislovodsk, Rina Zaslinskaya stressed:

> "Rumour has it that when the Germans came, the Karachays descended from the mountains, claiming they had lived there for 25 years and waited. They were traitors of the Motherland [. . .] The Germans would not have known who lived where, if not for them [. . .] They caused a lot of grief [. . .] And when Soviet power was restored, all the Karachay people were deported, every single one of them. It did not take very long. What had happened to the Jews, happened to the Karachay people."[26]

---

24 Survivors of the Holocaust in the North Caucasus were not interviewed for the principal oral history projects in Europe. In 1946, for example, Dr. David P. Boder, a psychology professor from Chicago's Illinois Institute of Technology travelled to Europe to record the stories of Holocaust survivors in their own words. During a three-month period, he visited refugee camps in France, Switzerland, Italy, and Germany, carrying a wire recorder and 200 spools of steel wire, upon which he was able to record over 90 hours of first-hand testimonies. These recordings represent the earliest known oral histories of the Holocaust, which are now available in the online archive. 22 interviews in Russian were conducted with Soviet Jews who survived the Holocaust in the ghettos of today's Baltic countries and the concentration camps of Eastern Europe. *The Voices of the Holocaust Collection at the Illinois Institute of Technology*, http://voices.iit.edu/ (accessed march 23, 2020); Beate Müller, "Translating Trauma: David Boder's 1946 Interviews with Holocaust Survivors," *Translation and Literature* 23, no. 2 (2014): 257–271.
25 "Svidetel'skoe pokazanie Fany Sklyar," 1975, in *YVA*, O3, no. 3934; "Svidetel'skoe pokazanie Vadim Maniker," 1975, in *YVA*, O3, no. 4108; "Svidetel'skoe pokazanie Rina Zaslinskaya," 1975, in *YVA*, O3, no. 4927.
26 "Svidetel'skoe pokazanie Rina Zaslinskaya," l. 2,10.

Stories of such collaboration and subsequent Stalinist deportations were silenced in the USSR for many years. Perhaps Holocaust survivors' interviews conducted in Israel in the 1970s are the earliest testimonies of these complex ethnic relations.

Since the 1970s the Oral History Division Archive at the Avraham Harman Institute of Contemporary Jewry at the Hebrew University of Jerusalem, hereafter the OHD, has been filled with the testimonies of Soviet Jews. A wide variety of thematic oral history projects has been conducted at the Institute since 1959.[27] The experience of Holocaust survivors in the North Caucasus was documented in several oral history projects, such as "The Mountain Jews of the USSR" in the mid-1970s, "Evacuation and Flight of Jews in the USSR during World War II" in 1987, "Non-Ashkenazi Jews in the USSR" at the end of the 1980s, and "Jewish Newcomers from the USSR" in the mid-1990s. Holocaust history was not the central topic of interviews in the majority of these oral history projects. Nonetheless narrators mentioned it, since the Holocaust played a key role in their lives. Oral interviews in the form of thematic conversations are typical for most oral history projects of the 1970s and 1980s. The interviewer asks detailed and clear questions in order to make the narrator's story more understandable for an outside listener/reader and to establish the factual accuracy of the story. In many ways, this method impedes the narrator's emotional freedom; we get interviews with the clearly expressed active position of the interviewer who often interrupts the narrator to specify or clarify misunderstandings.

The Holocaust Survivors Film Project for the Fortunoff Video Archive at the beginning of the 1980s, by contrast, is the first attempt to record video interviews in the form of free narrative. This method of conducting interviews gives the leading role to the narrator during the conversation. Interviewers ask questions to clarify the time and place mentioned in the narration, or to learn additional facts about the narrator's life. The narrators are the experts in their own life story, and the interviewers are there to listen, to learn, and to clarify.[28]

---

27 The OHD Archive contains the memories of individuals from various segments of Israeli and Jewish society throughout its modern history, providing generations of future researchers with an invaluable social history of the Yishuv, the State of Israel and Diaspora relations and Jewish communities in the Diaspora. OHD official website, http://www.hum.huji.ac.il/english/units.php?cat=4247&incat=4246 (accessed March 23, 2020).
28 "About the Fortunoff Video Archive," *Fortunoff Video Archive for Holocaust testimonies*, http://web.library.yale.edu/testimonies/about (accessed March 23, 2020). More about the method of conducting Video interviews for this project: Dori Laub, "Bearing Witness or the Vicissitudes of Listening," in Shoshana Felman and Dori Laub, ed., *Testimony: Crises of Witnessing in Literature, Psychoanalysis, and History* (New York, 1992), 71–74; Geoffrey H. Hartman, *The Longest Shadow: In the Aftermath of the Holocaust* (New York: Palgrave Macmillan, 1996), 133–150.

Nevertheless, this project is not of great importance for the regional Holocaust memory dispositif, since out of more than 4,400 interviews only one was with a Jew who survived the Holocaust in Stavropol Krai.[29]

In the mid-1990s, the USCSF carried out the largest project to gather oral testimonies from Holocaust survivors all over the world, including the countries of the former USSR. This project has also become the most important by number of interviews concerning the survival of Jews in the North Caucasus. The awarding of compensation payments to Holocaust survivors from different European foundations including the Claims Conference was an additional incentive for Holocaust survivors from the former Soviet Union to share their memories in the 1990s.[30] The goal of the whole USCSF Oral History project was to record the life experiences of Jews throughout the 20th century. An approximate questionnaire was developed, according to which 20 per cent of the interview time was devoted to discussing the pre-war period and 20 per cent concerned the post-war life of the narrator. The central topic was the war period, which was allotted up to 60 per cent of the interview time. According to Darlene Basch, the coach, therapist, social worker and trainer of interviewers at the Shoah Foundation, "an interview for the Foundation is not just a journalistic and historical interview. Its goal is to help the survivor give testimony the way she or he sees it and ask the right questions." Historian Igor' Kotler considers video testimonies for the Foundation to be ethnographic interviews.[31] Indeed, many interviews could become a significant source in reconstructing the everyday life of the Soviet Jewry throughout the 20th century.

The Oral History department at the United States Holocaust Memorial Museum (USHMM) has been conducting oral interviews on the territory of the former USSR since the 2000s.[32] Ukraine, Belarus, and Russia became the key states in the post-Soviet space for seeking Holocaust eyewitnesses and Holocaust survivors. Interviews for the USHMM are rather short; sometimes they are kind of

---

29 Interview with David H., born in 1921, in *Fortunoff Video Archive for Holocaust Testimonies at Yale University*, HVT-879, October 13, 1991.
30 For more about the Compensation Payment Programmes see: *The Claims Conference official website,* http://www.claimscon.org/what-we-do/compensation/ (accessed March 24, 2020).
31 "Interviewer training session in Moscow," August 1996, in VHA *USCSF*, Digital and remote storage, ALT03014, videocassette 2.
32 In addition to their own interviews, the USHMM's Jeff and Toby Herr Oral History Archive contains various collections of Holocaust testimonies from all over the world. Many of the testimonies, primarily those conducted by the museum are available online. More about this archive see: "The USHMM's Oral History collections," *USHMM official website,* https://www.ushmm.org/collections/the-museums-collections/about/oral-history (accessed March 24, 2020).

a monologue of the narrator about the wartime period in his or her life, and survival of the Holocaust.[33] A question-and-answer part of the interview follows the narrator's story and aims at clarifying the narrative or placing emphasis in the story. Perhaps, this form of thematic interview is a result of the preliminary work with potential narrators: "The interviewer meets a candidate for the interview and listens to the survivor's story. Then the project manager selects candidates for conducting an audio interview with them. After that the film crew comes and records a video interview".[34] It turns out that the narrator tells the story in front of the camera for the third time, which means that the story itself acquires a more refined and rehearsed character. Such interviews are important for fixing historical events in the life of the narrator. However, they are not an ideal source for the analysis of Holocaust memory. There was also a preliminary stage of conducting oral interviews for the USCSF when the interviewer only got acquainted with the Holocaust survivor and they together filled out a Pre-Interview Questionnaire. The completed survey was supposed to help the interviewer in his or her preparation for the interview (reading special literature, learning the Holocaust history of the region where the Holocaust survivor stayed during the war years, etc.).[35] Nevertheless, interviewers in many cases followed a standardised questionnaire during the interview and did not take the specifics of the narrator's personal story into account.

I conducted about 30 interviews with Holocaust survivors in the North Caucasus, Moscow, and Israel between 2013 and 2016. Quite a number of my narrators had already been interviewed for the USHMM or the USCSF Oral History projects. I usually conducted narrative interviews with an open general introductory question. As the narrator told the story I would ask questions to clarify the

---

[33] See, for example: Interview with Vyacheslav Bezshkurny, born in 1925, in *USHMMA*, courtesy of the Jeff and Toby Herr Foundation, RG-50.653*0006, May 12, 2010, the village of Ilinskaya, Krasnodar Krai; Interview with Irina Rubanova-Talant, born in 1933, in *USHMMA*, courtesy of the Jeff and Toby Herr Foundation, RG-50.653*0002, May 11, 2010, Krasnodar; Interview with Valentina Kolbaskina, born in 1931, in *USHMMA*, courtesy of the Jeff and Toby Herr Foundation, RG-50.575*0089, October 3, 2009, the village of Blagodarnoe, Stavropol Krai.
[34] For more about the interview method of the USHMM, see: Anatolii Kardash, "Katastrofa evreiev glazami neevreiev (svidetel'stva ochevidtsev okkupatsii na territorii byvshego SSSR)," in *Voina na unichtozhenie: natsistskaya politika genotsida na territorii vostochnoi Evropy: materialy mezhdunarodnoi nauchnoi konferentsii*, ed. Aleksandr Dyukov and Olesya Orlenko (Moscow: Fond "Istoricheskaya pamyat'", 2010), 448–449.
[35] "Interviewer Guidelines" and "Pre-Interview Questionnaire," *USCSF official website,* http://sfi.usc.edu/vha/collecting; "Oral History Interview Guidelines written by Oral History Staff of USHMM," *USHMM official website,* https://www.ushmm.org/m/pdfs/20121003-oral-history-interview-guide.pdf (accessed March 23, 2020).

details. The fact that I didn't know what had happened to the narrator during the war made it possible for him or her to develop the story in a meaningful way and allowed me just to follow the narrative. Subsequently, comparing several testimonies of one Holocaust survivor conducted in different years and by different interviewers can help trace the work of the narrator's memory on significant events of his or her life and the influence of time upon the story.[36]

The following offers a more or less general portrait of the narrators who survived the Holocaust in the North Caucasus and gave testimonies for various Oral History projects. He or she is a Soviet Jew, a refugee or an evacuee from the western regions of the USSR to the North Caucasus, who became a Holocaust survivor in childhood or adolescence. The majority of narrators were children of war, who were of retirement age at the time of giving testimony. A small number of interviews were conducted with Holocaust survivors who survived the occupation in their late 20s or early 30s. As a rule, they were women with young children. Their stories are more reflective than those of children of war. The story of how the lives of their children were saved becomes the most important part of such interviews. Another group of narrators are those born just before or during the war. They generally retell the wartime memories of their parents, since their personal memories are either missing or very fragmentary. The number of female narrators is about one-and-a-half times higher than the number of men. Men of military age obviously fought at the front, and mostly old people and women with children inhabited the occupied areas. Table 6 presents the statistical data of the narrators of the studied interviews. I consider the interviews for the USCSF and my own because they present the most significant group for my research.

Almost all narrators had higher education and relatively successful careers after the war. This success was strongly dependent on the way the post-war Soviet society accepted Jews. Some of them had to struggle against the system for quite a while, while others hid their Jewish origin. Some of the Holocaust survivors expatriated to Israel in 1970s; others left the Soviet Union after its collapse. Nonetheless, the majority decided to stay in the post-Soviet space. They considered themselves too old to start a new life elsewhere, regardless of how comfortable it might be abroad. At the same time the succeeding generations of almost

---

36 For example, Yakov Krut, a native of Rostov-on-Don, survived the occupation in Krasnodar Krai. Since the 1990s he has given several interviews to various journalists, film directors, and researchers. He also wrote his memoirs and published his stories in Israeli Russian-language newspapers and magazines. Krut, *Povest' o podarennoi zhizni*; Interview with Krut, in *author's archive*; Interview with Yakov Krut, born in 1928, in *VHA USCSF*, code 5009, May 10, 1995, Ashkelon, Israel; Grigorii Reikhman, "Zheltaya zvezda" iz kubanskoi stanitsy, *Vesti (Prilozhenie "Veteran i voin")*, May 4, 2011.

all the narrators had emigrated to the USA or Israel in the 1990s. The narratives of Jewish expatriates differ from the stories of those who stayed in the post-Soviet space. These differences are important when narrators speak of the postwar period, and especially the post-Soviet period. Stories about experiences of antisemitism in the USSR and their more fortunate life after expatriation contrast with the resignation felt among those who remained to the controversies of Soviet history and poverty in old age. Oral testimonies are always stories about the perception of the past from today's individual perspective. Since the majority of interviews, primarily those for the USCSF, were conducted in the 1990s during a serious economic crisis in the post-Soviet countries, Holocaust survivors who lived in the post-Soviet space spoke a lot about their current life conditions. At the same time, no significant differences in narratives about wartime experiences appeared between these two groups of narrators. Albeit subjective and fragmentary, the memory of children of war is an excellent historical source for social history and anthropology, shedding light on everyday life, permitting psychological and gender readings of the history of World War II. Oral testimonies are also an important element of the regional Holocaust memory dispositif, showing the presence and willingness of Holocaust survivors to share their stories with younger generations and not let Holocaust history be forgotten locally.

The testimonies of Holocaust survivors about their wartime experience can be contextually divided into two parts: the story of surviving the Holocaust, and self-reflection about what they had to overcome and how this period influenced narrators' future life.[37] The following general thematic lines emerge in survivors' testimonies when describing wartime in the North Caucasus:

1. *The beginning of the war and the decision to evacuate from the Western regions of the USSR.* The majority of oral testimonies on the Holocaust in the North Caucasus belong to evacuees and refugees rather than to local Jews. As already mentioned, there were few Jewish communities in the North Caucasus in the pre-war years. Additionally, local Ashkenazi Jews found it more difficult to survive the occupation, since their social circle was much wider than that of the evacuated Jews, and they therefore more at risk of betrayal. The stories told by refugees to the North Caucasus about their flight from the Nazis are less detailed, compared with their descriptions of life in a new place. Families that survived the war after having been evacuated to Central Asia or the Urals enjoyed more

---

[37] I shall analyse only the part of the interviews dealing with the wartime period and thoughts of the narrators about it. This permits me to compare the majority of accumulated oral testimonies about the Holocaust in the North Caucasus. Thus, parts of interviews about the pre-war life of Jews in the USSR, their cultural traditions and customs will not be analysed here.

favourable living conditions. They had to work hard in the rear, survived hunger, and lack of sleep. But they neither met the Nazis face to face, nor did they have to live in constant fear for their lives.

2. *Life as evacuees in the North Caucasus before the occupation.* This is the brightest part of the survivors' memories. Many narrators emphasise the hospitality of local residents and the absence of antisemitism in the region. The majority of villagers had no idea who Jews were, still less of the ideologies leading to their discrimination or derision:

> Well, for them [the local residents] this all seemed new. They had no idea who Jews were until a certain moment. I remember, I brought a toy with me, a motorboat. There were tubes and when you poured water with a syringe it could stay on the surface. It was a wonderous toy for any villager. There was no light, no radio, nothing in the village. They [local children] asked to play with it.[38]

Many evacuated Jews were richer than collective farmers in the rural areas of the North Caucasus. The difference in incomes was also obvious in the higher quality of evacuees clothing and children's toys, which the villagers could not even dream of. Holocaust survivors sometimes falsely associated childish envy of toys with manifestations of antisemitism in rural areas.

Almost all Holocaust survivors described the beautiful nature and warm climate of the North Caucasus. Maya Gaba, who was evacuated to the North Caucasus along with her whole orphanage, recalls:

> We had something like a collective farm. Kuban bread is big, there were lots of strawberries and we had our own beehives. There were also apricots, so incredibly sweet! The sweetest I ever tasted. The climate was so favourable, one could live as a cat or dog and eat whatever nature gave [i.e., live off local vegetables and fruits].[39]

Warm climate and fertile soils became an additional factor that helped many Jews to survive the Holocaust in the region. Perhaps, it was the hospitality of the region that played such a cruel trick on many evacuated Jews. The state did not conduct a timely re-evacuation from the North Caucasus in the summer of 1942. At the same time, many Jews did not hurry to leave their new homes. Thanks to their work in collective farms, evacuees to the countryside had food and some of them even had their own garden plots.

---

[38] Interview with Boris Gluzkin, born in 1931, in *VHA USCSF*, code 48066–27, July 3, 1998, Gomel, Belarus.
[39] Interview with Maya Gaba, born in 1930, in *VHA USCSF*, code 19180, August 21, 1996, Sderot, Israel.

3. *The beginning of the occupation and stories about the death of relatives and loved ones.* This is the most dramatic part of the oral interviews with Holocaust survivors. The loss of a close relative becomes the main traumatic event of the war period that affects the entire future life of the Holocaust survivor. Thus, having lost her father, Irina Rubanova experienced emotional shock that affected her six months of wandering and constant hiding with her mother in occupied Krasnodar.[40] She was young enough to remember the events of her father's arrest. Although she did not witness his killing, she learned about it from her mother and eyewitnesses. Rubanova looked for the official evidences of her father's killing in the post-war period. In the course of time her memory about this fact incorporated other testimonies and findings. For Evelina Ekonomidi, as well, her mother's killing in Zmievskaya Balka was the main narrative of her testimony.[41] In contrast, however, the loss of her mother became her personal tragedy and she was unable to speak about it for more than 60 years, even among family members. Neither did she visit the memorial complex at the killing site. There is no portrait of her mother among the family portraits displayed in the living room of her apartment. Losing her mother in childhood is a trauma, which she prefers to keep private. Perhaps, in this way the narrator feels her mother is still alive.

4. *Survival strategies during the occupation.* Following the local traditions and Orthodox rituals, wearing traditional Slavic or Caucasian clothes, changing names, destroying or falsifying documents were among the basic survival strategies of Jews during the Holocaust, according to their testimonies.[42] However, resorting to these survival strategies was often not enough. There were always local collaborators who could betray Jews. It was especially easy to identify evacuated Jews in the countryside, where everyone knew each other. Jews who did not attend the registration of the Jewish population in the first weeks of the occupation either had to hide or constantly move from place to place where no one knew them and where their new story and false names sounded believable. As Leonid Brik recalls: "We were afraid of people, kept hiding all the time. We had only a couple of bags, always ready to leave."[43]

---

40 Interview with Irina Rubanova, born in 1933, in *author's archive*, code 13-X-KK01, February 13, 2013, Krasnodar.
41 Interview with Evelina Ekonomidi, born in 1933, in *author's archive*, code 13-X-RO02, February 26, 2013, Rostov-on-Don.
42 For more on the survival strategies of Jews in the occupation in the North Caucasus, based on oral interviews see: Rebrova "Traumatische Kindheit."
43 Interview with Leonid Brik, born in 1935, in *VHA USCSF*, code 44537–55, May 17, 1998, Odessa, Ukraine.

Life among strangers was better because no one could betray them and their invented life stories were taken at face value. At the same time, Holocaust survivors could be saved only thanks to local residents, who hid them, shared food, or simply did not give them away to German authorities. These Soviet people, mostly farmers from the rural regions with basic education are very well described in oral interviews. Narrators almost always remember their names and thank them for saving their lives. There are accounts of Jewish children of war from evacuated orphanages who survived because Circassian families adopted Jewish children and raised them as their own.[44] These children did not identify themselves as Jews in the post-war period; they accepted the traditions of the new families and became full members of the local Islamic communities preferring not to speak about their traumatic past. There are also stories about survival in places in the North Caucasus where Jews were concentrated. Jews were typically held together in one building for several weeks in the last period of the Nazi occupation. Sometimes they were kept together with Soviet POWs. According to one testimony, Jews were guarded, fed twice a day, and had a 15-minute daily walk in the fresh air.[45]

Stories about the survival of Jews in the occupation are stories about daily life, specific events, situations and facts experienced by the narrators themselves. It is possible to reconstruct everyday life in the occupation, the attitude of locals to the evacuated Jews in the South of Russia, and inter-ethnic relations during the war using these testimonies. At the same time, many Holocaust survivors also sought to understand how they survived. Narrators rethink their wartime experience due to the influence of the official memory culture of the war. The emphasis in this part of the narrative is transferred from the private realm to the public sphere. A relatively short occupation period in the lives of Holocaust survivors becomes the subject of their thoughts during the whole post-war life. Narrators often refer to higher powers such as God or destiny, which prevents any rational explanation of the narrators' survival. Indeed, hundreds and thousands of other Jews, often their own relatives were killed right in front of them. Tankha Otershtein, who survived the occupation as a teenager together with his family, believes: "My family was saved by a miracle and this is thanks to God. I do not know anyone who survived the Holocaust among

---

[44] The most famous example is the case of the orphans from Leningrad who were adopted and raised in Beslenei aul. See: Auron and Okhtov, *Podvig milisediya*.
[45] Interview with Yulii Vaispapir, born in 1928, in *VHA USCSF*, code 37626–48, July 3, 1997, Zaporozhye, Ukraine.

those whom I personally knew."[46] The sociologist Reeve Brenner studied oral testimonies of Holocaust survivors in European countries and found that the faith of Jews was significantly strengthened in the post-war years.[47] Here we speak of faith in general, not in specific religions. My analysis of interviews with Soviet Holocaust survivors confirms this thesis.

Some narrators think that accidents were the main reason of their survival. For example, a village teacher warned the Yuzhelevskie family "to escape the registration because only Jews were being registered."[48] Semen Shlagin recalls the Germans who came to plunder the local population: "They took food, and one of them ordered the others to go, saying that he would catch up with them. They left, and he told us in proper Yiddish: 'Juden, why are you sitting here, don't you know what you are doing? You must run, hide. Otherwise you will be caught.' This shocked us so much."[49]

The stories about casual encounters in the narratives of Holocaust survivors are taken as signs of fate, thanks to which they were able to survive. However, Faina Babitskaya mentioned the main reason why Jews managed to survive the Holocaust in the North Caucasus. Probably it was "due solely to the fact that the occupation was only six months long and then we were liberated."[50]

Certainly, the wartime period was the key event in the life of the majority of Holocaust survivors, so thoughts about why they survived are the result of both interaction with the interviewer at a particular historical moment and a long process of self-reflection after the war. Nevertheless, not all narrators explained why they survived; the majority tended simply to describe their everyday experience of surviving. This is explained primarily by the semantic content of these two seemingly interdependent parts of the narrative. The story of survival is a sequence of events that led to a happy ending. Thoughts about its reasons require time distance from the events and the post-war social experience of the narrator. Level of education, cultural background and knowledge of World War II history allow the narrator reflect on his or her own destiny. Self-reflection is the way Holocaust survivors try to relate their own experience to the official

---

46 Interview with Tankha Otershtein, born in 1932, in *VHA USCSF*, code 31823, May 25, 1997, Taganrog, Russia.
47 Reeve R. Brenner, *The Faith and Doubt of Holocaust Survivors* (New York: Free Press, 2014).
48 Interview with Maya Yuzhelevskaya, born in 1937, in *VHA USCSF*, code 37730, January 4, 1998, Braintree, USA.
49 Interview with Semen Shlagin, born in 1926, in *VHA USCSF*, code 44006, May 6, 1998, Zaporozhye, Ukraine.
50 Interview with Babitskaya. The quotation from her testimony is used in the title of this subchapter as well.

state memory politics of the war and the Holocaust. The deeper the war trauma is the more they speak of their struggles "against windmills," or prohibitions of the state or other institutions, leaving their experience of the war untouched. Some narrators recount their significant achievements in the post-war period to emphasise that their survival had purpose for the victims. Instead of their survival story some narrators spoke of installing a monument to victims of the Holocaust, or the search for official documents about the killing of their relatives; others were looking for their rescuers, praising their heroism; the rest searched for documentary evidence that would confirm them as former "young prisoners of fascism" or Holocaust survivors and give them the right to receive reparations.

8.2 "The city is destroyed. All the Jews have been shot:" Eyewitness Narratives of the Holocaust in the North Caucasus

As noted in Chapter 3, the collection of evidence of the Holocaust began right after the liberation of each locality. Along with the work of the ChGK at the state level, other Soviet commissions were established to document the war from a personal perspective. The Commission on the History of the Great Patriotic War, hereafter the Mints Commission, was founded by the Academy of Sciences of the USSR. It sought to collect the stories of war participants and witnesses of the Great Patriotic War in the USSR.[51] It was formed at the initiative of the famous Soviet historian

---

51 For more about the activity of the Mints Commission, see: Aleksei Kurnosov, "Vospominaniya – interv'yu v fonde Komissii po istorii Velikoi Otechestvennoi voiny Akademii nauk SSSR (organizatsiya i metodika sobiraniya)," *Arkheograficheskii ezhegodnik za 1973 god* (Moscow: Nauka, 1974), 118–133; Nina Arkhangorodskaya and Aleksei Kurnosov, "O sozdanii Komissii po istorii Velikoi Otechestvennoi voiny AN SSSR i ee arkhiva (k 40-letiyu so dnya obrazovaniya)," *Arkheograficheskij ezhegodnik za 1981 god* (Moscow: Nauka, 1982), 219–229; Vitalii Tikhonov, "Materialy Komissii po istorii Velikoi Otechestvennoi voiny 1941–1945 gg. akademika I.I. Mintsa kak istoricheskii istochnik," in *Rossiiskoe gosudarstvo: istoki, sovremennost', perspektivy*, ed. Aleksandr Bessudnov, vol. 2 (Lipetsk: Lipetskii gosudarstvennyi pedagogicheskii universitet, 2012), 32–40; Sergei Zhuravlev, ed., *Vklad istorikov v sokhranenie istoricheskoi pamyati o Velikoi Otechestvennoi voine: na materialakh Komissii po istorii Velikoi Otechestvennoi voiny AN SSSR, 1941–1945 gg.* (Moscow: Institut Rossiiskoi akademii nauk, 2015). For more on publications on the Mints Commission's records, see: Jochen Hellbeck, *Die Stalingrad-Protokolle. Sowjetische Augenzeugen berichten aus der Schlacht* (Frankfurt am Main: Fischer Verlag GmbH, 2012); Idem, ed., *Stalingrad. The City That Defeated the Third Reich*, (New York: Public Affairs, 2015); Sergei Kudelko et al., ed., *Gorod i voina: Khar'kov v gody Velikoi Otechestvennoi voiny* (Saint-Petersburg: Aleteya, 2013); Andrei

Isaak Mints[52] in December 1941[53] in order to write a "chronicle of the defence of Moscow." One of the main activities of this Commission was to conduct "interviews with the heroes and participants of the Battle of Moscow and civilians affected by fascist atrocities."[54] Later on, members of the Mints Commission conducted interviews about other important battles of the war. Along with writing up the annals of the armies and biographies of war heroes, its staff conducted interviews on the work of enterprises in the rear and life of Soviet citizens under the occupation. By the end of 1942, staff of the Mints Commission had developed a manual, describing all stages of its work, from collecting evidence to editing manuscripts.[55] Each transcript began with brief biographical data followed by a detailed story about the struggle at the frontline or everyday life under the occupation. The opportunity to speak out was very important to many people: many narrators felt as if they were "a part of history" and experienced psychological relief. Despite the unique work of the Mints Commission, collecting the "living voices" of the war participants, one should beware the obligatory literary editing of the final transcript.[56]

Marchukov, *Geroi-pokryshkintsy o sebe i svoem komandire: pravda iz proshlogo. 1941–1945* (Moscow: Tsentrpoligraf, 2014).

52 Isaak Izrailevich Mints (1896–1991) was a doctor of historical sciences, Soviet historian, specialist on modern history, member of the USSR Academy of Sciences in history and philosophy, laureate of Prizes of Lenin (1974) and Stalin (1943, 1946). He headed the secretariat of the Chief Editorial Office of the multi-volume *History of the Civil War in the USSR* in 1931. He gave lectures at the front during the war and practically led the Commission's work on the History of the Great Patriotic War. "K 60-letiyu so dnya rozhdeniya akademika I.I. Mintsa," *Voprosy istorii* 3 (1956): 67.

53 The Presidium of the Academy of Sciences of the USSR approved the members of the Mints Commission August 27, 1942. It included the chairman Georgii Aleksandrov, the deputy chairman Isaak Mints, the academician Emelyan Yaroslavskii, the deputy chairman of the SNK Rosaliya Zemlyachka, the division commissar and deputy commander of GlavPU KA Iosif Shikin, the army commissar Ivan Rogov, the secretary of the TsK of the All-Union Leninist Young Communist League [*Vsesoyuznyi leuninskii kommunisticheskii soyuz molodezhi*] Nikolai Mikhailov, the editor-in-chief of the newspaper *Pravda*, Peter Pospelov, the head of the Union of State Book and Magazine Publishers [*Ob"edinenie gosudarstvennykh knizhno-zhurnal'nykh izdatel'stv*], hereafter OGIZ, Pavel Yudin, the member of the editorial board of the newspaper *Izvestia*, Lev Rovinskii, the secretary of the TsK VKP(b) for propaganda, Ekaterina Leontieva, and the professor Efim Gorodetskii. Zhuravlev, *Vklad istorikov*, 59.

54 Ibid, 55.

55 "Deloproizvodstvo Komissii po istorii Velikoi Otechestvennoi voiny," in the *Scientific Archive of the Institute of Russian History at the Russian Academy of Science* [*Nauchnyi arkhiv instituta Rossiiskoi istorii Rossiiskoi akademii nauk*], hereafter NA of the IRI RAN, f.2, section XIV, op.1, d. 7, l. 34–41.

56 There was no common opinion on the editing of transcripts among the staff of the Mints Commission. Some considered it unnecessary and wanted to preserve the narrator's speech as

A department named "The Occupation Regime and the Restoration of Soviet Power" was established inside the Mints Commission in 1944, directed by the historian Esfir' Genkina.[57] The staff of the Mints Commission visited many liberated regions of the USSR to write down stories about "life under the Nazis." According to many historians, the topic of the German occupation regime had not only scientific, but also political and propagandistic significance.[58] In early 1944 this topic was among the priorities of the Mints Commission as a result of the liberation of the Soviet territory and the restoration of Soviet power there. Many members of the Mints Commission recorded the experience of people in the occupied Belorussian and Ukrainian SSRs; some of them visited other regions of the country, including the North Caucasus. A senior researcher Grigorii Anpilogov conducted about 20 interviews with Soviet scientists, teachers, and agricultural workers who lived under occupation in Stavropol Krai between September and October 1945. As a rule, narrators talked a lot about what they did in the occupation and described the political views and fates of their colleagues. Almost every testimony has a story about the mass killing of Jews in general, and/or about the killing of the narrator's Jewish colleagues and acquaintances:

> The Jewish population lived in the village [Vorontsovo-Aleksandrovskoe, Stavropol Krai], some of them may be gone now, but I knew the tailor since the time I began working there. He did not run [when the Germans came] for some reason and he was shot together with his family, that's to say, he was taken away, and what was done with him is unknown, the man just disappeared. They [the German authorities] printed a leaflet, an Appeal to the Jewish Population. They wrote: 'A Jewish committee had been established in Vorontsovo-Aleksandrovskoe. All Jews had to show up there at certain time of the day'. Jews had to be moved to another place, so it was necessary to show up at certain location, carrying no more than 30 kg of luggage. The rest was to be handed over to the landlord. Jews gathered near the police office. Then people simply disappeared and that was it. This Appeal was immediately spread over the whole village, apparently printed somewhere in advance.[59]

---

it was, while others insisted on improving the quality of the transcripts, simplifying the text's further comprehension. Zhuravlev, *Vklad istorikov*, 109–111.
57 "Deloproizvodstvo Komissii po istorii Velikoi Otechestvennoi voiny," in *NA of the IRI RAN*, f.2, section XIV, op.1, d. 3, l. 14; d. 38, l. 117.
58 Zhuravlev, *Vklad istorikov*, 123.
59 "Transcript of a conversation with comrade Ivan Nikolaevich Zavgorodnev, the chief agronomist of the Stavropol Regional Agricultural Department, born in 1901," in *NA of the IRI RAN*, f.2, section VI, op. 23, d. 10, l. 5, October 2, 1945, Stavropol.

The narrators of the Mints Commission were generally literate, educated people. Therefore, they were able to give a general picture of everyday life under the occupation and to speak not just of their own personal experience. The stories about the rounding up of Jews and their subsequent "disappearance" are very important for the historical continuity of the Holocaust in a particular locality. The fact that the Appeal to the Jewish Population was displayed around the whole village implies that the entire local population was informed. Most narrators speak of their killed Jewish colleagues, stating the facts baldly:

> Professor Beshchestnaya, the radiologist, died. She was a Jew. She was lonely. Why did she go there [to the registration]? First, you could hide. Plenty of us did not go to the registration. Those who didn't survived for sure. Brailovskii died, the professor and pathologist. He lived in Pyatigorsk and someone betrayed him there. They [the Nazis] often used mobile gas vans to kill people. Vilenksii, the associate professor and psychiatrist, also died. Professor Revutskaya, the histologist survived, she worked in the laboratory and didn't go to the registration, hid all her documents and got some kind of false certificate.[60]

Despite the considerable stylistic and semantic editing of the transcripts, the stories about mass killing of the Jews were nevertheless preserved in the final versions.[61] Such stories are present in almost all transcripts of the interviews recorded in the North Caucasus. Interestingly, the records of the Mints Commission almost always identify the differences between Nazi victim groups and do not use the Soviet term "peaceful Soviet citizens." This can be attributed to the fact that during the war no generally accepted way of describing the losses under the occupation had been developed. Most of the Soviet clichés to describe the war entered the language of participants and witnesses much later. On the other hand, the majority of transcripts, especially those about the occupation regime, were not going to be printed, so people were not afraid to be misunderstood and say too much about someone or something. They had absolutely no reason to flatter someone and they did not think about speaking the 'right' way.[62] Most narrators recalled people they knew, primarily their colleagues, since the Nazis killed only Jews among the body of professors, scientists, and managers and workers of enterprises. The ethnicity of Holocaust victims is therefore emphasised in the records of the Mints Commission. It was important for the narrator to highlight that the Soviet Academy of Sciences,

---

60 "Transcript of a conversation with Professor Mikhail Varfolomeevich Donich, doctor of medical sciences, Head of the Department of Hygiene at Stavropol Medical Institute, born in 1898," in *NA of the IRI RAN*, f.2, section VI, op. 23, d. 7, l. 3, September 26, 1945, Stavropol.

61 The interviews were literarily edited before being placed into the archive. The depth of editing is now almost impossible to detect since the original records were recycled. Zhuravlev, *Vklad istorikov*, 133.

62 Marchukov, *Geroi-pokryshkintsy*, 9.

8.2 "The city is destroyed. All the Jews have been shot" — 271

as well as many enterprises, lost many professors and skilled workers only because they were Jews. At the same time, none of the narrators attempted to explain the causes of the Holocaust, just stating it as a fact.

The collection of transcripts in the Mints Commission named "The Occupation Regime" is the earliest set of thematic oral interviews with Holocaust eyewitnesses in the occupied Soviet territories. They were conducted right after the liberation, when the recent occupation period was easy enough to remember. Unfortunately, the questionnaire for conducting these conversations is missing from the Mints Commission's archive.[63] The presence of Holocaust-related topics in almost every transcript of the interview from the liberated regions shows either that there was a question about the fate of Jews in the questionnaire, or that the mass killing of Jews was so momentous and painful, that the narrators simply could not fail to address it in their stories.

The Presidium of the Academy of Sciences of the USSR dissolved the Mints Commission on November 15, 1945. A new center for the Study of the History of the Great Patriotic War was formed as a part of the Institute of History of the Academy of Sciences of the USSR under the direction of Isaak Mints.[64] The documents produced by the Mints Commission are currently preserved at the Scientific Archive of the Institute of Russian History at the Russian Academy of Sciences (RAS) and consist of about 20,000 files.[65] Created under the supervision of the Soviet authorities, these records contain the first evidence of Holocaust eyewitnesses about the mass killing of Jews in the USSR. Not published until now, these records reflect the duality of the Soviet Holocaust memory dispositif. On the one hand, the facts of the mass killing of Jews were well known to the public from the

---

[63] It is theoretically possible to reconstruct the initial questions when repeating the structure of the interviews, but it is extremely difficult to understand what suggestive questions and remarks were made (and whether they were made at all) during the conversation. Zhuravlev, *Vklad istorikov*, 133.

[64] Boris Levshin, "Deyatel'nost' Komissii po istorii Velikoi Otechestvennoi voiny, 1941–1945 gg.," in *Istoriya i istoriki: istoriograficheskii ezhegodnik*, 1974, ed. Melissa Nechkina (Moscow: AN SSSR, 1976), 317.

[65] After Stalin's death in 1953, the records of the Mints Commission were checked and "resystematised." The files on the history of military units collected during the war were transferred to the TsAMO in Podolsk, where they remain. Another 2,827 folders were transferred to various institutions, and 391 folders were destroyed in agreement with the General Directorate for the Protection of State Secrets in the Press under the Council of Ministers of the USSR [*Glavnoe upravlenie po okhrane gosudarstvennykh tain v pechati pri Sovete Ministrov SSSR*]. Some of the materials fell into the Special storage of the RAS. For the Catalogue of the Mints Commission data stored in the NA IRI RAN see: Zhuravlev, *Vklad istorikov*, 129–130, 364–382; *Official website of the Mints Commission*, http://komiswow.ru/ (accessed March 24, 2020).

very beginning and stories of the Holocaust were an essential part of early testimonies about everyday life under the occupation. On the other hand, these unpublished personal testimonies never found a place in the Soviet official memory of World War II.

The records of the Mints Commission have always been open to researchers. They have become the subject of scholarly reflection in the last decades, albeit on topics not related to Holocaust history on Soviet soil. However, other evidence of the Holocaust collected by Soviet judicial authorities has not been declassified. The testimonies of Holocaust eyewitnesses became important evidence in Soviet and international trials against Nazi perpetrators and Soviet collaborators. As noted in Chapter 5, the defendants were judged for their collaboration with Nazis in the majority of Soviet trials. The ethnicity of the victims was not emphasised in the Soviet open trials and in the mass media.[66]

After 1947, open trials were gradually replaced by closed trials, held by military tribunals of the Ministry of Internal Affairs of the USSR. They were held in the location where the defendant was detained, without eyewitness testimony or an adversarial defence, and resulted almost universally in the defendant being sentenced to up to 25 years in a corrective labour camp.[67] Nevertheless, Soviet secret police continued to collect evidence about the Nazi occupation until the 1960s and 1970s. Such testimony was widely used to hunt down Soviet collaborators and bring them to trial. These trials could be either open – although not widely broadcast in the Soviet mass media[68] – or closed. These latter remain classified. Testimonies collected for Soviet trials were also sent abroad, primarily to East and West Germany, and were used as additional evidence of Nazi crimes in the USSR in trials against the Nazi war criminals. Thus, the General Prosecutors Office of the

---

66 See a selection of articles covering the Soviet open trials of the 1940s: *"Soviet Nuremberg"* website, https://histrf.ru/biblioteka/Soviet-Nuremberg (accessed March 24, 2020).
67 "Order of the Ministry of Internal Affairs of the USSR, the Ministry of Justice of the USSR, the USSR Prosecutor's Office No. 739/18/15/31126 about the transfer of completed cases of POWs, participants in atrocities in the temporarily occupied Soviet territories for investigation in closed courts where perpetrators were held, Moscow, November 24, 1947," in *GARF*, f. 9401, op. 1, d. 837, l. 164–165. Ibid (document no. 7.14), in *Voennoplennye v SSSR. 1939–1956: dokumenty i materialy*, ed. Maksim Zagorul'ko (Moscow: Logos, 2000), 1007. According to the Russian historian Dmitrii Astashkin, there was a shortage of highly qualified investigators in the USSR after the war. The Soviet state also spent a lot of money on the open trials against Nazi perpetrators, held in 21 Soviet cities during 1943–1949, which essentially fulfilled their propaganda function. Astashkin, *Protsessy nad natsistskimi*, 7.
68 See, for example: Valentin Yakhnevich, "Vozmezdie: v Krasnodare zakonchilsya sud nad esesovtsami-karatelyami," *Pravda*, October 25, 1963; Idem, "Vozmezdie," *Pravda*, February 1, 1966.

USSR provided records against former servicemen of Einsatzgruppe D – Kurt Christmann, Heinrich Görz, Kurt Trimborn, and Walter Kehrer – in response to a request of the West-German embassy to the USSR in 1963.[69] The investigative cases of trials held in West and East Germany contain records sent by the Soviet Union such as ChGK files, records of interrogations with Soviet defendants and Holocaust eyewitnesses, and wartime photographs of killing sites in the liberated localities. Many of these Soviet files refer to Nazi crimes in the North Caucasus. Most of the testimonies of Soviet citizens were collected by Soviet law enforcement agencies in the 1960s. Because of the classified access to the former KGB files in Russia, one can only assume that the Soviet prosecutor's office conducted a special interrogation of civilians who survived the occupation. It is possible that these statements were subsequently used in Soviet trials against collaborators and then forwarded to their German colleagues. It is important to note that the testimonies of Soviet citizens of the 1960s and 1970s refer to specific groups of Nazi victims such as Jews, people with disabilities, Sinti and Roma, communists, and the term "peaceful Soviet citizens" was practically not used. Thus, the witness Antonina Ivanova, who lived near State Farm [*sovkhoz*] No.1 in Krasnodar, described the mass killing of Jews in August 1942 when she was interrogated on October 26, 1964:

> In August 1942, about five days after the Nazis captured Krasnodar, I went past a detached building, belonging to the All-Union Scientific Research Institute of Tobacco and Makhorka. As I approached it, I saw people in German military uniform and Jewish women, children, old people, and men in trucks, apparently brought from the city. I thought that they had been taken there for some work and just kept going. After a while, working in the vineyard, I heard people screaming, and then shooting. These sounds were audible for a long time. It became obvious that the Germans were shooting the people I had seen before. Indeed, later it turned out to be true. People of Jewish ethnicity were shot and buried near that building. I did not know exactly how many people were there, but I am sure quite a lot.[70]

It would appear that in interrogations of the 1960s, one of the main questions was the victims' ethnicity. Almost all eyewitnesses refer to the Jewish identity of the Nazi victims. If the eyewitnesses themselves did not know who was shot exactly, they referred to rumours circulating in the city: "I heard that the

---

[69] "Letter from Acting Prosecutor General of the USSR Mikhail Malyarov to the senior prosecutor at the Munich District Court I, December 2, 1964," in *StA München*, Staatsanwaltschaft, 35308/30, Bl. 1. See also: "Übersetzung der Note No. 22/3eo des Sowjetischen Außenministeriums vom 31.01.1973," in *StA München*, Staatsanwaltschaft, 21672/19, Bl. 451.

[70] "Record of interrogation with the eyewitness Antonina Ivanova from Krasnodar of October 26, 1964," in *StA München*, Staatsanwaltschaft, 35308/30, Bl. 102.

Germans allegedly shot the Jewish population in the first shootings [. . .] [in Krasnodar in August 1942]. But I cannot say whether it is true since I could not determine their ethnicity from the place where I was standing."[71] Eyewitness Maria Trufanova, born in 1904, living in the Vtoraya Zmeevka village near Rostov-on-Don also mentioned the killing of Jews, as told to her by fellow villagers, in her interrogation of October 8, 1964:

> I learned from local residents that a shooting of Soviet citizens of Jewish ethnicity was committed in a sandy quarry in the afternoon of August 11, 1942. Eyewitnesses, I do not remember now who they were exactly, said that Jews were brought in cars and on foot, then shot in a sandy quarry. I was convinced of the fact that the Jews were actually shot in a sandy quarry, as I, along with other citizens, repeatedly had to bury the corpses of Soviet citizens, which had opened and decomposed after the rains.[72]

Despite the widespread use of the term "peaceful Soviet citizens" in the official post-war Soviet media, Holocaust eyewitnesses testified to what they saw themselves and what the locals talked about. As in the original ChGK files, received from eyewitnesses who survived the occupation, the transcripts of witness interrogations twenty years later show that the memory of ordinary citizens was distinct from politics and ideology. Living in remote provincial towns and villages they preserved memories of the Holocaust and named certain groups of Nazi victims. Even in the presence of government officials in the face of MGB/KGB investigators, they used vocabulary reflecting real history in their testimonies. They saw, heard, and knew that the main Nazi victims were Jews, Sinti and Roma, handicapped persons, and communists.

The testimonies of Soviet Holocaust eyewitnesses also contain information on the methods of murdering the Jews: "I heard frequent shots," "I was in the garden with my husband and personally saw this execution," "I personally saw this car, a gas van and how the gendarmes dragged the corpses of people who had just been killed out and threw in the ditch."[73] Eyewitnesses speak less often about the number of victims, and almost never mentioned the names of the perpetrators. Eyewitnesses, mostly women who remained during the occupation, said that the killing site was far from where they saw it, so they obtained only a general picture and could not identify a specific victim or perpetrator. Testimonies of Holocaust

---

[71] "Record of interrogation with the eyewitness Irina Talashchenko from Krasnodar of September 14, 1964," in *StA München*, Staatsanwaltschaft, 35308/30, Bl. 98.
[72] "Record of interrogation with the eyewitness Maria Trufanova from Rostov-On-Don of October 8, 1964," in *StA München*, Staatsanwaltschaft, 35308/32, Bl. 24
[73] See: "Records of interrogations with Soviet Holocaust eyewitnesses in Russian and translated into German," in *StA München*, Staatsanwaltschaft, 35308/30–33; 35275/17–20; 21672/19–20.

eyewitnesses of the 1960s and 1970s are an integral element of the regional Holocaust memory dispositif. These statements did not become part of the official memory culture of victims of the Nazis in the USSR and then Russia, yet they clearly demonstrate that Holocaust eyewitnesses knew and remembered the main group of Nazi victims.

Holocaust eyewitnesses rarely devoted their written memoirs to this topic during the Soviet period. The exceptions are memoirs about Holocaust victims, written at someone's request. Ust-Labinsk schoolchildren collected evidence about the pioneer-hero Musya Pinkenson, whose story was mentioned in Chapter 1. Musya's former teachers and classmates wrote memoirs in his honour at the request of the founders of the school museum in the 1970s.[74] These memoirs present Musya as a talented violinist, a diligent schoolboy, and a "brave pioneer who was killed by the Nazi invaders." Holocaust history is completely silenced in such memoirs, which correspond to the Soviet ideology and memory politics. Such memoirs are a useful source of information about both individual acts of resistance by Jews during the Holocaust and general Holocaust remembrance in the USSR.

Despite the muting of Holocaust history in the Soviet period, Holocaust eyewitnesses remembered the history of the war as they experienced it. They were mainly villagers who rarely had higher education and were far removed from politics. More than half a century later they gave testimonies about the mass killing of different groups of Nazi victims. As confirmed by the results of Yahad-In Unum's expeditions to remote regions of the former USSR, memory at the individual level is not subject to change over time or to the influence of state ideology. Since 2006, teams of mainly young people from Yahad-In Unum have travelled the roads of Ukraine, Belarus, Russia, Lithuania, Romania, and Poland, videotaping interviews with non-Jewish eyewitnesses to shed new light on the Holocaust in Eastern Europe. Their main goal is to identify the mass grave sites of the genocide, not to hunt for the killers or to bring perpetrators to justice: "Instead, we are looking for victims, often cast aside and forgotten in the historical reconstruction of events. We patiently pull the threads of memory of the witnesses to establish crime scenes and identify the precise locations of the mass graves. We are indeed in a race, competing with time itself, trying to interview as many of these witnesses as possible, the majority of which are in their 80s or 90s."[75]

---

[74] See memories of Musya Pinkenson's former classmate Anna Bakieva, teachers Elena Sakhno and Nadezhda Bannova, and neighbour Maria Repeshchuk (written in 1970s?) in personal archive of Irina Svetlichnaya.
[75] Patrick Desbois, "The Holocaust by Bullets," *The Holocaust and the United Nations Outreach Programme. Discussion Papers Journal* 2 (New York, 2002): 81.

So far Yahad-In Unum teams have made 10 research trips to the North Caucasus out of 24 trips to Russia, and a total of 175 trips between 2004 and 2019.[76] They conducted 335 interviews in small towns and villages in Rostov Oblast and Krasnodar and Stavropol Krais, as well as in the Republics of Karachay-Cherkess, Kabardino-Balkaria, and Adygea. "Now-adult/then-child"[77] testimonies of ordinary villagers are an important source for reconstructing life during the war. Linguistically poor due to the relative inarticulacy of most narrators – many of whom have spent most of their lives performing heavy physical labour on a state farm in the middle of the Caucasus countryside – these testimonies are valuable for their facts. Many years later, narrators describe the events of the war in detail, they talk about what they saw as teenagers and what the rumours in the village were about. The memories of ordinary Holocaust eyewitnesses were only rarely sought in Russia at any time after World War II. That is why when the narrators recount the history of the mass killing of Jews in their village, it is usually for the first time.

Before the fieldtrip, the Yahad-In Unum team carefully study the primary sources, both Soviet – mainly ChGK files – and German, including trial cases against Nazi perpetrators. The task of the preparatory stage of the trip is to identify the killing sites, the approximate number of victims, and the details of the killing. The close study of primary sources allows the Yahad-In Unum team to reveal inaccuracies or discrepancies in different sources and to compile fragmentary information about a particular shooting, which then gets turned into a set of questions for Holocaust eyewitnesses during the field research.[78] There is no unified questionnaire developed by the Yahad-In Unum team for conducting interviews with the Holocaust eyewitnesses. The interview is essentially formed around what the witness has to say. The usual topics include the narrator's life before the war, the arrival of German armed forces, everyday life during the occupation, the policy towards the Jews, and the methods by which Jews were killed in a particular locality. Some of these topics may be altered depending on the region in which the research is conducted. For example, "it is important

---

[76] The Yahad-In Unum staff kindly provided this information at my request on February 18, 2020.
[77] This term was used by Paul Shapiro in the foreword to Patrick Desbois's book about Yahad-In Unum's work in Ukraine. Paul Shapiro, introduction to Desbois, *The Holocaust by Bullets*, xiii.
[78] For more details on the research method of the Yahad-In Unum team, using the example of fieldwork in Belarus, see: Andrej Umansky, "Metodika issledovanii, provodimykh tsentrom 'Yahad-In Unum,' i ikh rezul'taty na primere polevykh rabot v gorode Motol', Belorussiya," in Dyukov, *Voina na Unichtozhenie*, 462–470; Johanna Lehr and Patrice Bensimon, *Case Studies: Intersection of Archives and Fieldwork, Motol (Pinsk region, Belarus)*, https://www.yahadinum.org/case-studies-intersection-of-archives-and-fieldwork-motol-pinsk-region-belarus/ (accessed March 25, 2020).

to ask about the mobile gas vans, poisoning of children, and the evacuation of Jews in the North Caucasus, whereas the ghettos and life under the occupation are only covered briefly, since the Nazis didn't stay in this area long enough."[79] The topic of poisoning children comes up in oral narratives too. The regional ChGK files contain data on the poisoning of children "by lubricating the mucous membrane of the lips with a poisonous substance."[80] The results of the Yahad-In Unum trips to the North Caucasus and eastern Ukraine confirm this fact. Thus, Dmitrii G. from the village of Petropavlovskaya (Krasnodar Krai) recalls: "the Jewish children were lined up on the edge of the pit. A German came up to the children holding a stick under their noses and kicked them so that they fell into the pit."[81] The detailed testimony given by Nikolai T. from Gulkevichi (Krasnodar Krai) gives further details:

> Women, children, and men were separated at the edge of the anti-tank ditch and put into three groups. The children were to be killed first. Each child had to crouch above the anti-tank ditch. The German, holding a container in one hand, approached the child, put a stick in the container, and then applied it to the lips of the child who fell back in the ditch.[82]

The Yahad-In Unum team conducts mostly structured interviews in the form of direct questions and answers. The identity of the narrator as well as his or her personal wartime experience is not the core of the conversation, although interviews almost always tend to begin by establishing a brief biography of the narrator. Interviewers of Yahad-In Unum aspire to discover the circumstances in which Jews were killed and reconstruct every single execution that the narrators witnessed. The interview is often divided into two parts. First, the narrator answers the questions of the interviewer at home, and then the team leaves for the killing site in order to conduct a reconstruction of the events from the place where the eyewitness observed them. Detailed questions such as "Where were you during the shooting?," "Who was with you?," "Who led the Jews?," "How did you know they were Jews?," "Was there any guard and what did they do?," "What kind of uniform did the perpetrators wear?" aim to reconstruct the shooting of Jews in 1942 as accurately as possible. The interviewer usually offers the narrator a possible scenario to develop the story, giving several answers to his or her

---

79 The Yahad-In Unum staff kindly provides the information on my official request from January 3, 2018.
80 "ChGK reports for Rostov Oblast, Stavropol and Krasnodar Krais," in *GARO*, f. 3613, op. 1, d. 30, l. 1; *GARF*, f. 7021, op. 16, d. 4, l. 10; d. 7, l. 37; *GANISK*, f. 1, op. 2, d. 359. l. 13–15, 20. See also: Rowe-McCulloch, "Poison on the Lips."
81 Interview with Dmitrii G.
82 Interview with Nikolai T.

own question, as if the narrator should choose one of them.[83] Such a method of conducting interviews demonstrates the uniqueness of Holocaust eyewitnesses, not as people who survived the occupation and went through the war, but as sources and witnesses. He or she confirms, refutes, and more often clarifies the historical facts that the Yahad-In Unum team has already uncovered in the official documents. The interviewer asks questions based on his or her knowledge of the sources about the mass killing in the locality and in the neighbourhood, or mentions the names of already documented policemen or Nazi perpetrators, assuming that the narrator can know or recall these facts or these specific names. As a result, the interviews of Yahad-In Unum resemble an interrogation, restoring certain details of the Nazi crimes. This kind of "interrogation-interview" is conducted by sympathetic "interviewer-inquirer," whose purpose is to reconstruct historical facts; the narrator plays the key role in this reconstruction.

The arrival of strangers with cameras constitutes an important event in the daily lives of rural residents. The elderly narrators become the center of attention. They speak in a simple way, sometimes using foul language, unafraid of being misunderstood by the foreign guests. The constant translation of questions and answers (interviews are usually conducted in English or French) and long pauses in the speech of narrators during the translation do not affect the nature of the story. The kinds of interaction associated with these interviews – pauses in the story, distracting inputs by team members on the way to the killing site – mean they cannot count as classic oral interviews, where memories are just as important for *how* it is narrated as for the events they narrate. Interviews collected in Yahad-In Unum fieldtrips are nevertheless valuable as additional source materials for the reconstruction of Holocaust history, especially in the North Caucasus; a region almost completely lacking in German primary sources on the topic. These interviews play a significant role in the regional dimension of the Holocaust memory dispositif. They reveal Holocaust memory at the level of the ordinary person, whose wartime experience and memories of the occupation have remained uninfluenced by questions of state ideology.

The initiative to collect the testimonies of Holocaust eyewitnesses has often come from state authorities or international organisations. Eyewitness interviews were always conducted for practical purposes, for example, to gather evidence for trials against Soviet collaborators and Nazi perpetrators or in order to determine the location of a killing site. These testimonies are unique. They are an

---

[83] For example, in an interview with Dmitrii G. the interviewer "investigates" the history of Jews gathering at an assembly point, asking the following questions: "What was the day like when they [Jews] were taken out? Were they made to walk in a row or not? Did our [*politsai*] gather them or not?" and so on. Interview with Dmitrii G.

important source for micro-historians and local historians, helping to reconstruct the historical events of the war in a specific locality. But the initiative to collect Holocaust testimonies can come from individuals and civil society as well. For example, some non-Jewish citizens wrote diaries about everyday life under the occupation. These authors mention the mass killing of Jews by German occupying forces, although the events of the Holocaust are described in general terms and more attention is paid to describing the authors' own experiences. There are not many wartime diaries from occupied localities in the North Caucasus.[84] I found only one relevant diary. This is the diary of Nikolai Saenko, a factory worker who remained in occupied Taganrog. The diary was stored as a ChGK file in the GARF and in the GARO.[85] It became known after its publication by historians Pavel Polyan and Nikolai Pobol' in the collection of diaries and memoirs entitled *Banned from the Light* [*Nam zapretili belyi svet*] in the mid-2000s.[86] Everyday life in the occupied city is presented in this diary from the factory worker's point of view. The author wrote entries almost every day up to June 13, 1942, following which there are many gaps or summarised entries. The history of the mass killing of Jews in Taganrog is presented along with the description of other Nazi acts, such as the issuing of decrees, sending people to Germany as forced labourers, and the shootings of communists and Soviet POWs. Saenko's diary entry of September 29, 1941 mentioned that about two weeks after the occupation of Taganrog "all Jews were ordered to gather in the Red Square [the name of the square in the city center] for resettlement, but no one knew their final destination. The order was to take food and as many belongings as one can carry."[87] The entry for September 30, 1941 reports: "Jewish children and elderly people gathered outside the city behind the airfield of the Factory No.31 and were shot."[88]

Saenko was an eyewitness to these events and therefore they are reflected in his diary. At the same time, the main topic of his diary is the survival of a Soviet man under the occupation, physically and psychologically. The author

---

[84] As a rule, diary writing is a pastime of the educated intelligentsia. It is almost impossible to find diaries of rural residents about their life under the occupation.
[85] "Dnevnik Saenko," in *GARF*, f. 7021, op. 40, d. 632; *GARO*, f. P-3613, op. 1, d. 447.
[86] Nikolai Saenko, "Taganrogskii dnevnik (2 oktyabrya 1941 g. – 1 sentyabrya 1943 g.)," in *"Nam zapretili belyi svet . . . ": Al'manakh dnevnikov i vospominanii voennykh i poslevoennykh let*, ed. Pavel Polyan and Nikolai Pobol' (Moscow: ROSSPEN, 2006), 25–113. An excerpt from the diary was published by Pavel Polyan as an announcement of a future collection in the newspaper *Izvestiya*: Pavel Polyan "Taganrogskii dnevnik. Zhizn' i okkupatsiya glazami rabochego Nikolaya Saenko," *Izvestiya*, March 3, 2005.
[87] Saenko, "Taganrogskii dnevnik," 31.
[88] Ibid, 31.

dwells in detail on the everyday search for food and discusses prices at the markets.[89] He also wrote down many of his thoughts concerning the behaviour of Soviet people in the occupied city.[90] Saenko's diary is a kind of anthem to Soviet soldiers, for whom the residents of Taganrog waited during more than 22 months of occupation.[91] The long-awaited liberation of Taganrog happened only on August 30, 1943. The last entry of the diary, September 1, 1943 reports the happiness of liberation as well as with the grief of loss:

> The exhumation of victims shot by the Germans in Petrushina Balka began at 9 a.m. Barely sprinkled with earth, the corpses began to decompose from the heat during the removal. Many people fainted and were hysterical. Everyone had tears in their eyes. Many corpses had their hands tied back and were shot to the head, lying with their heads in one direction, naked, apparently ordered to lie down during the killing. Teenagers and children up to 15 years old were also among the victims.[92]

Petrushina Balka is the last address of Jews, communists, partisans, and POWs who were killed in Taganrog. Saenko did not care about the victims' ethnic or social status. They all were fellow citizens to him, deserving of remembrance and a worthy funeral.

Another important type of personal narrative of Holocaust eyewitnesses is letters to the relatives, friends, and colleagues of killed Jews. The Holocaust Center has collected wartime letters by Jews and letters about Jewish fates in the Holocaust for many years.[93] It has already published five volumes of the letters

---

89 For example, the diary contains the following entries: "The bread distribution began on new tickets. We did not have them so we could not get the bread. We have five numbers cut out on two tickets, because when the house committee filled in the lists for receiving those tickets they noted in the column how many bread products there are per family. So, being honest and precise, I said that we have 15 kg of grain. That's why they cut out five numbers" (entry of March 31, 1942). "A lot of fish appeared on the market. The price of fish (Zander) is still high; a two-kilo fish costs up to 70 roubles. And it cost up to 120 roubles earlier" (entry of April 26, 1942). Ibid, 70, 74.

90 For example, Saenko writes about the revival of religion in the city: "It is Sunday, the whole spiritual gang [svora] stopped hiding, threw off their sheepskin and already dressed in spiritual clothes – cassocks with wide sleeves and crosses on chains around their necks" (entry dated May 3, 1942), Ibid, 77.

91 The recording of almost every day contains information about the state of the front: "gunfire is audible from the front, all day and night there were air raids" or "nothing is heard from the front."

92 Ibid, 112.

93 For more on the Holocaust Center's collecting of military letters, see: Leonid Terushkin "Pis'ma i dnevniki kak istochnik po istorii Vtoroi mirovoi voiny i Kholokosta (po materialam arkhiva nauchno-prosvetitel'nogo tsentra "Kholokost")," in *Kholokost: novye issledovaniya i materialy: materialy XVIII mezhdunarodnoi ezhegodnoi konferentsii po iudaike*, ed. Viktoria

and diaries.⁹⁴ Again, the collection of letters about the fate of Jews in the North Caucasus is much smaller compared to collections from other regions of the USSR. This could be because the letters of relatives did not always reach their addressees, and the villagers renting a room to evacuated Jews were largely illiterate and could not respond to the letters. The few known letters that relatives of the Holocaust victims received during the war contain unique details about the mass killing of Jews. The North Caucasus was the last postal address of many evacuated Jews, whose husbands or children struggled at the frontline. Therefore, after the liberation of the North Caucasus, relatives who lost contact with their loved ones wrote letters to the place of their last address, hoping to receive at least some information about them. Relatives of Holocaust victims who learned about the shooting of Jews often retold these stories in letters to Il'ya Ehrenburg, who collected evidence for *The Black Book* during the war. The head of the military hospital Boris Eisenberg, for example, wrote a letter to Ehrenburg in June 1943, stating that his wife Galina Eisenberg and 10-year-old son Aleksandr were killed in Essentuki (Stavropol Krai): "During talks with the locals I learned that the naive Jews did not know for sure that the Germans would commit such an insidious and monstrous massacre of innocent people. Mass media and Soviet authorities did not warn the population about the danger for Jews to become Nazi victims in that region."⁹⁵

Correspondence with local residents who survived the occupation was often initiated at the official request of the Jewish relatives to the local authorities. Neighbours or representatives of local authorities reported details of the Holocaust in the locality back to the relatives. For example, a relative of the famous professor and doctor of medicine Ioann Baumgolts received a letter from

---

Mochalova, vol. IV (Moscow: Sefer, 2011), 94–104; Idem, "Pis'ma rodstvennikov, sosedei i druzei v evakuatsiyu – neizvestnye svidetel'stva o Kholokoste," in *Istoriya. Pamyat'. Lyudi: materialy VIII mezhdunarodnoi nauchno-prakticheskoi konferentsii*, ed. Maria Makarova (Almaty: Assotsiatsiya evreiskikh natsional'nykh organizatsii "Mitsva," 2017), 158–172.

94 The published collections contain letters from Jews evacuated to the North Caucasus. They are mostly about everyday life in a new place before the occupation. There are also the last letters to relatives, full of fear and sorrow. See: "Pis'mo Rozy Golub Synu Nikolayu, August 28, 1942," in *Sokhrani moi pis'ma . . .: sbornik pisem evreev perioda Velikoi Otechestvennoi voiny*, ed. Il'ya Al'tman and Leonid Terushkin, vol. 2, (Moscow: Tsentr i Fond "Kholokost", 2010), 142–144; "Pis'mo Naumu Fel'dmanu ot B.Sh. Gamarnika, March 11, 1942," *Ibid*, vol. 2 (Moscow: Tsentr i Fond "Kholokost," 2010), 105–107; "Pis'mo Sterny Freidson sestre Rivy i plemyannitse Base, November 1941," *Ibid*, vol. 3 (Moscow: Tsentr i Fond "Kholokost," 2013), 171–172.

95 "Pis'mo Borisa Aizenberga korrespondentu gazety "Krasnaya zvezda" Il'ye Eerenburgu, June 1943," in *Sovetskie evrei pishut Il'ye Erenburgu, 1943–1966*, ed. Mordechai Altshuler, Yitshak Arad, and Shmuel Krakowski (Jerusalem: Yad Vashem, 1993), 125–126. See also unpublished letters to Il'ya Erenburg in *YVA*, P.21 "Ilya Ehrenburg Collection."

a friend who visited a woman renting a room for the Baumgolts family in Kislovodsk and learned the details of the last days of their lives:

> Your mother and Melania, the nurse who serves Dr. Baumgolts, were taken out of the city with the rest of the Jewish population on the 4th or 5th of September 1942. The Gestapo published an order stating that the whole Jewish population of Kislovodsk should be resettled from Kislovodsk to another part of the country, but it was not specified where [. . .] On September 9 at 5 a.m., all Jews, except those of mixed marriages were to appear at the train station to be loaded into the carriages [. . .]. All of them were sent to Mineralnye Vody and from there, as reported, to a glass factory, which is about 6–7 km away from the city's main train station. That place was the last destination of the Jewish population from Kislovodsk.[96]

Most of the Jews arrived in the North Caucasus as early as the late summer of 1941, and so local residents lived and worked with the evacuees for almost a year before the occupation started. Hence, the locals got to know the newcomers. It should be noted that the fact of living together may have given Jews a false sense of security, leaving them unprepared when locals betrayed them to Nazis and sometimes participated in the mass shootings of Jews. However, the letters of Holocaust eyewitnesses do not contain such information.

Non-Jewish relatives of murdered Jews who survived the occupation were also eyewitnesses to the Holocaust and authors of wartime letters. For example, the letter of Zinaida Serpik (Kuklina) to her husband's relatives describes the killing of his mother, sister Berty (Beti), and son Dmitrii (Dima) from his first marriage: "this Plague [an expressive term for the Nazis] took many lives while raging in the occupied regions of our country. It took the lives of mother and Beti. Dima probably died too. All Jews, from the newly born to the extremely old, and all the Russian wives of Jews and their children were shot."[97]

Jewish soldiers of the Red Army who liberated Soviet localities also became eyewitnesses to the Holocaust. Moisei Mazur wrote in his letter: "the city is destroyed. All the Jews have been shot, youths have been taken to Germany. I feel sorry for Rostov."[98] The Mountain Jew Manuvakh Dadashev, who also liberated Rostov-on-Don, wrote to his sister, with more insight: "I am happy, and only

---

96 "Pis'mo P. Aksakova rodstvenniku doktora Baumgolts, April 12, 1943," in *Archive of the Holocaust Center*, f. 12, op. 1, d. 66, l. 1–3. See published correspondence about the fate of Dr. Baumgolts: Leonid Terushkin, "'I am the only witness who survived': fragments from the family correspondence," in *Materials of Conference in Riga 2015–2016*, ed. Menahems Barkahans (Riga: Society "Shamir," 2016), 525–528.
97 "Pis'mo Zinaidy Serpik (Kuklinoi) sem'e brata muzha Isaaku Serpiku, March 14, 1943," in *Archive of the Holocaust Center*, f. 5, op. 1, d. 159.
98 "Pis'mo Moiseya Mazura svoemu ottsu Izrailyu Mazuru," April 14, 1943, in Al'tman, *Sokhrani moi pi'sma*, vol. 2, (Moscow: Tsentr i Fond "Kholokost," 2010), 196–197. The quotation from this letter is used in the title of this subchapter as well.

the monstrous traces of the brutality of the Germans poison the joy of victory. They destroyed everything. There were 14,000 Jews shot in the zoological garden of Rostov [Zmievskaya Balka]. I looked at it, clenching my fists in hatred. There were women and children, children, children . . . "[99] Jewish soldiers conveyed the tragedy of their people more viscerally and always specified the ethnicity of the Holocaust victims in their wartime letters.

Another group of authors of letters are representatives of the intelligentsia who wrote about the occupation to their colleagues. The letter of a Stavropol historian and journalist Leonid Polskii (Leonidov-Polskii) to the local historian German Belikov offers a detailed account of everyday life in occupied Stavropol.[100] Notable is the level of education of the author, which allows him to describe the period of occupation in broad terms, and to judge those who were, in his opinion, responsible for the Holocaust:

> Our Soviet authorities in Kislovodsk and Pyatigorsk could have let the Jews know that they cannot stay in one place waiting for the Germans. And then, instead of being shot in anti-tank ditches near Mineralnye Vody, they could have walked through the Zolskii fields to Nalchik, where the Germans stamped around on the spot for a month, modifying the railway, and not conducting military operations. All this somewhat makes one think about another sinister crime of Stalin, who did absolutely nothing to save Jews who were condemned to die![101]

The letter of Polskii is unique as a reflection of a literate person. He offers independent judgement of the war, the Soviet authorities, and the fate of a small person in the flow of history.

Thus, ego-documents of Holocaust eyewitnesses are an important source for reconstructing Holocaust history in a specific locality. They become more relevant in cases where the facts are not well documented in official wartime sources. Personal narratives of Holocaust eyewitnesses though, unlike those of Holocaust survivors, can hardly be used to analyse the work of memory of the narrators. These are stories about events that people heard or saw during the occupation, but did not experience personally. Holocaust eyewitnesses generally shared their stories using simple language, full of details yet lacking in many cases any personal reflection of the witnessed events.

---

99 "Pis'mo Manuvakha Dadasheva kollegam po redaktsii tatskoi gazety 'Zakhmethesh' na adres sestry Milko Dadashevoi, February 18, 1943," in Al'tman, *Sokhrani moi pi'sma*, vol. 3, 203–204.
100 "Pis'mo Leonida Leonidova-Pol'skogo Germanu Belikovu," in *Pod nemtsami: vospominaniya, svidetel'stva, dokumenty: istoriko-dokumental'nyi sbornik*, ed. Kirill Aleksandrov (St. Petersburg: Scriptorium, 2011), 430–465.
101 Ibid, 436.

***

When examining the core information in any kind of personal narrative, one observes that each piece tells the story of a single person in actions involving many others. Each person speaks in his or her own individual voice, which is neither the voice of history nor of collective memory. As Samuel Hynes notes, the span of time in a narrative is not historical but personal, because very few personal narratives begin with the start of the war and end when it ends. They rather tell a part of the whole story, which is that person's individual experience of the war or period of occupation, in the cases of Holocaust survivors and eyewitnesses.[102] The liberation was a meaningful event for Holocaust survivors, yet they relate it to their personal story and not to official history. People only exist, act, and sometimes die in personal narrations. When they die, they do it personally, in private. That is why the language of the narratives, both of Holocaust survivors and Holocaust eyewitnesses is outside the influence of state ideology. It is totally free of the clichés of Soviet officialdom. The war narrative is a personal story, a part of the personal history happening within the historical reality of World War II. Post-war reflections on events that they endured during the war, on the contrary, were formed in the person's mind under the influence of the state propaganda, self-education, imagery, and mass media. Personal narratives are an opportunity for Holocaust survivors and Holocaust eyewitnesses to speak out, which became possible only decades later. The Holocaust survivor Aleksandra Rosina, summarising her testimony for the USCSF, stated: "my memories oppressed me all the time. I had to speak out so that people knew what it was like to survive [the Holocaust] as a pure-blood Jew!"[103]

This important observation is pertinent to most personal narrations. Since Holocaust-related topics were not discussed in the official Soviet memory policy of the war, one was not supposed to speak about it publicly. The untold and unprocessed experience of the majority of Holocaust survivors and eyewitnesses was a traumatic experience, which could be overcome by writing memoirs or giving interviews.[104] Even though personal narratives, primarily oral testimonies, of the Holocaust in the North Caucasus have been collected during past decades, they are not at present widely used at regional level. Most written memoirs and oral testimonies are part of the archival collections of international Holocaust

---

102 Hynes, "Personal Narratives," 219.
103 Interview with Aleksandra Rosina, born 1929, in *VHA USCSF*, code 47599, September 10, 1998, Carmel, Israel.
104 There was no post-war rehabilitation programme in the Soviet Union for Red Army soldiers or Soviet citizens who survived the occupation and the loss of their family.

museums and research centers. With very few exceptions in the online database, Russians have almost no access to them. The written memoirs of Holocaust survivors are usually published in small editions and circulate among members of Jewish communities abroad. Testimonies of Holocaust eyewitnesses are interesting primarily to foreign scholars and organisations. Few survivors and eyewitnesses of the Holocaust attend thematic school and other commemorative events. They pass their memories to the new generations only in the localities where Holocaust memory already exists. Hopefully, personal narratives on the Holocaust in the North Caucasus will one day assume a worthy place in the regional Holocaust memory dispositif.

9 Conclusion. Key Features of the Holocaust Memory Dispositif

In December 2017, the Claims Conference Kagan Fellowship in Advanced Shoah Studies launched a competition for a funded teaching position in Holocaust studies in universities that do not offer Holocaust-related courses in Hungary, Lithuania, Latvia, Ukraine, and Russia. In light of my research on Holocaust remembrance in the North Caucasus, I could appreciate the efforts that remembering communities are making to commemorate victims of Holocaust in the region. The work of representatives of Jewish communities, non-Jewish activists and local historians to preserve remembrance of the Holocaust is visible in every big city and small village of the North Caucasus that I visited during my research trips between 2013 and 2016. However, the regional dimension of the Holocaust memory dispositif, unlike at the state level, is characterised by more complex interactions among its various actors, a diverse composition of remembering communities, and a disjuncture between articulated and unspoken experience. Here the conflict between officially permitted actions, discursive practices and nonverbal experience, on the one hand, and practices of unspoken but implicit action and the semantic meaning of quite formal commemorative events, on the other, is more obvious. At the same time, as my analysis of school curricula and teaching activities demonstrates, other than at the initiative of individual teachers, Holocaust history is almost completely excluded from the classroom. The hegemonic institutional and political memory of World War II differs from the experience of Holocaust survivors and eyewitnesses and the knowledge of local historians.

Encouraged by my research I proposed to the Dean of the History Faculty at a university in southern Russia to participate in the Claims Conference programme and offer courses on Holocaust history in the classroom. Courses on the Holocaust have only recently been introduced at a few central Russian universities.[1] The absolute majority of regional universities continue to implement the Soviet approach

---

[1] For example, Il'ya Al'tman has taught a special class "The Holocaust in the USSR: sources, historiography, memorialisation, and denial" [*Kholokost na territorii SSSR: istochniki, istoriografiya, memorializatsiya, otritsanie*] at the Russian State Humanitarian University for the last several years. The Higher School of Economics offered a class "The Holocaust: the destruction of European Jewry" [*Kholokost: unichtozhenie evropeiskogo evreistva*] in the 2016–2017 academic year. A class "Historical memory in the XX–XXI centuries in Europe: cultural, social, and mental dimensions" was given by professor Thomas Sniegon at the Faculty of History of Moscow State University in the 2017–2018 academic year. "Kholokost:

to studying the history of the "Great Patriotic War." There are no special courses on Holocaust history. Moreover, it is often omitted in study programmes on the general history of World War II. As I was corresponding with the Dean of the History Faculty, I learned that Holocaust history was "largely politicised." Its systematic study, in his opinion, contradicts the fundamental principle of historical practice, which, coupled with a systematic approach, seeks to study the past in its full variety. A separate course on Holocaust history would violate this systematic approach to the study of the history of the war as a whole: "Of course, the Holocaust happened as a historical fact, no one denies it. The death of six million people is terrible. But why is a special term for the death of Jews propagated everywhere? What about the Poles and Soviet people? After all, the Germans killed Jews and Communists. Some were killed for their ethnicity, others for their convictions. Either of these is murder. They are both crimes. But no one speaks about Communists or Gypsies. They only talk about the Jews!"[2]

I considered it inappropriate to pursue this discussion further, since this statement by an experienced academic, teacher, historian, and citizen is more likely to reflect the official position on remembrance of the "Great Patriotic War" in Russia, whose mouthpieces in the provinces are, above all, in the educational system. Moreover, this statement was not to be taken as an invitation to discussion, but as an indisputable assertion, as fact. The story of the Karatygina sisters, which received a huge response in the media, had already demonstrated the weakness of aspects of Russian historical education several years ago. Nevertheless, as the Dean's statements reveal, there has been little visible change in the content of historical education in the regions of the country. The judgements of such experienced historians only strengthen the distance between real experience and ideologically endowed knowledge.

I visited the village of Arzgir, Stavropol Krai in May 2016. This village is located on the eastern outskirts of the North Caucasus almost 200 km from its regional capital, Stavropol. It was founded and populated in 1876 by settlers

---

unichtozhenie evropeiskogo evreistva," *Higher School of Economics official website*, https://www.hse.ru/edu/courses/206066592 (accessed March 24, 2020); "Il'ya Al'tman," *Russian State Humanitarian University official website*, https://www.rsuh.ru/who_is_who/detail.php?ID=119993 (accessed March 24, 2020); "News of the Historical Faculty of Moscow State University," *Moscow State University official website*, http://www.hist.msu.ru/about/gen_news/37943/?sphrase_id=67381 (accessed March 24, 2020).
2 For ethical reasons I do not mention the name and affiliation of the historian. This quotation is given verbatim and taken from my private correspondence with this person in December 2017.

from the central Russian Empire, mostly peasants freed from serfdom in the 1861 reforms.³ Several collective farms were established in the village during the early years of Soviet power. Arzgir became a shelter for evacuees and refugees from the western regions of the country at the beginning of the Great Patriotic War; many of them were Jews. The village was occupied and put under German administration for about five months from August 15, 1942 to January 13, 1943.⁴ The ChGK reported that "in September 1942, according to the order of the German command and gendarmerie, 695 people were killed in Arzgirskaya Balka [. . .] 675 of them were Jewish, 15 Russians (four partisans and the rest civilians), and five Moldovans."⁵ Almost all the evacuated Jews who remained in the occupied village became Holocaust victims. The village remained ethnically Russian, both before and after the war, and no Jewish community has ever existed here.⁶ According to the commemorative tradition of the post-war Soviet state a monument to "peaceful Soviet citizens" was installed at the killing site in Arzgirskaya Balka in 1968. Members of the regional branch of the Komsomol – the political youth organisation of the USSR – raised funds for the monument, which was officially opened on the 50th anniversary of its founding on October 29, 1968.⁷ Since that time, young Pioneers and village Komsomol members held mourning events, called Torchlight Processions, at the monument on the eve of each Victory Day. Yet in addition to the solemn words said in memory of "peaceful Soviet citizens brutally tortured by fascist invaders," the villagers would recall this monument as a Jewish one.⁸ This unofficial knowledge was passed on to younger generations, and those Komsomol members who participated in raising funds for the

---

3 Aleksandr Korobeinikov, ed., *Ocherki istorii Stavropol'skogo kraya*, vol. 1, *S drevneyshikh vremen do 1917 g.* (Stavropol: Knizhnoe izdatel'stvo, 1984), 237.

4 Yurii Shiyanov, ed., *Arz – imya sobstvennoe* (Budenovsk: Budenovskaya tipografiya, 2004), 286.

5 "ChGK report for Arzgir village of July 1, 1943," in *GARF*, f. 7021, op. 17, d. 9, l. 49.

6 According to the results of the 2010 Census, out of almost 15,000 residents of Arzgir, only nine people had stated their Jewish ethnicity. *Upravlenie federal'noi sluzhby gosudarstvennoi statistiki po Stavropol'skomu krayu, KChR i KBR, itogi Vserossiiskoi perepisi naseleniya 2010 goda*, vol. 3 book 1, *Natsional'nyi sostav i vladenie yazykami, grazhdanstvo*, http://stavstat.gks.ru/wps/wcm/connect/rosstat_ts/stavstat/ru/census_and_researching/census/national_census_2010/score_2010/score_2010_default (accessed March 24, 2020).

7 Certificate No. 06-22/04 for the monument "Victims of Fascism, shot by the occupiers in 1942" of January 11, 2013. Provided by the staff of the Arzgir Local History Museum.

8 Interview with Karnaukh; Video recording in the museum "Origins" of Secondary School No.3.

monument's construction later became representatives of the local administration and teachers who led commemorative events at the "Jewish monument" every year. The annual Torchlight Procession continued in the village until the collapse of the USSR.

Anatolii Karnaukh, a native of Arzgir, started conducting historical research after his retirement in 2010. Beginning by seeking his distant relatives, who had participated in World War II, he expanded his activity to identifying the names of all those fellow villagers who had never returned from the battlefield.[9] While working in the GASK he found a ChGK report on the mass killing of the Jews in Arzgirskaya Balka, right at the place where the Komsomol committee had organised Torchlight Processions under his leadership during the Soviet years. This document gave him a new impulse for his search. Karnaukh began to collect the testimony of fellow villagers about the Nazi occupation of the village and the fates of the Jews who had taken refuge there. Cooperation with the Holocaust Center and participation in its educational programmes enabled him to understand Holocaust history better. He was gradually able to restore the names of eight victims of the Holocaust in Arzgirskaya Balka. In 2012, on the 70th anniversary of the tragedy, a memorial plaque inscribed with these names was installed on the Soviet monument.

Being a member of the Russian Union of Journalists, Anatolii Karnaukh frequently publishes articles about the results of his research in the regional press. In 2014, he published a story about the rescue of an eight-year-old Jewish boy by the Palaguta family who raised the boy as their own son. Karnaukh presented this story to Aleksei Palaguta, the head of the municipal administration in Arzgir, since his father was the rescued boy, the protagonist of the story.[10] Anatolii still continues his search for the names of the Holocaust victims in Arzgir and its neighbouring villages. His activity is supported by the local administration and funded by the RJC through the Restore Dignity project. In 2016 a new memorial in Arzgirskaya Balka was opened. Its inscription bears the names of the 19 Holocaust victims identified so far, and states that Arzgirskaya Balka is a Holocaust site.[11] Representatives of the local administration, honorary guests from the Holocaust Center, and schoolchildren attended the opening of the memorial in September 2016. The latter are now participants in both the Torchlight Procession – the tradition was revived in 2015 – and

---

9 Aleksandr Zagainov, "Anatolii Kanaukh iz sela Arzgir zanimaetsya vosstanovleniem imen zemlyakov, pogibshikh i propavshikh vo vremya Velikoi Otechestvennoi," *Stavropol'skaya pravda*, July 17, 2015.
10 Interview with Karnaukh.
11 Klimchenko, "Budem pomnit'! Pomnit' vsegda!"

the March of the Living held on the anniversary of the mass killing of Jews in Arzgirskaya Balka.¹² Anatolii Karnaukh's commemorative activity has now spread throughout Stavropol Krai. By mid-2018, 12 memorials of Holocaust victims had been erected on his initiative, seven of which were designed by him.¹³ According to Karnaukh's data, there are more than 80 mass grave sites of Jews in the Stavropol Krai, only 20 of which are marked with a commemorative sign.¹⁴ The activity of this singular individual from a remote village is highly supported by the regional administration, and together they have done much to commemorate Holocaust victims throughout Stavropol Krai over the past few years.

Anatolii Karnaukh is a regular participant of the thematic evenings with schoolchildren dedicated to the remembrance of Nazi victims. His findings are exhibited in the Arzgir local history museum and the "Origins" museum of the village Secondary School No.3. Several research works were written by schoolchildren based on his findings, and one of them was dedicated to the analysis of his activity as a local historian preserving the memory of Holocaust victims.¹⁵ Meanwhile, Karnaukh continues to conduct research in his native village and designs memorials to be installed at other killing sites in the region. In 2019 a documentary was made about the mass killing of Jews in Arzgirskaya Balka,¹⁶ and Anatolii became one of the main consultants for the film.

These two contrasting stories show the complexity and ambiguity of Holocaust memory in modern Russia. Its presence throughout the post-war Soviet period as experience unspoken in commemorative practices, the offset of semantic meanings in ritual activities, the need to read between the lines of the Soviet press, all resulted in there been no language with which to commemorate Holocaust victims in modern Russia. The term "peaceful citizens" is still present in the inscriptions of memorials at Holocaust killing sites, although some of the names listed clearly evidence the ethnicity of the Nazi victims. The term Holocaust entered Russian discourse only after the collapse of the USSR. The establishment of an eponymous scientific

---

12 "V Rossii prokhodyat 'marshi zhivykh' v chest' ubitykh evreev," *Vesti: Izrael po-russki*, August 11, 2017, https://www.vesty.co.il/articles/0,7340,L-5001586,00.html (accessed March 24, 2020).
13 "Otkrytie memorial'nykh plit," *Novosti Arzgirskogo munitsipal'nogo raiona*, http://arzgiradmin.ru/news.htm (accessed March 24, 2020).
14 "A letter to the governor of the Stavropol Krai from the co-chairman of the Holocaust Center Il'ya Al'tman, January 28, 2016, no. 21," in *personal archive of Anatolii Karnaukh*.
15 Mitenko, "Tragediya v Arzgirskoi Balke"; Evgeniya Pisarenko, "Zvuchi pamyati nabat! (po materialam issledovanii kraeveda s. Arzgir Anatoliya Karnaukha)." The work is provided by Sara Konovalova.
16 *695*, produced by Sergei Miroshnichenko (Studiya Ostrov, 2019), 44 min.

and educational center in Moscow served as an impetus for the intelligentsia, and above all scholars, to use this term in their speeches and research. It took about 20 years for this term to be recognised in the regions of Russia, while many Russians, like the Karatygina sisters, still do not know its meaning. In May 2016, I attended the official opening of a memorial plaque to Holocaust victims appended to the Soviet monument on Mount Koltso near Kislovodsk. I went there along with a delegation of local Jewish community members, elderly Jews whose relatives had become Holocaust victims there or in other places. While speaking to one of the elderly women, I learned a lot about her family and personal history during the war. But when I asked if she considered herself a Holocaust survivor, she simply did not understand the question, since she did not know or use this term. In Russia those Jews who survived the war rarely call themselves Holocaust survivors, increasingly using the more familiar Soviet terminology such as "war veterans," "rear workers," or "children of war." This again shows the complexity and ambiguity of memory: many of these people have received material compensation from numerous European foundations since the late 1990s, precisely for being Holocaust survivors, yet they prefer to use Soviet terminology to identify themselves. These are just some examples of the strong influence of the Soviet ideologisation of the memory of the war in modern Russia, with the canonisation of its heroes and homogenisation of its victims into "Soviet citizens." The appropriate commemorative language for specific Nazi victim groups is only beginning to emerge in Russia.

Analysis of the regional dimension of the Holocaust memory dispositif shows that the relations among its various elements are complex and constantly subject to change under the influence of external and internal factors. They constantly acquire different forms and expressions, being visible or hidden against the background of annual celebrations of Victory Day. If one looked only at official practices of remembering the history of the war in the USSR and Russia, this book would have no subject. It is only in looking "from below" at the informal practices of local people that one can trace the memory of Holocaust victims throughout the Soviet era. The official model of remembering the war was centred on the hero. Commemorative events were held around Graves of the Unknown Soldier and gained all-Union significance since the celebration of the 20th anniversary of the Victory. The construction of memorial complexes to war heroes and "peaceful Soviet citizens," the Eternal Flame, the minute's silence, the Victory Parade on Red Square and in every single Russian locality, the honouring of war veterans, the St. George ribbon, the phrase "Grandfather's victory is my victory" [*Pobeda deda – moya pobeda*], and the recent all-Russian annual "Immortal Regiment" [*Bessmertnyi polk*] campaign have become the dominant symbols of Victory Day for modern generations. They willingly participate in such events, taking them as part of what it means to be

Russian, but sometimes do not even know which victory in which war they are celebrating. Such a state-patriotic model of war remembrance excludes other types of memory and "unpatriotic" stories of the war which lie outwith the official war narrative. The concept of memory dispositif, on the other hand, makes it possible to explore these official narratives as only part, even if a significant and hegemonic part, of the story. It permits one to trace memory from below, to examine connections that arise among the various elements of the dispositif, and which show remembrance of the war as dynamic, fluid, and evolving through time. The ambiguity of the Holocaust memory dispositif is expressed in its coexistence and sometimes in its response to the dominant Soviet memory dispositif of the Great Patriotic War. Holocaust memory dispositif thus becomes a kind of contra-memory with a more developed and complex structure.

Key Features of the Holocaust Memory Dispositif

As this study affirms, the elements of the regional Holocaust memory dispositif are interdependent and influenced not only by internal regional factors, but also reflect the national state's position on Holocaust remembrance and respond to the international commemorative environment. Even studying the 'life of remembrance' from the bottom up, power players, media, and the internationally recognised position on the remembrance of Holocaust victims will inevitably influence the way Holocaust history is remembered at the regional level. Therefore, when identifying key features of the regional dimension of the Holocaust memory dispositif, it is necessary also to describe how it relates to the broader Russian national level. The North Caucasus is geographically, economically, and politically part of modern Russia, just as it was a part of the RSFSR during the Soviet era, so the development of the regional Holocaust memory dispositif occurs within this nationwide memory dispositif.

The main feature of the all-Russian Holocaust memory dispositif is the parallel coexistence of two or many dispositifs. Among them are the grand-memory dispositif of the "Great Patriotic War" with its articulation of the heroism of the Soviet soldier as its main distinctive feature and the memory dispositif of its war victims as "peaceful Soviet citizens," standing in for Holocaust victims, Soviet POWs, Communists and partisans, people with disabilities, victims of Stalinist deportations, and so forth. This grand-memory dispositif has persistently sought to exclude other memory dispositifs throughout its history. An ideological language through which the war could be remembered was formed during the Soviet era, which sought to displace any forms of contra-memory. The heroic war, albeit with tears in the eyes, was the officially remembered one. Nevertheless, other memory

9 Conclusion. Key Features of the Holocaust Memory Dispositif — 293

dispositifs still found a place alongside this official memory dispositif, but always in its shadows. Based on the analysis of various elements of the regional Holocaust memory dispositif in the North Caucasus in the chapters of this work it is possible to single out the main stages of the development of a Holocaust memory dispositif in the USSR and Russia.

During the war years, when the official Soviet language of describing the Great Victory and its main hero the Soviet soldier had not yet formed, the memory of different groups of Nazi victims had already begun to find articulation. The Soviet wartime press did not often use the term "peaceful Soviet citizens," and those who survived the occupation wrote statements and gave testimonies to ChGK members in which they clearly distinguished the Nazi victims, specifically singling out the Jews. Illiterate peasants whose families sheltered many refugees and evacuees, most of whom were Jews, in 1941 became involuntary witnesses to the Holocaust under the occupation. The Nazi policy against the Jews was more than obvious for them, as they reported to the representatives of the liberating Soviet forces in 1943. The relatives of Holocaust victims and local citizens installed modest monuments at the Holocaust sites on their own initiative. Most of the articles and stories in the wartime Soviet media, especially in Yiddish, contained information about mass killings of the Jews. *The Black Book* was prepared and supposed to reflect the fate of the Jews in the occupied regions of the Soviet Union. The famous slogan "Never again!" was the unspoken symbol of this activity to commemorate Jews killed during the war years. This reservoir of lived experience and personal memory would form part of the regional memory dispositif, providing the basis for a contra-memory at odds with the official narrative, through to the present day.

By the end of the war, an official Soviet language for its memorialisation had already been formed. In the context of increasing state antisemitism in the late 1940s, and considering the Soviet interest in homogenising ethnic difference in the name of a Soviet people, it was not acceptable to consider Jews as the main victim group of the Nazis. In this war for the very existence of the Soviet Union, the Soviet people were united as victims, and eventual overthrowers, of the "barbarous German-fascist invaders." *The Black Book* was never published in Russian in the USSR, the JAC was later closed down, and many of its active members became victims of the Stalinist terror. All groups of war victims were united under the universal term "peaceful Soviet citizens." Monumental memorial complexes began to be installed in their honour since the mid 1960s, their size and grandeur giving way only to the memorial complexes to the war's heroes. The Eternal Flame and minute of silence in memory of all the dead, the honour guard and solemn speeches on the significance of the Great Victory became official and integral elements of the celebration of the Victory Day throughout

(Soviet) Russia. Holocaust victims were remembered only by representatives of Jewish communities, often behind the scenes of the main celebration. It was mainly Jews who visited modest monuments or killing sites unidentified throughout the entire Soviet era, not during the Victory celebrations, but on the anniversaries of the particular killing actions. In the 1950s Jews were forbidden from holding commemorative events at Holocaust sites, therefore religious rituals in memory of Holocaust victims were conducted only in the synagogue in most regions. Evgenii Evtushenko's poem Babi Yar, published in *Literaturnaya gazeta* in 1961, provoked wide public discussion, which clearly demonstrated the coexistence of two memory dispositifs in the USSR for the first time: the state victorious grand-memory dispositif and the Holocaust memory dispositif. However the grand-memory dispositif prevailed, and the commemorative activity of Jewish communities to remember Holocaust victims paled in comparison to the increasing military patriotic work, searching for the graves of unknown soldiers, involvement of veterans in the upbringing of younger generations, and the idea that this Victory was the main achievement of the Soviet people during the 20th century. The role of the socialist state in the downfall of National Socialism became more significant during the Cold War, especially in the 1950 and 1960s.

The Holocaust memory dispositif has gradually been incorporated into the official state war narrative since the Perestroika period of the mid-1980s. The subsequent collapse of the USSR made it possible to openly criticise the blunders of the Soviet leadership and research previously silenced facts of Soviet history. Since the early 1990s, the term Holocaust has entered scholarly, and then gradually everyday discourse in Russia. Russia began to present itself at international meetings on the preservation of Holocaust remembrance. It also facilitated the conducting of several thousand interviews with Holocaust survivors in post-Soviet countries for USCSF in the 1990s. Russian scholars have researched the history of the Holocaust on Soviet soil since that time. Revived Jewish communities are active in commemorating the Holocaust. The term "Holocaust by Bullets," introduced in the late 2000s, has begun to be used by the scholarly community to characterise the mass killing of Jews on Soviet soil, with Babi Yar as its main symbol. International Holocaust Remembrance Day, established in 2005 by the General Assembly of the United Nations on the initiative of Russia, alongside other countries, has been widely commemorated throughout the country in the past few years. The term "Holocaust" is mentioned in school history textbooks in the section on the "Great Patriotic War." The internationalisation of Holocaust remembrance gradually acquires a national and local character. The Jewish Museum and Tolerance Center opened in Moscow in 2012. One of the central halls of the permanent exhibition represents the participation of Jews in the war and the fate of Holocaust victims on Soviet soil. A number of documentaries in Russian, and above all the project "Holocaust in the USSR" by Israeli director Boris Maftsir, show the history of the Holocaust on Soviet soil. These are important milestones that demonstrate the gradual incorporation of a Holocaust memory dispositif into the grand-memory dispositif of the Great Patriotic War in modern Russia. However, Holocaust remembrance remains far from an inalienable and equal element of the all-Russian state grand-memory dispositif of World War II.

Another important feature of the Soviet and Russian Holocaust memory dispositif is the overwriting of the historical fact of the mass killings of Jews during the war years in a manner commodious to Soviet officialdom. The very act of the mass killing of Soviet Jews by the Nazis has never been denied in the postwar Russia. Moreover, its memory always existed, but it was encoded in the Soviet commemorative act of remembering all "peaceful Soviet citizens." The functioning of the Soviet judicial system clearly demonstrates the overwriting of the meaning of the historical fact. Dozens of open and closed trials of Soviet collaborators and Nazi perpetrators were held throughout the Soviet era since July 1943; hundreds of Soviet collaborators who served on the side of Nazi Germany during the war years were judged in Soviet courts. They participated in the capture of Jews, escorted them to the execution sites, and some individuals personally shot their victims. However, Soviet courts judged the defendants not for the killing of Jews and other Nazi victims under the occupation, but for their betrayal of the Motherland. The numerous testimonies of Holocaust eyewitnesses – forming the backbone of the prosecution's evidence and attesting directly to the participation of many defendants in the Holocaust – became only trivial considerations when it came to sentencing. As the records of most Soviet trials against Soviet collaborators remain unavailable, this thesis is confirmed by the analysis of the Soviet literary works about open trials. Their authors, as shown in Chapter 5, depict the mass killings of Jews in their pieces, the facts of which were presented as evidence of the defendants' crimes. However, their participation in the killing actions is secondary to the central crime of betraying the Soviet Motherland. Scenes detailing the defendants' participation in the mass killing of Jews in a variety of stories written by Soviet authors merely underline the "beastly nature" of Nazism and individual Soviet collaborators, who defected to the Nazi side and were justly punished by the Soviet judicial system.

Examples of the slip in meaning that occurs when remembering historical fact "in a Soviet way" are also noticeable in commemorative activity and museum exhibitions to the Great Patriotic War. A widespread tendency to install memorial complexes to commemorate the memory of peaceful Soviet citizens in centrally located city locations in the 1960s and 1970s went hand-in-hand with the installing of modest monuments or even commemorative stones at Holocaust sites. Such modest obelisks, following the example of large-scale memorial complexes, were almost always dedicated to "peaceful Soviet citizens" in Soviet Russia. At the same time the residents of each locality knew exactly who was buried at the site. For example, villagers of Arzgir unofficially referred to the Soviet monument "To Victims of Fascism" as the "Jewish monument." Almost every museum exhibition on the history of the Great Patriotic War presents Holocaust artefacts in the section about Nazi occupation, e.g., a copy of the Appeal to the Jewish Population, death certificates of Nazi victims with Jewish names, and Soviet photos of the exhumation of corpses in mass graves. However, these showpieces are silent witnesses of the occupation, rarely entering the guide's narrative. The stories of the heroic battles, partisan movement, and underground leaders figures more prominently in military-patriotic work with young generations than stories of the fates of hundreds and thousands of Holocaust victims. Thus, the Holocaust as historical fact moves to the background of the Soviet grand-memory dispositif of the war, with no scope for critical reflection on its ideological basis and the implications thereof. The very reason for the commemoration of Holocaust victims becomes subordinate to Soviet ideological practices of remembering "peaceful Soviet citizens." The Holocaust gains a different semantic meaning under the influence of Soviet ideology. Ordinary citizens who survived the war referred to it in the Soviet way, yet experienced differently. They could read about "peaceful Soviet citizens" on memorial plaques, in fiction, and the media, but they would still place flowers on the "Jewish monument" and give testimonies about the mass killings of Jews in their native locality.

Continuous usage of Soviet language in Holocaust commemorative events is another distinctive feature of the Russian Holocaust memory dispositif of the post-Soviet period. Inscriptions in Hebrew and Yiddish can coexist with the mention of "peaceful Soviet citizens" on many newly erected memorials at Holocaust sites in modern Russia. This term, coined by Soviet authorities, speaks about all Nazi victim groups without giving priority to any of them. It also does not necessitate a conscious reflection on the causes of the mass killing of any group of victims, especially Holocaust victims. Such a position was beneficial for the Soviet social system, not singling out any victim group according to its social position, class, or ethnicity. Modern Russia prefers to support the heroic concept of war remembrance, which lacks a worthy place for the victims of this war. Otherwise, it would be necessary to revise the heroic grand-memory dispositif of the war and include non-patriotic facts in it, such as the participation of Soviet perpetrators and collaborators in the Holocaust, Stalinist deportations of People of Caucasus, and so on. Most modern history textbooks of the Great Patriotic War pay little attention to the Holocaust compared to the description of other Nazi crimes. The significance of the Holocaust, therefore, often cannot be adequately understood by schoolchildren. One can still hear speeches commemorating "peaceful Soviet citizens" at opening events for new monuments in post-Soviet Russia in the presence of Jewish community members. This happens even if it is already documented that this is the site of the mass killing of people targeted specifically because they were Jews. Many activists in modern Russia grew up in Soviet times. For many of them the use of Soviet language in Holocaust commemorations is a natural and unreflected action.

In today's Russia even Holocaust remembrance attempts to enter the heroic register. This is most obvious in the example of the Pecherskii cult in Rostov-on-Don, and in Russia more widely. Aleksandr Pecherskii, the leader of a successful uprising and mass escape of Jews from Sobibor extermination camp, lived his whole post-war life in Rostov-on-Don.[17] He died a forgotten figure in 1990. During the Soviet period, far from being a hero, he was simply a former POW and prisoner of a camp. Yet this changed in 2016, when President Vladimir Putin posthumously awarded Pecherskii the Order of Courage. A monument devoted to the "Hero of Sobibor" was opened in 2018. A feature film about the uprising, with famous Russian actor Konstantin Khabenskii in the lead role, was made and shown in all cinemas in Russia in 2018, streets in Rostov-on-Don and Moscow and Rostov Secondary School No.52 now bear his name.[18] This is an important milestone on the way to recognising and accept the history of the

---

[17] More about Pecherskii's deeds, see: Semen Vilenskii, Grigorii Gorbovitskii, and Leonid Terushkin, ed., *Sobibor. Vosstanie v lagere smerti* (Moscow: Vozvrashchenie, 2010); Lev Simkin, *Poltora chasa vozmezdiy*a (Moscow: Zebra E, 2013); Selma Leydesdorff, *Sasha Pechersky: Holocaust Hero, Sobibor Resistance Leader, and Hostage of History* (New York and London: Routledge, 2017); Il'ya Vasil'ev and Nikolai Svanidze, *Sobibor. Vozvrashchenie podviga Aleksandra Pecherskogo* (Moscow: Eksmo, 2018).

[18] "Aleksandr Pecherskii nagrazhden ordenom muzhestva," *Holocaust Center official website*, March 1, 2016, http://holocf.ru/ александр-печерский-награжден-орден/ (accessed March 26, 2020); Sergei Kopylov, "Odna iz ulits Moskvy nazvana v chest' Aleksandra Pecherskogo," *Russian Historical Society official website*, https://historyrussia.org/proekty/75-letie-vosstaniya-v-lagere-smerti-sobibor/odna-iz-ulits-moskvy-nazvana-v-chest-aleksandra-pecherskogo.html (accessed March 26, 2020); "V Rostove-na-Donu otkryt pamyatnik Aleksandru Pecherskomu," *News on the official website of the Russian Military-Historical Society*, April 24, 2018, https://rvio.histrf.ru/activities/news/item-4853 (accessed March 26, 2020); *Sobibor*, directed by Konstantin Khabenskii (Russia, 2018), 112 min.

Holocaust in Russia. At the same time, it is again a demonstration of how the Russian leadership continue in the heroic tradition of remembrance established during the Soviet period. Pecherskii is now presented as an heroic Soviet soldier who instigated a successful uprising from a death camp. The Holocaust is thus externalised: shown as somewhere in Europe, far from Russia; and the Soviet solder is presented as trying to prevent it. Speaking about Pecherskii himself, his Jewishness is, if not silenced, at least secondary to his other, military-patriotic attributes. This hero cult fulfils all the contemporary requirements for official remembrance of the "Great Patriotic War." Ironically, its personification in the appropriated figure of Aleksandr Pecherskii permits the history of the Holocaust to be skipped over, and the story of the hundreds of thousands of its victims on Soviet soil and, more important, in Russia, to be ignored.

Along with the main features of the Soviet and Russian Holocaust memory dispositif, the features of its regional dimension can also be highlighted. The North Caucasus is an ethnically diverse region, although one which has never had a significant Jewish presence. Mountain Jews always lived compactly and more or less apart, and Ashkenazi Jews settled in large cities and worked largely in the professions and intelligentsia. During the war, all major Jewish communities were almost completely destroyed, along with Jews who sought refuge or were evacuated to the North Caucasus before it fell under Nazi occupation. After the war, local Jewish communities were either not re-established or were weak. Non-Jewish citizens became the main actors in the regional dimension of the Holocaust memory dispositif. The first modest stones and obelisks were often installed at the initiative of local historians rather than Jewish representatives, who had returned from the evacuation to their home cities and villages. There are enthusiastic local historians in almost every North Caucasian region who, like Anatolii Karnaukh, reconstruct Holocaust history in their particular locality and do their best to commemorate Holocaust victims. Thanks to regional authors like Svetlana Chechulina from the village of Bratskii, memory of the war is filled with new content and meaning, including original and moving literary responses to the Holocaust. Regional school history teachers in cooperation with the Holocaust Center, such as Irina Svetlichnaya from Ust-Labinsk, teach about the Holocaust in the classroom. Thus, not only Jews, but also regional activists and local historians are among the main actors of the regional Holocaust memory dispositif. With their activities they underline the universality of the Holocaust in its human dimension.

Another important feature of the regional dimension of the Holocaust memory dispositif is the strong influence of the Soviet and Russian Orthodox model of remembering different groups of Nazi victims in comparison with the other Russian regions. Even unveiling ceremonies of memorials with Jewish symbolism are often proclaimed as secular and do not respect Jewish religious traditions. This may be a consequence of the low Jewish presence in the North Caucasus. Moreover, Soviet Jews were largely nonreligious, and according to Inna Kalabukhova, wearing a Jewish *yarmulke* [*kippah*] and reading prayers is more a trend than an expression of religious faith. It is important to bear in mind that nearly all of the mass graves of Jews were also the last address for Soviet POWs, communists, Sinti and Roma, and/or people with disabilities. So, speaking of Holocaust sites on Soviet soil without mentioning other groups of Nazi victims does not correspond to the historical reality. Therefore, representatives of Jewish communities installing a monument on the site of mass killing of Jews do sometimes indicate that "Soviet citizens" are buried here. It is possible to hear the prayers of a Russian Orthodox priest more often than those of a Jewish rabbi during commemorative events at Holocaust sites.

The ethnic diversity of the North Caucasus influences the development of a regional Holocaust memory dispositif. The history of the participation of representatives of different ethnicities in the Great Patriotic War is more significant in the region than Holocaust history, of which the victims were mainly newcomers, but not locals. However, if Holocaust memory is officially accepted at the state level, many tragic episodes in the history of the Peoples of the Caucasus and the Cossacks during the Soviet and wartime periods are yet to be acknowledged and re-evaluated. For many Peoples of the Caucasus deported during the war it is still important to pay attention to the heroic side of their participation in the war, which does not contradict, but rather seeks to achieve recognition within, the state grand-memory dispositif of the Great Patriotic War. At the same time, nurturing the memory of war victims – including that of victims of the Stalinist terror – is not a priority for the Peoples of the Caucasus. As long as they live in the Russian Federation, they need to relate their memory policy to the official state position. The idea of the North Caucasus as a multi-ethnic region and the supportive friendship between ethnicities shapes regional policy, which is reflected in the development of the regional Holocaust memory dispositif. On the one hand, there are activists and local historians who support Holocaust remembrance in the region. On the other, regional authorities often prevent such initiatives, trying not to accord remembrance of the victims of war on an ethnic basis. Prevention of interethnic tension is one of the priorities of regional domestic policy, which sometimes exaggerates its real significance. During the lawsuit on the illegal removal of the plaque from the Zmievskaya Balka memorial in Rostov-on-Don, representatives of the city's ethnic communities highlighted that not only Jews were buried there.

Unwillingness to accept Holocaust history among regional authorities is noticeable in those regions of the North Caucasus where antisemitism was historically strong. For instance, there is still no memorial at Holocaust sites in Krasnodar Krai with an inscription dedicated to Jews. Local historians do research and find evidence of the Holocaust in their native localities, but they would still commemorate the memory of "peaceful Soviet citizens." The unwillingness of the region's authorities to accept the Holocaust as a historical fact leads to such extreme manifestations as the installation of Orthodox crosses in sites where only Jews were murdered in Novopokrovskii District, Krasnodar Krai. Most Holocaust memorials installed in the post-Soviet period are not under the protection of the relevant regional or city organisations; therefore, no one is responsible for them. Memorials with Jewish symbols or inscriptions are constantly subjected to acts of vandalism both in the North Caucasus and throughout Russia. In this case regional authorities use mainly the strategy of preventing any initiative that draws attention to ethnically marked historical facts instead of educating the younger generation about tolerance. As a result of such a hostile environment, Yurii Teitelbaum, the former chairman of the Krasnodar Jewish community, has expatriated to Israel, and the Dean of the History Faculty, mentioned above, does not let the Holocaust enter the classroom. Nevertheless, the defensive and intransigent position of local authorities cannot completely eliminate Holocaust remembrance. It is well known in Russia that "water wears away stones" [*voda kamen' tochit*]. Anatolii Karnaukh's initiative to commemorate Holocaust victims in one village, for instance, spread across the whole of Stavropol Krai five years later, and a previously wary regional administration supported the erection of a renewed monument near Urochishche Stolbik in Stavropol in 2017.

Another distinguishing feature of the regional dimension of the Holocaust memory dispositif is that despite a wealth of personal narratives of Holocaust survivors and Holocaust eyewitnesses, they are rarely used in educational programmes and commemorative events in the region. They form part of the international archives of Holocaust museums and scientific centers and remain almost inaccessible in the North Caucasus. The voices of Holocaust survivors are the most important evidence of the Holocaust, and are widely used in research and educational programmes in the Western world. Practically all major oral history projects with Holocaust survivors and Holocaust eyewitnesses were conducted by international research centers, such as USCSF, USHMM, and Yahad-In Unum, and are not part of the Russian commemorative landscape. Likewise, written memoirs of Holocaust survivors circulate among representatives of Jewish communities in exile, remaining unknown to regional (non-Jewish) readers. In order to study the regional Holocaust memory dispositif, one needs not only to come to the region, but also to work in international archives. Such an outward orientation of the regional Holocaust memory dispositif makes Holocaust history in the North Caucasus available for study by the international scholarly community, but at the same time limits its development within the region itself.

Thus, the existence of Holocaust remembrance in Soviet and present-day Russia cannot be reduced to an analysis of vertical relations of power and society or horizontal relations among remembering communities. It is made more complex by the competition over the memorialisation of heroes and victims of the war and between different groups of victims. In the North Caucasus, this competition is complicated by the unprocessed experience of other significant traumas of the history of the war and of the 20th century in general. In this sense, Holocaust remembrance in the region is a very mobile dispositif, with a complex arrangement of actors. It has been shaped throughout the post-war period, influenced by both internal (state ideology, personal memory of Holocaust survivors and Holocaust eyewitnesses, the research of local historians, and members of Jewish communities) and external factors (adoption of international laws on commemorating Holocaust victims, activities of international organisations).

Unlike most countries of the Western world and Israel, Holocaust remembrance in Russia is being formed afresh, re-constructed through grassroots engagement. While still hegemonic, the top-down schema of official memory culture are set into dialogue with the on-the-ground efforts of people to establish facts, preserve remembrance, and tell diverse and different stories. This is why the study of the internal connections between various actors in this process, and the decoding of individual verbal and nonverbal acts of commemoration that would otherwise remain invisible to the outsider, requires a broad theoretical toolkit. The concept of "memory dispositif" is a useful addition to this toolkit, helping to understand the nature of Holocaust remembrance in Soviet and post-Soviet Russia. It makes it possible to trace the complex of extensive links among its individual elements in an historical perspective, to trace the development, evolution, and gradual emergence into the limelight of Holocaust memory as a counter-history, the elements of which have at times been variously included or excluded from the larger state-authorised grand-memory dispositif of the "Great Patriotic War." In short, it allows for the re-constructing of Holocaust memory in the Russia in general, and the North Caucasus in particular.

Bibliography

Unpublished Sources

Archive of Kislovodsk "Fortress" Local History Museum
    f. 108 – Monuments in Kislovodsk.
Archive of Rostov-on-Don VOOPIiK.
Archive of the school competition "Chelovek v istorii. Rossiya – XX vek" of the society "Memorial," Moscow, Russia.
Author's archive, Berlin, Germany
    Project MEM – Oral history interviews with the members of Jewish communities, schoolteachers, activists, and scholars in the North Caucasus;
    Project X – Oral history interviews with Holocaust survivors in the North Caucasus;
    Project F – Oral history interviews with artists.
BAL, Ludwigsburg, Germany
    B 162 – Zentrale Stelle der Landesjustizverwaltungen.
Fortunoff Video Archive for Holocaust testimonies at the Yale University, USA
    Interviews with Holocaust survivors.
GAKK, Krasnodar, Russia
    f. R-487 – German city govermnent in Tikhoretsk (1942–1943);
    f. R-687 – Executive Committee of the Krasnodar Regional Council of Peoples' Deputies;
    f. R-897– ChGK records for Krasnodar Krai.
GANISK, Stavropol, Russia
    f. 1 – Stavropol Krai Committee of the VKP(b).
GARF, Moscow, Russia
    f. A-327 – The main resettlement administration under the Council of Ministers of the RSFSR and its predecessors (1942–1956);
    f. R-6991– Council on religions at the council of the ministers of the USSR, 1943–1991;
    f. R-7021 – ChGK records for Russian Federation:
        op. 12 – for North Ossetian ASSR,
        op. 16 – for Krasnodar Krai,
        op. 17 – for Stavropol Krai,
        op. 40 – for Rostov Oblast,
        op. 116 – ChGK paperwork,
        op. 148 – German captured records;
    f. R-8114 – JAC;
    f. R-8131 – Prosecutor's office of the USSR.
GARO, Rostov-on-Don, Russia
    f. R-3613 – ChGK records for Rostov Oblast;
    f. R-4096 – City Administration of Culture of the Rostov-on-Don Executive Committee of the City Council of Peoples' Deputies, Rostov-on-Don, 1945–1989.
GASK, Stavropol, Russia
    f. R-1368 – ChGK records for Stavropol Krai.
Holocaust Center Archive, Moscow, Russia
    f. 5 – Collection of letters;

f. 12 – Doctor Baumgolt's personal collection;
f. 46 – Arthur Khavkin's personal collection.

NA of the IRI RAN, Moscow, Russia
f. 2, section VI – Occupation period in the North Caucasus;
f. 2, section XIV – Papers of the Commission for the History of the Great Patriotic War.

OHD, Jerusalem, Israel
106 – The Mountain Jews of the USSR;
199 – Non-Ashkenazi Jews in the USSR;
202 – Evacuation and flight of Jews in the USSR during World War II;
217 – Jewish newcomers from the USSR.

Personal archives of
Anatolii Karnaukh, Arzgir village, Russia;
Irina Svetlichnaya, Ust-Labinsk, Russia;
Nadezhda Suvorova, Krasnodar, Russia;
Sergei Shaitanov, Simferopol, Russia;
Yurii Teitelbaum, Bat-Yam, Israel.

RGAE, Moscow, Russia
f. 15A – The national composition of the population in the republics, krais and oblasts of the USSR;
f. 1562, op. 336 – Summary tables of the main results of the 1939 census.

RGAKFD, Krasnogorsk, Russia
Wartime chronicle and newsreels about the occupation and liberation of the North Caucasian localities.

RGASPI, Moscow, Russia
f. 17 – TsK CPSU.

StA München, Munich, Germany
Staatsanwaltschaft, 21672 – 22Js 201/61- Investigations against members of SK 10a;
Staatsanwaltschaft, 35275 – 114Js 117/64 – Investigations against members of SK 10a;
Staatsanwaltschaft, 35280 – 22Js 206/61- Investigations against members of EK 12;
Staatsanwaltschaft, 35308 – 22Js 202/61- Investigations against members of SK 10a.

TsAMO, Podolsk, Russia
f. 32 – GlavPU KA collection.

TsDNIKK, Krasnodar, Russia
f. 1072 – Krasnodar City Committee of the VKP(b);
f. 1774-A – Krasnodar Krai Committee of the CPSU.

TsDNIRO, Rostov-on-Don, Russia
f. R-9 – Rostov-on-Don City Committee of the VKP(b):
f. R-1886 – Rostov Oblast Committee of the CPSU.

USHMMA, Washington D.C., USA
RG-06.025 – Central Archives of the Federal Security Services (FSB, former KGB) of the Russian Federation records relating to war crime trials in the Soviet Union, 1939–1992 (bulk dates 1945–1947);
RG-22.002M – Extraordinary State Commission to Investigate German-Fascist Crimes Committed on Soviet Territory from the USSR, 1941–1945;
RG-22.016 – Reports and investigative materials compiled by the Military Commissions of the Red (Soviet) Army related to the crimes committed by the Nazis and their

collaborators on the occupied territories of the Soviet Union and Eastern Europe during WWII, 1942–1945;

RG-22.028M – Records of the Jewish Anti-Fascist Committee from the State Archives of the Russian Federation (GARF), Fond 8114, opis 1, 1942–1948;

RG-50 – The Jeff and Toby Herr Testimony Initiative, a multi-year project to record the testimonies of non-Jewish witnesses to the Holocaust.

VHA USCSF, Los Angeles, USA

Collection of oral interviews with Holocaust survivors and witnesses "The European Holocaust";

Digital and remote storage.

YVA, Jerusalem, Israel

M.33 – Records of the Extraordinary State Commission to Investigate German-Fascist Crimes Committed on Soviet Territory;

M.40 – Central Archives in Russia;

O.3 – Yad Vashem Collection of Testimonies;

O.33 – Collection of various testimonies, diaries and memoirs;

O.75 – Letters and Postcards Collection;

P.21 – Ilya Ehrenburg Collection.

Yahad-In Unum Archive, Paris, France

Oral testimonies with Holocaust eyewitnesses in the North Caucasus.

Primary Published Sources

Aleksandrov, Kirill, ed. *Pod nemtsami: vospominaniya, svidetel'stva, dokumenty: istoriko-dokumental'nyi sbornik.* St. Petersburg: Skriptorium, 2011.

Al'tman, Il'ya, ed. *Kholokost na territorii SSSR: entsiklopediya.* Moscow: ROSSPEN, Nauchno-prosvetitel'nyi Tsentr "Kholokost," 2009.

Al'tman, Il'ya, ed. *Neizvestnaya "Chernaya kniga:" materialy k "Chernoi knige."* Moscow: AST: CORPUS, 2015.

Al'tman, Il'ya, and Terushkin, Leonid, ed. *Sokhrani moi pis'ma . . .: sbornik pisem evreev perioda Velikoi Otechestvennoi voiny.* 5 vol. Moscow: Tsentr i Fond "Kholokost," 2010–2019.

Altshuler, Mordechai, Arad, Yitshak, and Krakowski, Shmuel, ed. *Sovetskie evrei pishut Il'e Ehrenburgu, 1943–1966.* Jerusalem: Yad Vashem, 1993.

Angrick, Andrei, et al., ed. *Deutsche Besatzungsherrschaft in der UdSSR 1941–1945, Dokumente der Einsatzgruppen in der Sowjetunion.* Vol. 2. Darmstadt: Wissenschaftliche Buchgesellschaft, 2013.

Arad, Yitzhak, Krakowski, Shmuel, and Spector, Shmuel. *The Einsatzgruppen Reports: Selections from the Dispatches of the Nazi Death Squads' Campaign against the Jews in Occupied Territories of the Soviet Union July 1941 – January 1943.* New York: Holocaust Library, 1989.

Arad, Yitzhak, ed. *Unichtozhenie evreev SSSR v gody nemetskoi okkupatsii (1941–1944): sbornik dokumentov i materialov.* Jerusalem: Yad Vashem, 1992.

Arad, Yitzhak, ed. *Neizvestnaya "Chernaya kniga."* Moscow – Jerusalem: Tekst, 1993.

Arad, Yitzhak, ed. *Unknown Black Book: Evidence of Eyewitness Accounts of the Catastrophe of Soviet Jews (1941–1944)*. Jerusalem – Moscow: Yad Vashem, 1993.

Avramenko, Anatolii, et al. *Ekaterinodar-Krasnodar: dva veka goroda v datakh, sobytiyakh, vospominaniyakh [1793–1993]: materialy k letopisi*. Krasnodar: Knizhnoe izdatel'stvo, 1993.

Belokon', Violetta, et al., ed. *Golosa iz provintsii: zhiteli Stavropol'ya v 1941–1964 godakh: sbornik dokumentov*. Stavropol: Komitet Stavropol'skogo kraya po delam arkhivov, 2011.

Belyaev, Aleksandr, et al., ed. *Krasnodarskii krai v 1937–1941 gg.: dokumenty i materialy*. Krasnodar: "Edvi," 1997.

Belyaev, Aleksandr, and Bondar', Irina, ed. *Kuban' v gody Velikoi Otechestvennoi voiny: 1941–1945: rassekrechennye dokumenty: khronika sobytii*. Vol. 1. *Khronika sobytii 1941–1942 gg*. Krasnodar: Sovetskaya Kuban', 2000.

Bloknot agitatora Krasnoi armii. Moscow: Voenizdat, 1942–1945.

Bloknot agitatora voenno-morskogo flota SSSR. Moscow: Voenmorizdat, 1942–1945.

Boiko, Sergei, ed. *Stavropol'e v Velikoi Otechestvennoi voine 1941–1945 gg: sbornik dokumentov i materialov*. Stavropol: Knizhnoe izdatel'stvo, 1962.

Die Verfolgung und Ermordung der Europäischen Juden durch das Nationalsozialistische Deutschland 1933–1945. Vol. 7. Hoppe, Bert, and Glass, Hildrun, ed. *Sowjetunion mit Annektierten Gebieten I. Besetzte Sowjetische Gebiete unter Deutscher Militärverwaltung, Baltikum und Transnistrien*. München: Oldenbourg Verlag 2011.

Dokumenty obvinyayut: sbornik dokumentov o chudovishchnykh zverstvakh germanskikh vlastei na vremenno zakhvachennykh imi sovetskikh territoriyakh. 2 vols. Moscow: Gospolitizdat, 1943, 1945.

Grossman, Vasilii, and Erenburg, Il'ya, ed., *Chernaya kniga*. Moscow: Ast: Korpus, 2015.

Kononenko, Elena. *Pered sudom naroda*. Moscow: OGIZ, Gospolitizdat, 1943.

Kulaev, Chermen, ed. *Narody Karachaevo-Cherkessii v gody Velikoi Otechestvennoi voiny (1941–1945 gg.): sbornik dokumentov i materialov*. Cherkessk: Stavropol'skoe knizhnoe izdatel'stvo, Karachaevo-Cherkesskoe otdelenie, 1990.

Levendorskaya, Lyudmila, ed. *Zaveshchano pomnit' . . . donskie arkhivy – 70-letiyu Velikoi Pobedy*. Rostov-on-Don: Al'tair, 2015.

Maiorov, Sergei, ed. *Vneshnyaya politika Sovetskogo Soyuza v period otechestvennoi voiny*. Vol. 1. *22 iyunya 1941 g. – 31 dekabrya 1943 g.* Moscow: Gospolitizdat, 1946.

Mallmann, Klaus-Michael, et al., ed. *Deutsche Berichte aus dem Osten. Dokumente der Einsatzgruppen in der Sowjetunion*. Vol. 3. Darmstadt: Wissenschaftliche Buchgesellschaft, 2014.

Naselenie SSSR: 1973: statisticheskii sbornik. Moscow: Statistika, 1975.

Pamyatka agitatora. Moscow: Gospolitizdat, 1943.

Robinson, Jacob, and Sachs, Henry, ed. *The Holocaust: The Nuremberg Evidence; Digest, Index and Chronological Tables*. Jerusalem: Yad Vashem, 1976.

Rubenstein, Joshua, and Altman, Ilya, ed. *The Unknown Black Book: The Holocaust in the German-Occupied Soviet Territories*. Bloomington and Indianapolis: Indiana University Press, 2008.

Rudenko, Roman, ed. *Nyurnbergskii protsess nad glavnymi nemetskimi voennymi prestupnikami: sbornik materialov*. 7 vols. Moscow: Gosyurizdat, 1957–1961.

Shekikhachev, Mukhamed, ed. *Kabardino-Balkariya v gody Velikoi Otechestvennoi voiny 1941–1945 gg.: sbornik dokumentov i materialov*. Nalchik: Elbrus, 1975.

Sklyarova, Valentina, ed. *Kuban', opalennaya voinoi (o zhertvakh i zlodeyaniyakh zakhvatchikov na territorii Krasnodarskogo kraya, vremenno okkupirovannoi v 1942–1943 gg.)*. Krasnodar: Periodika Kubani, 2005.
SSSR. *Administrativno-territorial'noe delenie soyuznykh respublik na 1 maya 1940 goda*. Moscow: Izdatel'stvo vedomostei verkhovnogo soveta RSFSR, 1940.
SSSR. *Chrezvychainaya gosudarstvennaya komissiya po ustanovleniyu i rassledovaniyu zlodeyanii nemetsko-fashistskikh zakhvatchikov: sbornik soobshchenii chrezvychainoi gosudarstvennoi komissii o zlodeyaniyakh nemetsko-fashistskikh zakhvatchikov*. Moscow: Gospolitizdat, 1946.
Sudebnyi protsess po delu o zverstvakh nemetsko-fashistskikh zakhvatchikov i ikh posobnikov na territorii g. Krasnodara i Krasnodarskogo kraya v period ikh vremennoi okkupatsii. Moscow: Gospolitizdat, 1943.
Sverdlov, Fedor. *Dokumenty obvinyayut: Kholokost: svidetel'stva Krasnoi armii*. Moscow: Tsentr "Kholokost," 1996.
The People's Verdict: A Full Report of the Proceeding at the Krasnodar and Kharkov Nazi Atrocity Trials. London – New York – Melbourne: Hutchinson & Co, 1944.
Veselov, Vladimir, ed. *Zverstva nemetsko-fashistskikh zakhvatchikov: dokumenty*. Vol. 2,6,7. Moscow: Voenizdat, 1942–1943.
Vodolazhskaya, Valeriya, et al., ed. *Stavropol'e v period nemetsko-fashistskoi okkupatsii (avgust 1942 – yanvar' 1943 gg.): dokumenty i materialy*. Stavropol: Knizhnoe izdatel'stvo, 2000.
Yudin, Ivan. *Sledy fashistskogo zverya na Kubani*. Moscow: Gospolitizdat, 1943.
Zagorul'ko, Maksim, ed. *Voennoplennye v SSSR. 1939–1956: Dokumenty i materialy*. Moscow: Logos, 2000.

Diaries and memoirs

Benjamin, Walter. *Moskauer Tagebuch*. Frankfurt am Main: Suhrkamp Verlag, 1980.
Borovoi, Saul. *Vospominaniya*. Moscow-Jerusalem: Hebrew University in Moscow, 1993.
Danilova, Svetlana, ed. *Deti voiny: Vospominaniya*. New-York: Design-ER, 2015.
"Dnevnik Sheyny Gram (22 iyunya – 8 avgusta 1941 goda)". In *Evrei v menyayushchemsya mire: materialy 8-i mezhdunarodnoi konferentsii, Riga, 30 iyunya – 4 iyulya 2011*, edited by German Branover and Ruvin Ferber, 90–96. Riga: Latviiskii universitet, 2015.
Dimitrov, Georgii. *Selected Works*. Vol. 2. Sofia: Sofia Press, 1972.
Fridman, Filipp. *Gibel' l'vovskikh evreev*, http://samlib.ru/h/hohulin_aleksandr_wasilxewich/gibelxlxwowskihewreew.shtml.
Khveister, Aleksei, ed. *"Kto vyzhivet – rasskazhet": vospominaniya poslednikh svidetelei Kholokosta*. Wismar: Verlag Koch und Raum Wismar, 2015.
Lazerson-Rostovskaya, Tamara. *Zapiski iz Kaunasskogo getto (katastrofa skvoz' prizmu detskikh dnevnikov)*. Moscow: Vremya, 2011.
Linkimer, Kalman. *Nineteen Months in a Cellar: How 11 Jews Eluded Hitler's Henchmen: The Diary of Kalman Linkimer (1913–1988)*. Translated by Rebecca Margolis. Riga: Jewish Community of Riga: Museum "Jews in Latvia," 2008.
Poppe, Nicholas. *Reminiscences*, edited by Henry G. Schwarz. Bellingham, Wash.: Western Washington University, 1983.
Rolkiteite, Maria. *Ya dolzhna rasskazat'*. Leningrad: Soviet writer, 1976.

Rolnikaite, Masha. *Ikh muz dertseyln*. Warsaw-Moscow: Yidish buch, 1965.
Rozenfeld, Lyubov', ed. *Voennoe detstvo: sbornik rasskazov detei, nynche zhitelei Ashkelona (Izrail'), perezhivshikh Velikuyu Otechestvennuyu voinu*. Ashkelon – Tel-Aviv, 2015.
Schramm, Percy E., ed. Kriegstagebuch des Oberkommandos der Wehrmacht (Wehrmachtführungsstab), 1 Januar 1942 – 31 Dezember 1942. Holocaust Impiety in Literature: Bernard und Graefe, 1982.
Saenko, Nikolai. "Taganrogskii dnevnik (2 oktyabrya 1941 g. – 1 sentyabrya 1943 g.)." In *"Nam zapretili belyi svet . . . ": Al'manakh dnevnikov i vospominanii voennykh i poslevoennykh let*, edited by Pavel Polyan and Nikolai Pobol', 25–113. Moscow: ROSSPEN, 2006.
Vyzhit' i pomnit': vospominaniya byvshikh uznikov getto i kontslagerei – zhitelei g. Karmielya, Izrail'. Carmiel, 2014.

Literary Works

Amiramov, Efrem. *U obeliska*. https://utices.com/page-text.php?id=20199.
Belyaev, Aleksandr, et al., ed. *Bez grifa "sekretno": iz istorii organov gosbezopasnosti na Kubani: ocherki, stat'i, dokumental'nye povesti*. Krasnodar: Sovetskaya Kuban', 2005.
Bodalov, Roman. *Andzhi-name: rasskazy o gorskikh evreyakh*. Tel-Aviv, 2004.
Burakovskii, Yurii. *Vremya dozhdyu razlivat'sya rekoi: stikhi*. Tel-Aviv: Dfus Shakhaf, 2004.
Chechulina, Svetlana. *Okkupatsiya: dokumental'naya povest' v stikhakh o sobytiyakh 1941–1943 gg. v khutore Bratskom Ust-Labinskogo raiona i ego okrestnostyakh*. https://www.stihi.ru/2012/08/15/7764.
Chistyakov, Nikolai. *Pamyat' serdtsa: leningradskie cherkesy Besleneya*. Saratov: OOO "Privolzhskoe izdatel'stvo," 2012.
Erenburg, Il'ya. *Burya: roman*. Magadan: Sovetskaya Kolyma, 1947.
Evtushenko, Evgenii. "Babii Yar". *Literaturnaya gazeta*. September 19, 1961.
Evtushenko, Evgenii. "Storozh Zmievskoi Balki". *Komsomolskaya pravda*. December 14, 2014.
Ginzburg, Lev. *Bezdna. Povestvovanie, osnovannoe na dokumentakh*. Moscow: Sovetskii Pisatel', 1967.
Glebova, Lina. *Kogda ya snova budu . . .* Moscow: Molodaya gvardiya, 1970.
Gneushev, Vladimir. "Klavdiya Il'inichna: dokumental'naya povest'." In *Vo imya zhizni: o podvigakh Stavropol'tsev na fronte i v tylu Velikoi Otechestvennoi voiny*, 254–298. Stavropol: Knizhnoe izdatel'stvo, 1987.
Gofman, Genrikh. "Dos iz Geshen in Taganrog" [in Yiddish]. Translated by Moisei Itkovich. *Sovetish Heymland* 2 (1966): 119–124.
Gofman, Genrikh. *Geroi Taganroga: dokumental'naya povest'*. Moscow: Molodaya Gvardiya, 1966.
Grossman, Vasilii. *Treblinskii ad*. Moscow: Voennoe izdatel'stvo, 1945.
Grossman, Vasilii. *Zhizn' i sud'ba: roman*. Kuybyshev: Knizhnoe izdatel'stvo, 1990.
Guzenko-Vesnin, Gennadii. *Ukradennaya iz ada*. https://www.proza.ru/2015/06/23/1286.
Itskovich, Saul. *Musya Pinkenson*. Moscow: Malysh, 1981.
Kalabukhova, Inna. *V proshchan'i i v proshchen'i*. Moscow: Dom evreiskoi knigi, 2008.
Karnaukh, Anatolii. *Neveroyatnoe spasenie mal'chika Vani*. http://www.proza.ru/2014/08/21/2040.
Khentov, Igor'. *Vzglyad: stikhi rasnykh let i novelly*. Vol. 1. Rostov-on-Don: Lunkina N.V., 2011.
Khentov, Igor'. *Rodoslovnaya*. Kyiv: Izdatel'stvo Dmitriya Yuferova, 2017.

Kolyabin, Vladimir. *Predatel'stvo*. https://www.proza.ru/2014/11/30/182.
Krut, Yakov. *Povest' o podarennoi zhizni*. Petakh-Tikva, 2009.
Kuznetsov, Anatolii. *Babii Yar: roman-dokument*. Moscow: Molodaya gvardiya, 1967.
Rozhdestvenskii, Robert. "Requiem". Translated by Walter May. *The Anglo-Soviet Journal*. September (1968): 11–14.
Rubin, Rivke. *Jidishe Froyen* [in Yiddish]. Moscow: Ogiz, 1943.
Rybakov, Anatolii. *Tyazhelyi pesok: roman*. Moscow: Sovetskii pisatel', 1979.
Selvinskii, Il'ya. "Ya eto videl!" *Bol'shevik*. January 23, 1942.
Selvinskii, Il'ya. *Krym: Kavkaz: Kuban': stikhi*. Moscow: Sovetskii pisatel', 1947.
Shklovskii, Viktor." Nemtsy v Kislovodske." In *Chernaya kniga*, edited by Vasilii Grossman and Il'ya Erenburg, 293–297. Moscow: AST: CORPUS, 2015.
Smirnov, Vladislav. *Rostov pod ten'yu svastiki*. Rostov-on-Don: OAO "Rostovkniga," 2006.
Velikanov, Vasilii. "Ranenaya skripka." In *Put' otvazhnykh: rasskazy*, edited by Sergei Baruzdin, 101–110. Moscow: Detgiz, 1962.

Films (Feature Films, Documentaries, and Television News)

Beslenei. Pravo na zhizn'. Directed by Vyacheslav Davydov. "TONAP" film studio, 2008.
Deti nashego aula. Directed by Yurii Gavrilov. Pyatigorskoe TV, 1970.
Glebova, Anna, and Yakovlev, Konstantin. "'Zabytye" – Vystavka pod takim nazvaniem otkrylas' v Rostove-na-Donu. Don Pravoslavnyi TV. Aired August 23, 2017.
Holocaust: The Eastern Front. Directed by Boris Maftsir, 2016.
Judenfrei: Svobodno ot evreev. Directed by Yurii Kalugin. "Dontelefil'm," 1992.
Kholokost – Klei dlya oboev? Directed by Mumin Shakirov. Bakteria-film, 2013.
Nepokorennye. Directed by Mark Donskoi. Kievskaya kinostudiya, 1945.
Obyknovennyi fashizm. Directed by Mikhail Romm. Kinostudiya "Mosfil'm," 1965.
Odna rodnya. Directed by Marina Sasikova. ORTK Nalchik, 2005.
Otomstim. Directed by Nikolai Karamzinskii. Tsentral'naya kinostudiya dokumental'nykh fil'mov, 1942.
Po sledam fashistskogo zver'ya. Directed by Shalva Chagunava. Tbilisskaya kinostudiya, tsentral'naya kinostudiya dokumental'nykh fil'mov, 1943.
Potana, Elena. "Otkrytie muzeya Sabiny Shpil'rein." *REN TV Rostov*. Aired November 11, 2015.
Prigovor naroda. Shot by the film crew of the North Caucasus Front. Tsentral'naya studiya kinokhroniki, July 1943.
Prigovorennye k zabveniyu. Directed by Marat Duraev and Anton Stepanenko. 2019.
Proskurina, Aleksandra. "Istoriki Ust-Labinska rasskazali o podvige evreiskogo mal'chika-skripacha." *TV channel "Kuban' 24."* Aired April 16, 2015.
Rai. Directed by Andrei Konchalovskii. DRIFE Productions and Prodyuserskii tsentr Andreya Konchalovskogo, 2016.
Semestr, kotorogo ne bylo. Okkupatsiya. Directed by Andrei and Igor' Kartashevy. Stavropol: tvorcheskaya masterskaya "Brat'ya Kartashevy," 2018.
Skripka pionera. Directed by Boris Stepantsev, written by Yurii Yakovlev, camera Mikhail Druyan. Soyuzmul'tfil'm, 1971.
Sobibor. Directed by Konstantin Khabenskii. Russia, 2018.
"Sobytiya. Vremya mestnoe." *Kanevskaya Televisiya*. Aired May 13, 2015.

Soyuzkinozhurnal no. 9. Directed by Irina Setkina, Z. Dembovskaya. Tsentral'naya kinostudiya dokumental'nykh fil'mov, 1943.
Soyuzkinozhurnal no. 114. Directed by Arkadii Levitan, Georgii Popov, and Andrei Sologubov. Tsentral'naya kinostudiya dokumental'nykh fil'mov, 1941.
Stradanie pamyati. Directed by Yurii Kalugin. St. Petersburg, 2013.
Tufel'ki. Directed by Konstantin Fam. Costa Fan Production, 2012.
Tyazhelyi pesok: TV series. Directed by Anton and Dmitrii Barshchevskii. Kinokompaniya "Risk," 2008.
Voina i mir Karachaya. Directed by Vyacheslav Davydov. Kinostudiya Tonap, 2011.
Ya rodom iz detstva: dokumental'nyi fil'm. Directed by Nadezhda Popova. Studio "S.S.S.R.," 2013.
Zhertvy gitlerovtsev v Pyatigorske. Cameramen Andrei Sologubov and Jakov Avdeenko. Tsentral'naya kinostudiya dokumental'nykh fil'mov, 1943.

Music

Klebanov, Dmitrii. *Symphony no. 1 "Pamyati muchenikov Bab'ego Yara."* 1945.
Levin, Igor'. *Poem "Bol' Zemli for symphony orchestra, mixed and children's choruses."* 2014.
Shostakovich, Dmitrii. *Symphony no. 13*. 1962.
Stankovich, Evgenii. *Kadish-rekviem "Babii Yar."* 1991.

History Textbooks and Teaching Manuals

Achmiz, Kazbek, ed. *Istoriya Adygei: uchebnoe posobie dlya 9 klassa obshcheobrazovatel'nykh uchrezhdenii*. Maykop: Adygeiskoe respublicanskoe knizhnoe izdatel'stvo, 2002.
Al'tman, Il'ya, Gerder, Alla, and Poltorak, David. *Istoriya Kholokosta na territorii SSSR: uchebnoe posobie dlya 9 klassa*. Moscow: Fond "Kholokost," 2001.
Al'tman, Il'ya, Gerber, Alla, and Prokudin, Dmitrii, ed. *Kholokost v russkoi literature: sbornik urokov i metodicheskikh rekomendatsii*. Moscow: Tsentr i fond "Kholokost," 2006.
Al'tman, Il'ya, ed. *Prepodavanie temy Kholokosta v XXI veke*. Moscow: Fond "Kholokost," 2000.
Al'tman, Il'ya, ed. *Kholokost i evreiskoe soprotivlenie na okkupirovannoi territorii SSSR: uchebnoe posobie dlya studentov vysshikh uchebnykh zavedenii*. Moscow: Fond "Kholokost," 2002.
Al'tman, Il'ya, ed. *Kholokost: vzglyad uchitelya: metodicheskoe posobie – sbornik rabot pedagogov Rossii*. Moscow: Tsentr i Fond "Kholokost," 2006.
Al'tman, Il'ya, ed. *Tema Kholokosta v shkol'nykh uchebnikakh: posobie dlya uchitelya*. Moscow: Tsentr i Fond "Kholokost," 2010.
Anisina, Natalia, ed. *Uroki Kholokosta i narushenie prav cheloveka v sovremennoi Rossii: pedagogicheskii aspekt*. Moscow: Tsentr i Fond "Kholokost," 2014.
Bekaldiev, Muhamed. *Istoriya Kabardino-Balkarii, 8–9 klassy: uchebnik dlya obshcheobrazovatel'nykh uchrezhdenii*. 4th ed. Nalchik: Izdatel'stvo "Elbrus," 2013.
Danilov, Aleksandr, and Kosulina, Lyudmila. *Istoriya Rossii. XX Vek: uchebnik dlya 9 klassa obshcheobrazovatel'nykh uchrezhdenii*. Moscow: Prosveschenie, 1995.

Danilov, Aleksandr, Kosulina, Lyudmila, and Brandt, Maksim. *Istoriya Rossii, XX – nachalo XXI veka: 9 klass: uchebnik dlya obshcheobrazovatel'nykh uchrezhdenii.* Moscow: Prosveschevie, 2013.
Dmitrenko, Vladimir, Esakov, Vladimir, and Shestakov, Vladimir. *Istoriya otechestva. XX vek: uchebnik dlya 11 klassa obshcheobrazovatel'nykh uchebnykh zavedenii.* Moscow: Drofa, 1995.
Kim, Maksim, ed. *Istoriya SSSR (1938–1972): uchebnik dlya 10 klassa.* 4th ed. Moscow: Prosveshchenie, 1975.
Kim, Maksim, ed. *Istoriya SSSR (1938–1976): uchebnik dlya 10 klassa.* 7th ed. Moscow: Prosveshchenie, 1978.
Kim, Maksim, ed. *Istoriya SSSR (1938–1978): uchebnik dlya 10 klassa.* 9th ed. Moscow: Prosveshchenie, 1980.
Kislitsyn, Sergei, and Kislitsyna, Irina. *Istoriya Donskogo kraya: uchebnik dlya 9 klassa obshcheobrazovatel'nykh uchrezhdenii.* Rostov-on-Don: Donskoi izdatel'skii dom, 2004.
Klokova, Galina. *Istoriya Kholokosta na territorii SSSR v gody Velikoi Otechestvennoi voiny (1941–1945): posobie dlya uchitelya.* Moscow: Tsentr "Kholokost," 1995.
Krugov, Aleksei. *Stavropol'skii krai v istorii Rossii (konets XVII – nachalo XXI v.): regional'nii uchebnik dlya 10–11 klassov obshcheobrazovatel'nykh uchrezhdenii.* Moscow: Russkoe slovo, 2006.
Kuchiev, Vasilii. *Istoriya Osetii. XX vek: uchebnik dlya starshih klassov obshcheobrazovatel'nykh shkol.* Vladikavkaz: Ir, 2011.
Kudrina, Taisiya, ed. *Muzei i shkola: posobie dlya uchitelya.* Moscow: Prosveshchenie, 1985.
Levandovskii, Andrei, and Shchetinov, Yurii. *Rossiya v XX veke: uchebnik dlya 10–11 klassa obshcheobrazovatel'nykh uchebnykh zavedenii.* Moscow: Prosveschenie, 1997.
Levandovskii, Andrei, Shchetinov, Yurii, and Mironenko, Sergei. *Istoriya Rossii XX – nachalo XXI veka: uchebnik dlya 11 klassa obshcheobrazovatel'nykh uchrezhdenii: Bazovii uroven'.* Moscow: Prosveschenie, 2013.
Luchina, Ol'ga. Tolerantnost' – usvoenie urokov Kholokosta: urok pamyati, posvyashchennyi mezhdunarodnomu dnyu pamyati zhertv Kholokosta (9–11 klassy). *Obrazovanie v Kirovskoi oblasti* 2 (2012): 54–58.
My ne mozhem molchat': Shkol'niki i studenty o Kholokoste. 15 vols. Moscow: Fond "Kholokost," 2005–2018.
Nakhushev, Vladimir, ed. *Narody Karachaevo-Cherkessii: istoriya i kul'tura: uchebnoe posobie dlya 10–11 klassov obshcheobrazovatel'nykh uchrezhdenii.* Cherkessk: Poligrafist, 1998.
Ostrovskii, Valerii, and Utkin, Aleksei. *Istoriya Rossii. XX vek: 11 klass: uchebnik dlya obshcheobrazovatel'nykh uchebnykh zavedenii.* Moscow: Drofa, 1995.
Pankratova, Anna, ed. *Istoriya SSSR: uchebnik dlya srednei shkoly.* 11th ed. Moscow: Prosveshchenie, 1952.
Shestakov, Vladimir, Gorinov, Mikhail, and Vyazemskii, Evgenii. *Istoriya Rossii XX – nachalo XXI veka: uchebnik dlya 9 klassa obshcheobrazovatel'nykh uchrezhdenii.* 7th ed. Moscow: Prosveschevie, 2011.
Shestakov, Vladimir. *Istoriya Rossii XX – nachalo XXI veka: uchebnik dlya 11 klassa obshcheobrazovatel'nykh uchrezhdenii. Profil'nyi uroven'.* 5th ed. Moscow: Prosveschenie, 2012.
Studenikin, Mikhail. *Metodika prepodavaniya istorii v shkole: uchebnik dlya studentov vyschikh uchebnykh zavedenii.* Moscow: VLADOS, 2003.

Svetlichnaya, Irina. "Ne stat' ravnodushnymi nablyudatelyami: elektivnyi kurs 'Istoriya Kholokosta.'" *Uchitel'skaya gazeta*, October 26, 2010.

Tyulyaeva, Tatyana, ed. *Istoriya 10–11 klassy: programmno-metodicheskie materialy*. Moscow: Drofa, 1999.

Tyulyaeva, Tatyana, ed. *Istoriya. 5–9 klassy: programmno-metodicheskie materialy*. Moscow: Drofa, 1999.

Volobuev, Oleg, Klokov, Valerii, Ponomarev, Mikhail, and Rogozhkin, Vasilii. *Istoriya. Rossiya i mir: 11 klass: bazovii uroven': uchebnik dlya obshcheobrazovatel'nykh uchrezhdenii*. 12th ed. Moscow: Drofa, 2013.

Vyazemskii, Evgenii. *Kak prepodavat' istoriyu v shkole: posobie dlya uchitelya*. Moscow: Prosveschenie, 2000.

Zagladin, Nikita, Kozlenko, Sergei, Minakov, Sergei, and Petrov, Yurii. *Istoriya Rossii XX – nachalo XXI veka: uchebnik dlya 11 klassa obshcheobrazovatel'nykh uchrezhdenii*. 5th ed. Moscow: Russkoe slovo, 2007.

Zagladin, Nikita, ed. *Istoriya Rossii. XX – nachalo XXI veka: uchebnik dlya 9 klassa obshcheobrazovatel'nykh uchrezhdenii*. 11th ed. Moscow: Russkoe slovo, 2013.

Zaitsev, Andrei, ed. *Kubanovedenie: uchebnoe posobie dlya 9 klassa obshcheobrazovatel'nykh uchrezhdenii*. Krasnodar: Perspektivy obrazovaniya, 2012.

Encyclopedias, Catalogues, and Guidebooks

Agranovskii, Genrikh, and Gusenberg, Irina. *Po sledam litovskogo Ierusalima: pamyatnye mesta evreiskoi istorii i kul'tury: putevoditel'*. Vilnius: Pavilniai, 2011.

Al'tman, Il'ya, et al., ed. *Kholokost: unichtozhenie, osvobozhdenie, spasenie: buklet istoriko-dokumental'noi vystavki*. Moscow: Tsentr "Kholokost," 2017.

Astashkin, Dmitrii, ed. *Protsessy nad natsistskimi prestupnikami na territorii SSSR v 1943–1949 gg.: katalog vystavki*. Moscow: Gosudarstvennyi tsentral'nyi muzei sovremennoi istorii Rossii, 2015.

Baumann, Ulrich, Oppermann, Paula, and Schmittwilken, Christian. *Mass Shootings. The Holocaust from the Baltic to the Black Sea 1941–1944: Exhibition Catalogue*. Berlin: Stiftung Topographie des Terrors, Stiftung Denkmal für die ermordeten Juden Europas, 2016.

Belen'kii, Gennadii, and Red'kov, Nikolai. *Pamyatniki monumental'nogo iskusstva goroda Rostova-na-Donu*. Rostov-on-Don: Donskoi izdatel'skii dom, 2016.

Bol'shaya sovetskaya entsiklopediya. Vol. 47. Moscow: Gosudarstvennoe nauchnoe izdatel'stvo, 1957.

Grechko, Andrei, et al., ed. *Istoriya Vtoroi mirovoi voiny. 1939–1945*. 12 vols. Moscow: Voenizdat, 1973–1982.

Itskhak, Oren (Nadel'), and Zand, Mikhael', ed. *Kratkaya evreiskaya entsiklopediya*. Vol. 2, 4, 7. Jerusalem: Keter, 1982, 1988, 1994.

Jahn, Peter, ed. *Mordfelder: Orte der Vernichtung im Krieg gegen die Sowjetunion: Ausstellungskatalog*. Berlin: Deutsch-Russisches Museum Berlin-Karlshorst, 1999.

Kartashev, Andrei, ed. *Stavropol'skii meditsinskii: Dorogami voiny: Biograficheskii spravochnik*. Stavropol: StGMU, 2015.

Klee, Ernst. *Das Personenlexikon zum Dritten Reich: wer war was vor und nach 1945?* Frankfurt am Main: Fischer, 2003.

Kosven, Mark, et al., ed. *Narody Kavkaza.* Vol. 1. Moscow: Akademiya Nauk SSSR, 1960.
Lenskis, Ilja, ed. *Holocaust Commemoration in Latvia in the Course of Time, 1945–2015: Exhibition Catalogue.* Riga: Muzejs "Ebreji Latvijā", 2017.
Mak, Irina. *The Jewish Museum and Tolerance Center: Its History, Exposition and Foundations.* Translated by Ben McGarr. Moscow: August Borg, 2017.
Megargee, Geoffrey P., ed. *Encyclopaedia of Camps and Ghettos.* 2 vols. Bloomington: Indiana University Press, 2009, 2014.
Neumärker, Uwe, ed. *Holocaust. Der Ort der Information des Denkmals für die ermordetet Juden Europas: Katalog.* Berlin: Stiftung Denkmal für die ermordeten Juden Europas, 2015.
Pionery-geroi: al'bom-vystavka. Moscow: Malysh, 1969.
Pospelov, Petr, et al., ed. *Istoriya Velikoi Otechestvennoi voiny Sovetskogo Soyuza. 1941–1945.* 6 vols. Moscow: Voenizdat, 1961–1965.
Pozner, Valérie, ed. *Filmer la guerre 1941 – 1946: les Soviétiques face à la Shoah: Mémorial de la Shoah: à l'occasion de l'Exposition "Filmer la guerre: les Soviétiques face à la Shoah 1941 – 1946" 9 janvier – 27 septembre 2015, conçue par le Mémorial de la Shoah.* Paris: Mémorial de la Shoah, 2015.
Rebrova, Irina, ed. *"Pomni o nas . . . ": katalog vystavki, posvyashchennoi pamyati patsientov psikhiatricheskikh klinik, detei-invalidov i vrachei-evreev, ubitykh v period natsistskoi okkupatsii Severnogo Kavkaza.* Krasnodar: Tipografiya "EdArt print," 2019.
Shvartsman, Inna, ed. *Kniga pamyati: martirolog zhertv Kholokosta: Rostov-na-Donu, Zmievskaya Balka, 1942 god.* Rostov-on-Don: Rostovskaya evreiskaya obshchina, 2014.
Sokolova, Ol'ga, ed. *Muzei evreiskogo naslediya i Kholokosta: buklet.* Moscow: RJC, 1999.
Ushakov, Dmitrii. *Bol'shoi tolkovyi slovar' russkogo yazyka: sovremennaya redaktsiya.* Moscow: Dom slavyanskoi knigi, 2008.
Weinberg, Jeshayahu, and Elieli, Rina. *The Holocaust Museum in Washington.* New York: Rizzoli, 1995.

Newspapers

Argumenty i fakty
Bol'shevik
Gazeta evreev Severnogo Kavkaza
Gorod N
Izvestiya
Komsomolets Kubani
Komsomolskaya pravda
Krasnodarskie izvestiya
Maykopskie novosti
Molodezh' Moldavii
Molot
Moskovskii komsomolets na Donu
Neues Leben
Pravda
Rossiiskaya gazeta
Rostov ofitsial'nyi

Sel'skaya nov' (town of Ust-Labinsk)
Sovetskaya Kuban'
Stavropol'skaya pravda
Taganrogskaya pravda
The New York Times
Uchitel'skaya gazeta
Vechernii Kishinev
Vechernii Rostov
Vesti (prilozhenie "Veteran i voin")
Vesti: Izrael' po-russki
Vestnik Arzgirskogo raiona
Vremya evreev
Vremya: gazeta goroda Mineralnye Vody
Zarya: obshchestvenno-politicheskaya gazeta Arzgirskogo raiona Stavropol'skogo kraya

Websites

http://273-фз.рф/ – Implementation of the Federal Law "On Education in the Russian Federation";
http://38ddt1.ru/ – Dom detskogo tvorchestva [House of Children's Creativity] No. 1, Irkutsk;
http://arzgiradmin.ru/site_pk/index.php – Administration of Arzgir Municipality, Stavropol Krai;
http://babiyar.org/en – BYHMC;
http://db.saur.de/DGO/login.jsf;jsessionid=07f014bec31763a4fdfc84e42a0e– "Nationalsozialismus, Holocaust, Widerstand und Exil" Online Database;
http://eajc.org/ – Euro-Asian Jewish Congress;
http://holocaust.su/ – Holocaust. Rostov-on-Don. 1942;
http://holocf.ru/ – Holocaust Center;
http://jewish.ru/ – Global Jewish Online Center;
http://komiswow.ru/ – Mints Commission;
http://kremlin.ru/ – Official Website of the President of the Russian Federation;
http://m-academ.centerstart.ru/ – "Small Academy;"
http://moypolk.ru/ – Bessmertyi Polk [Immortal Regiment];
http://old.hum.huji.ac.il/english/units.php?cat=4246 – OHD at the Hebrew University in Jerusalem;
http://samlib.ru/ – "Samizdat" Magazine;
http://sfi.usc.edu/ – USCSF;
http://southru.info/ – SouthRu.info News;
http://staviropk.ru/index.php – SKIRO PK and PRO;
http://stavstat.gks.ru/wps/wcm/connect/rosstat_ts/stavstat/ru/ – Department of the Federal State Statistics Service for the North Caucasus Federal District;
http://turcentrrf.ru/ – Federal Center for Children and Youth Tourism and Local History;
http://unpo.org/ – Unrepresented Nations and Peoples Organisation;
http://urokiistorii.ru/ – Uroki istorii XX vek [20th Century History Lessons];
http://victorymuseum.ru/ – Victory Museum in Moscow;
http://voices.iit.edu/ – Voices of the Holocaust project at Illinois Institute of Technology;

http://www.aen.ru/ – Agenstvo evreiskikh novostei [Jewish News Agency];
http://www.claimscon.org/ – Conference on Jewish Material Claims against Germany;
http://www.demoscope.ru/weekly/2018/0765/index.php – Demoscope Weekly;
http://www.dspl.ru/ – Don State Public Library in Rostov-on-Don;
http://www.gorby.ru/gorbi_fund/about/ – Gorbachev Foundation;
http://www.hist.msu.ru/about/gen_news/ – History Faculty of Moscow State University;
http://www.holocaustinussr.com/ – Documentary Project by Boris Maftsir;
http://www.karaims.ru/index.php?cod=ru – Moscow Karaite Society;
http://www.minobrkuban.ru/ – The Ministry of Education, Science and Youth Policy of Krasnodar Krai;
http://www.rjc.ru/ – RJC;
http://www.stavarhiv.ru/ – GASK;
http://www.stavmuseum.ru/ – SGMZ;
http://www.stolpersteine.eu/start/ – Stolpersteine Project;
http://www.taglib.ru/index.html – Taganrog Central City Public Library;
http://www.vha.fu-berlin.de/en/index.html – VHA at Freie Universität Berlin;
http://www.yahadinunum.org/ – Yahad-In-Unum;
http://yadvashem.org/ – Yad Vashem;
http://yurist-online.com/ – Legal Services in the Russian Federation;
https://1sept.ru/ – "Pervoe sentyabrya" [First September] Publishing House;
https://45parallel.net/ – Classical and Contemporary Russian Poetry;
https://7x7-journal.ru/ – "7x7: novisti, mneniya, blogi" [News, Opinions, Blogs] Magazine;
https://dic.academic.ru/ – Academic Dictionaries and Encyclopaedias;
https://en.unesco.org/ – UNESCO;
https://felicina.ru/ – Felistyn Museum in Krasnodar;
https://historyrussia.org/ – Russian Historical Society;
https://histrf.ru/biblioteka/Soviet-Nuremberg – Project "Soviet Nuremberg;"
https://lechaim.ru/ – Lekhaim Jewish Community Journal;
https://motl.org/ – International March of the Living;
https://nsvictims.ru/ – "Pomni o Nas" [Remember Us] Travelling Exhibition;
https://planeta.ru/ – Planeta.ru Crowdfunding Website;
https://proza.ru/ – Proza.ru literary Portal;
https://ria.ru/ – RIA Novosti;
https://rvio.histrf.ru/ – Russian Military-Historical Society;
https://sites.google.com/site/school2historyproject/ – "Kislovodskaya Valley 1942–1943: Memory, Pain and Mercy" School Project;
https://stihi.ru/ – Stihi.ru Literary Portal;
https://stmegi.com/ – STMEGI International Charitable Foundation;
https://web.library.yale.edu/testimonies – Fortunoff Video Archive for Holocaust Testimonies at Yale University;
https://www.consultant.ru/ – KonsultantPlus Legal/Financial Consulting Online Platform;
https://www.hse.ru/en/ – Higher School of Economics;
https://www.memo.ru/ru-ru/ – Society "Memorial;"
https://www.memorialmuseums.org/pages/home – International Portal to European Sites of Remembrance;
https://www.opengaz.ru/ – "Otkrytaya gazeta" [Open Newspaper] Newspaper;

https://www.ort.org/ – World ORT Jewish Global Education Network;
https://www.rsuh.ru/ – Russian State University for the Humanities;
https://www.stav.kp.ru/ – "Komsomolskaya pravda" [Komsomol's Truth] Newspaper;
https://www.stiftung-denkmal.de/denkmaeler/denkmal-fuer-die-ermordeten-juden-europas.html
 – Memorial to the Murdered Jews of Europe and Informational Center in Berlin;
https://www.svoboda.org/ – Radio svoboda [Radio Liberty];
https://www.ushmm.org/ – USHMM;
https://www.war.ekimovka.ru/index.php – "Zhivy navsegda. Stavropol'e pomnit Pobedu"
 [Alive Forever. Stavropol Remembers Victory] Project;
https://www.yuga.ru/ – Yuga.ru Internet Portal;
https://минобрнауки.рф/ – Ministry of Education and Science of the Russian Federation.

PhD Theses

Bochkareva, Zoya. "Okkupatsionnaya politika fashistskoi Germanii na Severnom Kavkaze." Candidate of science diss. in history: Kuban State University, 1992.

Bulgakova, Natalia. "Sel'skoe naselenie Stavropol'ya vo vtoroi polovine 20-kh – nachale 30-kh gg. XX veka: izmeneniya v demograficheskom, khozyaystvennom i kul'turnom oblike." Candidate of science diss. in history: Stavropol State University, 2003.

Epifanov, Aleksandr. "Organizatsiya i deyatel'nost' chrezvychainoi gosudarstvennoi komissii po ustanovleniyu i rassledovaniyu natsistskikh zlodeyanii." Candidate of science diss. in law, Military Academy of Ministry of Interior Affairs, 1996.

Ershova, Irina. "Evolyutsiya metodov obucheniya istorii v sovetskoi shkole, 1930-kh – nachalo 1990-kh." Candidate of science diss. in pedagogy, Institute of Secondary School of Russian Academy of Education, 1994.

Glukhov, Vasilii. "Adygeya v dni Velikoi Otechestvennoi voiny." Candidate of science diss. in history, Maykop State Institute, 1948.

Khubova, Dariya. "Ustnaya istoriya i arkhivy: zarubezhnye kontseptsii i opyt." Candidate of science diss. in history, Russian State University for the Humanities, 1992.

Linets, Sergei. "Severnyi Kavkaz nakanune i v period nemetsko-fashistkoi okkupatsii: sostoyanie i osobennosti razvitiya, iyul' 1942 – oktyabr' 1943 g." Doctor of science diss. in history, Pyatigorsk State Technological University, 2003.

Malakhova, Galina. "Stanovlenie i razvitie rossiiskogo gosudarstvennogo upravleniya na Severnom Kavkaze v XVIII – XIX vv." Doctor of science diss. in history, Russian Academy of Public Administration under the President of the Russian Federation, 2001.

Mirzoyan, Elena. "Sibirskaya i kavkazskaya ssylka dekabristov, 1826–1856 gg.: opyt sravnitel'nogo issledovaniya." Candidate of science diss. in history, Siberian branch of the RAN, 2002.

Rakachev, Vadim. "Natsional'nyi sostav naseleniya Kubani v XX veke: istoriko-demograficheskii aspekt." Candidate of science diss. in history, Kuban State University, 2003.

Rebrova, Irina. "Velikaya Otechestvennaya voina v memuarakh: istoriko-psikhologicheskii aspekt." Candidate of science diss. in history, North-West Academy of Public Administration, 2005.

Rowe-McCulloch, Maris. "The Holocaust and Mass Violence in the German-Occupied City of Rostov-on-Don, 1941–1943." PhD diss., Department of History at the University of Toronto, 2020.
Rudneva, Larisa. "Evolyutsiya shkol'nogo uchebnika istorii v 1940–80-e gody 20-go veka." Candidate of science diss. in history, Kursk State Unoversity, 2005.
Sanders, Marian R. "Extraordinary Crimes in Ukraine: An Examination of Evidence Collection by the Extraordinary State Commission of the USSR, 1942–1946." PhD diss., Ohio University, 1995.
Stepanenko, Sergei. "Deyatel'nost' Chrezvychainoi gosudarstvennoi komissii SSSR po vyyavleniyu voennykh prestuplenii fashistkoi Germanii na territorii Krasnodarskogo kraya." Candidate of science diss. in history, Maykop State University, 2010.
Tsifanova, Irina. "Pol'skie pereselentsy na Severnom Kavkaze v XIX veke: osobennosti protsessa adaptatsii." Candidate of science diss. in history, Stavropol State University, 2002.
Umansky, Andrej. "L'extermination des juifs dans le Caucase du nord pendant la seconde guerre mondiale (1942–1943)." PhD diss., Doctoral school on Human and Social Sciences of the University of Picardie Jules Verne, 2016.
Voitenko, Elena. "Kholokost na yuge Rossii v period Velikoi Otechestvennoi voiny (1941–1943 gg.)." Candidate of science diss. in history, Stavropol State University, 2005.
Winkler, Christina. "The Holocaust in Rostov-on-Don: official Russian Holocaust remembrance versus a local case study." PhD diss., School of History at the University of Leicester, 2015.

Secondary sources

Abram, Ido, and Matthias, Heyl. *Thema Holocaust: Ein Buch für die Schule*. Reinbek bei Hamburg: Rowohlt, 1996.
Adams, Jenni, ed. *The Bloomsbury Companion to Holocaust Literature*. London: Bloomsbury, 2014.
Agamben, Giorgio. *Remnants of Auschwitz: The Witness and the Archive*. Translated by Daniel Heller-Roazen. New York: Zone books, 1999.
Agamben, Giorgio. *What is an Apparatus? And Other Essays*. Stanford: Stanford University Press, 2009.
Aharony, Michal, and Rosenfeld, Gavriel D. Holocaust Commemoration: New Trends in Museums and Memorials. *Dapim: Studies on the Holocaust* 30, no. 3 (2016): 162–165.
Chernogorov, Aleksandr, ed. *Krai nash Stavropol'e: ocherki istorii*. Stavropol: Shat-Gora, 1999.
Alexander, Jeffrey C. *Remembering the Holocaust: A Debate*. New York: Oxford University Press, 2009.
Aliskerov, Islam. "Vliyanie religioznykh i etnicheskikh faktorov na voenno-politicheskuyu obstanovku na Severnom Kavkaze v gody Velikoi Otechestvennoi voiny 1941–1945 gg." In *Religioznye organizatsii Sovetskogo Soyuza v gody Velikoi Otechestvennoi voiny 1941–1945 gg.: materialy 'kruglogo stola', posvyashch. 50-letiyu pobedy, 13 aprelya 1995 g.*, edited by Nikolai Trofimchuk, 78–94. Moscow: Rossiiskaya akademiya gosudarstvennoi sluzhby, 1995.

Allen, Barbara. "Re-creating the Past: The Narrator's Perspective in Oral History." *Oral History Review* 12 (1984): 1–12.

Al'tman, Il'ya. *Dnevnik Lyusi Kaliki "Odessa. 820 dnei v podzemel'e" kak istoricheskii istochnik.* http://www.holocf.ru/Editor/assets2/dnevnik_lusy_kaliki.pdf.

Al'tman, Il'ya. *Zhertvy nenavisti. Kholokost v SSSR 1941–1945.* Moscow: Kollektsiya "Soversheno sekretno," 2002.

Al'tman, Il'ya. "Memorializatsiya Kholokosta v Rossii: istoriya, sovremennost', perspektivy." *Neprikosnovennyi zapas*, 2–3 (2005): 252–263.

Al'tman, Il'ya. "Krasnaya armiya i Kholokost: k postanovke problem." In *Kholokost na territorii SSSR: materialy XIX mezhdunarodnoi ezhegodnoi konferentsii po iudaike*, edited by Arkadi Zeltser. Vol. 1, 86–93. Moscow: Sefer, 2012.

Al'tman, Il'ya. "Mesto Kholokosta v rossiiskoi istoricheskoi pamyati." In *Pamyat' o Kholokoste: problemy memorializatsii: materialy 6-i mezhdunarodnoi konferentsii "Uroki Kholokosta i sovremennaya Rossiya,"* edited by Il'ya Al'tman, 15–21. Moscow: Tsentr i Fond "Kholokost," 2012.

Altshuler, Mordechai. "Escape and Evacuation of the Soviet Jews at the Time of the Nazi Invasion." In *Holocaust in the Soviet Union: Studies and Sources on the Destruction of the Jews in the Nazi-Occupied USSR, 1941–1945*, edited by Lucjan Dobroszyski, 77–105. Armonk: Sharpe, 1993.

Altshuler, Mordechai. "Jewish Holocaust Commemoration Activity in the USSR under Stalin." *Yad Vashem Studies* 30 (2002): 271–295.

Altshuler, Mordechai. "Evacuation and Escape During the Course of the Soviet-German War." *Dapim: Studies on the Holocaust* 28, no. 2 (2014): 57–73.

Angrick, Andrej. "Die Einsatzgruppe D." In *Die Einsatzgruppen in der besetzten Sowjetunion 1941/42: die Tätigkeits- und Lageberichte des Chefs der Sicherheitspolizei und des SD*, edited by Peter Klein and Andrej Angrick, 88–110. Berlin: Ed. Hentrich, 1997.

Angrick, Andrej. *Besatzungspolitik und Massenmord. Die Einsatzgruppe D in der Südlichen Sowjetunion 1941–1943.* Hamburg: Hamburger Edition, 2003.

Anisina, Natalia. "Muzei kak mesto pamyati Kholokosta v Rossii." In *Pamyat' o Kholokoste: problemy memorializatsii: materialy 6-i mezhdunarodnoi konferentsii "Uroki Kholokosta i sovremennaya Rossiya,"* edited by Il'ya Al'tman, 43–46. Moscow: Tsentr i Fond "Kholokost," 2012.

Arad, Yitzhak. *The Holocaust in the Soviet Union, The Comprehensive History of the Holocaust.* Jerusalem: Yad Vashem, 2009.

Arkhangorodskaya, Nina, and Kurnosov, Aleksei. "O sozdanii komissii po istorii Velikoi Otechestvennoi voiny AN SSSR i ee arkhiva (k 40-letiyu so dnya obrazovaniya)." *Arheograficheskii ezhegodnik za 1981 god*, 219–229. Moscow: Nauka, 1982.

Assmann, Aleida. *Mnemosyne: Formen und Funktionen der kulturellen Erinnerung.* Frankfurt am Main: Fischer-Taschenbuch-Verl, 1991.

Assmann, Aleida. *Cultural Memory and Western Civilisation: Functions, Media, Archives.* New York: Cambridge University Press, 2011.

Assmann, Jan, and Hölscher, Tonio, ed. *Kultur und Gedächtnis.* Frankfurt am Main: Suhrkamp, 1988.

Assmann, Jan. *Das kulturelle Gedächtnis.* München: C.H. Beck, 2007.

Auron, Yair, and Okhtov, Aleksandr. *Podvig milisediya: spasenie detei is blokadnogo Leningrada v cherkesskom aule Beslenei.* Moscow-Cherkessk-Jerusalem, 2018.

Averyanova, Irina. "Prepodavanie temy 'Kholokost' na urokakh istorii." In *Rossiya v perelomnye periody istorii: nauchnye problemy i voprosy grazhdansko-patrioticheskogo vospitaniya molodezhi*, edited by Marina Dmitrieva, 341–347. Ivanovo: Institut razvitiya obrazovaniya Ivanovskoi oblasti, 2012.
Avital, Moshe. "The Role of Songs and Music during the Holocaust." *Journal of Jewish Music and Liturgy* 31 (2011–2012): 51–60.
Azamatov, Kamil', et al. *Cherkesskaya tragediya*. Nalchik: El'brus, 1994.
Bajohr, Frank, and Steinbacher, Sybille, ed. ". . . Zeugnis ablegen bis zum letzten:" *Tagebücher und persönliche Zeugnisse aus der Zeit des Nationalsozialismus und des Holocaust*. Göttingen: Wallstein Verlag, 2015.
Bar, Doron. "Holocaust and Heroism in the Process of Establishing Yad Vashem (1942–1970)." *Dapim: Studies on the Holocaust* 30, no. 3 (2016): 166–190.
Barber, John, and Dzeniskevich, Andrei, ed. *Life and Death in Besieged Leningrad: 1941–44*. Houndmills: Palgrave Macmillan, 2005.
Baron, Lawrence. *Projecting the Holocaust into the Present: The Changing Focus of Contemporary Holocaust Cinema*. Lanham: Rowman and Littlefield, 2005.
Baron, Lawrence. "The First Wave of American "Holocaust" Films, 1945–1959." *American Historical Review* 115, no. 1 (2010): 90–114.
Bartov, Omer. *Murder in Our Midst: The Holocaust, Industrial Killing, and Representation*. New York: Oxford University Press, 1996.
Basu, Laura. *Ned Kelly as Memory Dispositif: Media, Time, Power, and the Development of Australian Identities*. Berlin-Boston: Walter de Gruyter, 2012.
Baudry, Jean-Luis. "The Apparatus: Metapsychological Approaches to the Impression of Reality in the Cinema." In *Narrative, Apparatus, Ideology: A Film Theory Reader*, edited by Philip Rosen, 299–318. New York: Columbia University Press, 1986.
Beer, Mathias." Die Entwicklung der Gaswagen beim Mord an den Juden." *Vierteljahrshefte für Zeitgeschichte* 3 (1987): 403–413.
Belikov, German. *Okkupatsiya, Stavropol', avgust 1942 – yanvar' 1943*. Stavropol: Fond dukhovnogo prosveshcheniya, 1998.
Belikov, Mikhail. *Severnyi Kavkaz: realii sotsial'no-ekonomicheskoi sfery na poroge tysyacheletiya: geograficheskii aspekt*. Krasnodar: Kubanskii gosudarstvennyi universitet, 2002.
Belonosov, Ivan. "Evakuatsiya naseleniya iz prifrontovoi polosy v 1941–1942 gg." In *Eshelony idut na vostok: Iz istorii perebazirovaniya proizvoditel'nykh sil SSSR v 1941–1942 gg.: sbornik statei i vospominanii*, edited by Yurii Polyakov, 15–30. Moscow: "Nauka," 1966.
Belting, Hans. "Place of Reflection or Place of Sensation?" In *The Discursive Museum*, edited by Peter Noever, 72–82. Vienna: MK, 2001.
Bennett, Tony. *The Birth of the Museum: History, Theory, Politics*. London: Routledge, 1995.
Beorn, Waitman Wade. *Marching into Darkness: The Wehrmacht and the Holocaust in Belarus*. Cambridge, London: Harvard University Press, 2014.
Berger, Alan L., and Berger, Naomi. *Second Generation Voices: Reflections by Children of Holocaust Survivors and Perpetrators*. Syracuse: Syracuse University Press, 2001.
Berger, Ronald J. *The Holocaust, Religion, and Politics of Collective Memory*. New Brunswick: Transaction Publishers, 2012.
Berkhoff, Karel C. *Harvest of Despair: Life and Death in Ukraine under Nazi Rule*. Cambridge-London: The Belknap press of Harvard University Press, 2004.

Berkhoff, Karel C. "Total Annihilation of the Jewish Population: The Holocaust in the Soviet Media, 1941–1945." *Kritika: Explorations in Russian and Eurasian History* 10, no. 1 (2009): 61–105.
Berkhoff, Karel C. *Motherland in Danger: Soviet Propaganda during World War II*. Cambridge, MA: Harvard University Press, 2012.
Berkhoff, Karel C. "The Dispersal and Oblivion of the Ashes and Bones of Babi Yar." In *Lessons and Legacies XII: New Directions in Holocaust Research and Education*, edited by Wendy Lower and Lauren Faulkner Rossi, 256–276. Evanston: Northwestern University Press, 2017.
Bezymenskii, Lev. "Informatsiya po-sovetski." *Znamya* 5 (1998), http://magazines.russ.ru/zna mia/1998/5/bezym.html.
Bliev, Mark. *Kavkazskaya voina (1817–1864)*. Moscow: TOO "Roset," 1994.
Blouin, Francis X., and Rosenberg, William G., ed. *Archives, Documentation, and Institutions of Social Memory: Essays from the Sawyer Seminar*. Ann Arbor: University of Michigan Press, 2007.
Bobrovnikov, Vladimir, and Babich, Irina, ed. *Severnyi Kavkaz v sostave Rossiiskoi Imperii*. Moscow: Novoe literaturnoe obozrenie, 2007.
Bogachkova, Anna. *Istoriya Izobil'nenskogo raiona*. Stavropol: Knizhnoe izdatel'stvo, 1994.
Bogolyubov, Aleksandr. *Polyaki na severnom Kavkaze v XIX–XX vv*. Krasnodar: Kubanskii gosudarstvennyi universitet, 2008.
Bolotina, Tatyana, and Prigodich, Elena. "Istoriya Kholokosta i problemy prepodavaniya etoi temy v shkole." *Methodist* 5 (2013): 16–18.
Bondar', Lev. *Detyam detei rasskazhite . . .* Bel'tsy: Assotsiatsiya evreiskikh organizatsiya goroda Bel'tsy, 2006.
Boswell, Matthew. *Holocaust Impiety in Literature, Popular Music and Film*. Basingstoke: Palgrave Macmillan, 2012.
Boiko, Ivan. "Opyt i problemy memorializatsii zhertv Kholokosta na raionnom urovne." In *Pamyat' o Kholokoste: problemy memorializatsii: materialy 6-i mezhdunarodnoi konferentsii "Uroki Kholokosta i sovremennaya Rossiya,"* edited by Il'ya Al'tman, 60–64. Moscow: Tsentr i Fond "Kholokost," 2012.
Boiko, Ivan. *Etapy bol'shogo puti: K 90-letiyu Novopokrovskogo raiona*. Krasnodar: Kniga, 2015.
Brenner, Rachel F. "Teaching the Holocaust in Academia: Educational Mission(s) and Pedagogical Approaches." *The Journal of Holocaust Education*, 8 no. 2 (1999): 1–26.
Brenner, Reeve R. *The Faith and Doubt of Holocaust Survivors*. New York: Free Press, 2014.
Brinkmöller-Becker, Heinrich. "Kino und die Wahrnehmung von Filmen: Das Kino-Dispositiv in 100-järiger Entwicklung." *Medien und Erziehung: Zeitschrift für Medienpädagogik* 38, no. 6 (1994): 327–332.
Broxup, Marie, ed. *The North Caucasus Barrier: The Russian Advance towards the Muslim World*. London: Hurst, 1992.
Bougai, Nikolai. *The Deportation of Peoples in the Soviet Union*. New York: Nova Science, 1996.
Bühmann, Andrea D. "Die Normalisierung der Geschlechter in Geschlechterdispositiven." In *Das Geschlecht der Moderne*, edited by Hannelore Bublitz, 71–94. Frankfurt am Main: Campus, 1998.
Bührmann, Andrea D., and Schneider, Werner. "Mehr als nur diskursive Praxis? – Konzeptionelle Grundlagen und methodische Aspekte der Dispositivanalyse." *Forum*

Qualitative Sozialforschung 8, no. 2 (2007), http://www.qualitative-research.net/index. php/fqs/article/view/237/526.
Bührmann, Andrea D., and Schneider, Werner. *Vom Diskurs zum Dispositiv. Eine Einführung in die Dispositivanalyse*. Bielefeld: Transcript Verlag, 2008.
Bührmann, Andrea D., and Schneider, Werner. "Vom 'discursive turn' zum 'dispositive turn'? Folgerungen, Herausforderungen und Perspektiven für die Forschungspraxis." In *Verortungen des Dispositiv-Begriffs: analytische Einsätze zu Raum, Bildung, Politik*, edited by Joannah Caborn Wengler, Britta Hoffarth, and Lukasz Kumiega, 24–28. Wiesbaden: Springer VS, 2013.
Carden-Coine, Ana. "The Ethics of Representation in Holocaust Museums." In *Writing the Holocaust*, edited by Jean-Marc Dreyfus and Daniel Langton, 167–184. London: Bloomsbury Academic, 2011.
Chaffee, Daniel and Lemert, Charles. "Structuralism and Poststructuralism." In *The New Blackwell Companion to Social Theory*, edited by Turner Bryan, 133–139. Chichester: Wiley-Blackwell, 2009.
Charnyi, Semen. "Rol' evreiskikh obshchin yuga Rossii v sokhranenii i memorializatsii pamyati o Kholokoste (na primere Rostova-na-Donu)." In *Istoriya Kholokosta na Severnom Kavkaze i sud'by evreiskoi intelligentsii v gody Vtoroi mirovoi voiny: materialy 7-i mezhdunarodnoi konferentsii "Uroki Kholokosta i sovremennaya Rossiya,"* edited by Kiril Feferman, 110–117. Moscow: Tsentr i Fond "Kholokost," 2013.
Chashchukhin, Aleksandr. "Uchebnye teksty i professional'no-politicheskaya sotsializatsya shkol'nykh uchitelei 1950-kh gg." In *Dorogoi drug. Sotsial'nye modeli i normy v uchebnoi literature 1900–2000 godov: istoriko-pedagogicheskoe issledovanie*, edited by Vitalii Bezrogov, Tatyana Markarova, and Anatolii Tsapenko, 35–48. Moscow: Pamyatniki istoricheskoi mysli, 2016.
Chekeres, Ol'ga. "Sanatorium Ordzhonikidze during the Great Patriotic War (1941–1945 years)." *Gardarika* 4, no. 3 (2015): 91–100.
Cherkasov, Aleksandr. "K nekotorym aspektam raboty sochinskoi gospital'noi bazy (1941–1945 gg.): periodizatsiya i effektivnost'." *Bylye gody* 2 (2008): 19–28.
Cherniy, Moisei. "Figura umolchaniya, ili 'Yudenfrai'." *Evreiskaya shkola* 2–3 (1995): 59–66.
Chevelya, Yanina, ed. *Zmievskaya Balka: vopreki*. Rostov-on-Don: Feniks, 2013.
Clowes, Edith W. "Constructing the Memory of the Holocaust: The Ambiguous Treatment of Babii Yar in Soviet Literature." *Partial Answers* 3, no. 2 (2005): 153–182.
Crownshaw, Rick. "Photography and Memory in Holocaust Museums." *Morality* 12, no. 2 (2007): 176–191.
Crysler, Greig C. "Violence and Empathy: National Museums and the Spectacle of the Society." *Traditional Dwellings and Settlements Review* 17, no. 2 (2006): 19–38.
Cubitt, Geoffrey. *History and Memory*. Manchester: Manchester University Press, 2007.
Dallin, Alexander. *German Rule in Russia 1941–1945. A Study of Occupation Policies*. New York: Palgrave, 1981.
Danilova, Svetlana, ed. *Gorskie evrei v Kabardino-Balkarii*. Nalchik: El'brus, 1997.
Danilova, Svetlana, ed. *Iskhod gorskikh evreev: razrushenie garmonii mirov*. Nalchik: Poligrafservis i T, 2000.
Danilova, Svetlana. "Kholokost na Severnom Kavkaze i iskazhenie faktov v kontekste sovremennykh sobytii v dannom regione." *Zametki po evreiskoi istorii* 2 (2012), http://ber kovich-zametki.com/2012/Zametki/Nomer2/Danilova1.php.

David-Fox, Michael, Holquist, Peter, and Martin, Alexander M., ed. *Holocaust in the East. Local Perpetrators and Soviet Responses*. Pittsburgh: University of Pittsburgh Press, 2014.

Dekel-Chen, Jonathan L. *Farming the Red Land: Jewish Agricultural Colonization and Local Soviet Power, 1924–1941*. New Haven: Yale University Press, 2005.

Deleuze, Gilles. "What is a Dispositif?" In *Michel Foucault Philosopher: Essays*, edited and translated by Timothy J. Armstrong. New York: Routledge, 1992.

Desbois, Patrick. "The Holocaust by Bullets." *The Holocaust and the United Nations Outreach Programme*: *Discussion Papers Journal* 2 (2002): 77–86.

Desbois, Patrick. *The Holocaust by Bullets: A Priest's Journey to Uncover the Truth Behind the Murder of 1.5 Million Jews*. New York: Palgrave Macmillan, 2008.

Desbois, Patrick, and Husson, Edouard. "Neue Ergebnisse zur Geschichte des Holocaust in der Ukraine: Das 'Oral History'-Projekt von Yahad-In Unum und seine Wissenschaftliche Bewertung." In *Besatzung, Kollaboration, Holocaust. Neue Studien zur Verfolgung und Ermordung der europäischen Juden*, edited by Johannes Hürter und Jürgen Zaruskiy, 177–188. München: Oldenbourg Verlag, 2008.

Desbois, Patrick, and Husson, Edouard. *In Broad Daylight: The Secret Procedures behind the Holocaust by Bullets*. New York: Arcade Publishing, 2018.

Deutsch, Nathaniel. *The Jewish Dark Continent: Life and Death in the Russian Pale of Settlement*. Cambridge: Harvard University Press, 2011.

Deinevich, Aleksandr. *Tam za povorotom*. Maykop: Polofraf Yug, 2012.

Dreesen, Philipp, Kumięga, Łukasz, and Spie, Constanze, ed. *Mediendiskursanalyse. Diskurse – Dispositive – Medien – Macht*. Wiesbaden: Springer VS, 2011.

Dubin, Boris. "'Krovavaya' voina i 'Velikaya' Pobeda." *Otechestvennye zapiski* 5 (2004): 86–84.

Dubossarskaya, Maya. "Svoi-chuzhoi-drugoi: k postanovke problemy." *Vestnik Stavropol'skogo gosudarstvennogo universiteta* 54 (2008): 167–174.

Dubson, Vadim. "Toward a Central Database of Evacuated Soviet Jews' Names, for the Study of the Holocaust in the Occupied Soviet Territories." *Holocaust and Genocide Studies* 26, no. 1 (2012): 95–119.

Dvuzhil'naya, Inessa. *Tema Kholokosta v akademicheskoi muzyke*. Grodno: Grodnenskii gosudarstvennyi universitet: 2016.

Dymshits, Valerii, ed. *Gorskie evrei: istoriya, etnografiya, kul'tura*. Jerusalem, Moscow: DAAT/Znanie, 1999.

Dzeranov, Timur. "Etnokonfessional'nye razlichiya naseleniya severnogo Kavkaza." *Fundamental'nye issledovaniya* 3–4, (2014): 861–865.

Edgre, Henrik, ed. *Looking at the Onlookers and Bystanders. Interdisciplinary Approaches to the Causes and Consequences of Passivity*. Stockholm: The Living History Forum, 2012.

Education on the Holocaust and Anti-Semitism: An Overview and Analysis of Educational Approaches. Warsaw: OBCE Office for Democratic Institutions and Human Rights, 2006.

Emonds, Friederike B. "Revisiting the Memory Industry: Robert Thalheim's 'Am Ende kommen Touristen'." *Colloquia Germanica* 44, no. 1 (2011): 55–78.

Endlich, Stefanie. *Wege zur Erinnerung: Gedenkstätten und -orte für die Opfer des Nationalsozialismus in Berlin und Brandenburg*. Berlin: Metropol, 2007.

Engelhardt, Isabelle. *A Topography of Memory: Representations of the Holocaust at Dachau and Buchenwald in Comparison with Auschwitz, Yad Vashem and Washington, DC*. Brussells: Lang, 2001.

Epifanov, Aleksandr. *Otvetstvennost' za voennye prestupleniya, sovershennye na territorii SSSR v gody Velikoi Otechestvennoi voiny. 1941–1956 gg*. Volgograd: Volgogradskaya akademiya MVD Rossii, 2005.

Epelboin, Annie, and Kovriguina, Assia. *La littérature des ravins: écrire sur la Shoah en URSS*. Paris: Robert Laffont, 2013.

Erll, Astrid. "Literatur und kulturelles Gedächtnis: Zur Begriffs- und Forschungsgeschichte, zum Leistungsvermögen und zur literaturwissenschaftlichen Relevanz eines neuen Paradigmas der Kulturwissenschaft." *Literaturwissenschaftliches Jahrbuch* 43 (2002): 249–276.

Erll, Astrid, and Nünning, Ansgar, ed. *A Companion to Cultural Memory Studies*. Berlin and New York: Walter de Gruyter, 2010.

Erokhin, Igor'. *Kubanskoe kazachestvo v emigratsii XX v*. Ekaterinburg: Izdatel'stvo knizhnyi perestok, 2013.

Evstaf'eva, Tatyana and Nakhmanovich, Vitalii, ed. *Babii Yar: chelovek, vlast', istoriya: dokumenty i materialy*. Vol. 1. *Istoricheskaya topografiya i khronologiya sobytii*. Kyiv: Vneshtorgizdat, 2004.

Evstaf'eva, Tatyana. "Babii Yar: Poslevoennaya istoriya mestnosti." In *Babyn Yar: masove ubyvstvo i pam"yat' pro n'oho*, edited by Vitalii Nakhmanovych, 21–31. Kyiv: Ukrainskii tsentr vyvchennya istoriyi Holokostu, 2012.

Fallace, Thomas D. "The Origins of Holocaust Education in American Public Schools." *Holocaust and Genocide Studies* 20, no. 1 (2006): 80–102.

Feferman, Kiril. "Nazi Germany and the Mountain Jews: Was There a Policy?" *Holocaust and Genocide Studies* 21, no. 1 (2007): 96–114.

Feferman, Kiril. "Soviet Investigation of Nazi Crimes in the USSR: Documenting the Holocaust." *Journal of Genocide Research* 5, no. 4. (2003): 587–602.

Feferman, Kiril. "A Soviet Humanitarian Action? Center, Periphery and the Evacuation of Refugees to the North Caucasus, 1941–1942." *Europe-Asia Studies* 61, no. 5 (2009): 813–831.

Feferman, Kiril. "Pamyat' o voine i Kholokoste v sovetskom i postsovetskom kollektivnom soznanii." In *Istoricheskaya pamyat': protivodeistvie otritsaniyu Kholokosta: materialy 5-i mezhdunarodnoi konferentsii "Uroki Kholokosta i sovremennaya Rossiya,"* edited Il'ya Al'tman, 76–87. Moscow: MIK, 2010.

Feferman, Kiril. "Nazi Germany and the Karaites in 1938–1944: Between Racial Theory and Realpolitik." *Nationalities Papers* 39, no. 2 (2011): 77–94.

Feferman, Kiril. "The Fate of the Karaites in the Crimea during the Holocaust." In *Eastern European Karaites in the Last Generations*, edited by Dan Shapira and Daniel Lasker, 171–191. Jerusalem: Benzvi Institute, 2011.

Feferman, Kiril. "Preaching to the Converted? Teaching about the Holocaust in Modern-Day Russia." In *Holocaust Education in the 21st Century*, edited by Eva Matthes and Elizabeth Meilhammer, 230–239. Bad Heilbrunn: Verlag Julius Klinkhardt, 2015.

Feferman, Kiril. *The Holocaust in the Crimea and the North Caucasus*. Jerusalem: Yad Vashem, 2016.

Fentress, James. *Social Memory*. Oxford: Blackwell, 1992.

Ferro, Mark. *Kak rasskazyvayut istoriyu detyam v raznykh stranakh mira*. Translated by Elena Lebedeva. Moscow: Vyshaya shkola, 1992.

Fish, Daniil. "Evrei v shkol'nom kurse vsemirnoi i russkoi istorii (chto chitayut sovetskie shkol'niki o roli evreev v istorii)." *Evreiskii samizdat. Ierusalim, evreiskii universitet v Ierusalime*, 15 (1978): 313–352.
Foucault, Michel. *The Archaeology of Knowledge and the Discourse on Language*. New York: Pantheon Books, 1972.
Foucault, Michel. *Discipline and Punish: The Birth of the Prison*. Translated by Alan Sheridan. New York: Pantheon Books, 1977.
Foucault, Michel. *The History of Sexuality. Vol. 1. An Introduction*. Translated by Robert Hurley. New York: Pantheon Books, 1978.
Foucault, Michel. "Nietzsche, Genealogy, History." In *Language, Counter-Memory, Practice: Selected Essays and Interviews*, edited by Donald F. Bouchard, 139–164. Ithaca: Cornell University Press, 1980.
Foucault, Michel. *Power/Knowledge: Selected Interviews and Other Writings 1972–1977*, edited and translated by Colin Gordon. New York: Pantheon Books, 1980.
Foucault, Michel. "Different Spaces." In *Essential Works of Foucault 1954–1984*. Translated by Robert Hurley. Vol. 2, 175–185. London: Penguin, 1998.
Fraser, Nancy. "Transnationalizing the Public Sphere: on the Legitimacy and Efficacy of Public Opinion in a Post-Westphalian World." *Theory, Culture and Society* 24, no. 4 (2007): 7–30.
Friedlander, Saul, ed. *Probing the Limits of Representation: Nazism and the "Final Solution."* Cambridge: Harvard University Press, 1992.
Friedlander, Saul, ed. *Holocaust Literature: A Handbook of Critical, Historical, and Literary Writings*. Westport: Greenwood Press, 1993.
Friedlander, Saul, ed. *Memory, History, and the Extermination of the Jews of Europe*. Bloomington: Indiana University Press, 1993.
Friedman, Philip. "The Karaites under Nazi Rule." In *On the Track of Tyranny. Essays Published by the Wiener Library to Leonard G. Montefiore, on the Occasion of his Seventieth Birthday*, edited by Max Beloff, 97–123. London: The Wiener Library, 1960.
Gabowitsch, Mischa, Gdaniec, Cordula, and Makhotina, Ekaterina, ed. *Kriegsgedenken als Event: Der 9. Mai 2015 im postsozialistischen Europa*. Paderborn: Ferdinand Schöningh, 2017.
Gammer, Moshe. *Muslim Resistance to the Tsar: Shamil and the Conquest of Chechnia and Daghestan*. London: Frank Cass, 1994.
Gapurov, Shakhrudin, Bugaev, Abdula, and Chernous, Viktor. K 150-letiyu okonchaniya Kavkazskoi voiny: o khronologii, prichinakh i soderzhanii. *Nauchnaya mysl' Kavkaza* 4 (2014): 90–100.
Garbarini, Alexandra. *Numbered Days: Diaries and the Holocaust*. New Haven: Yale University Press, 2006.
Garrard, John. "The Nazi Holocaust in the Soviet Union: Interpreting Newly Opened Russian Archives." *East European Jewish Affairs* 25, no. 2 (1995): 3–40.
Gedi, Noa, and Yigal Elam. "Collective Memory – What is It?" *History and Memory* 8, no. 2 (1996): 30–50.
Genkina, Marina. "Jewish Artists in the Soviet Union in the 1960s–1970s." In *Jews of Struggle: The Jewish National Movement in the USSR, 1967–1989*, edited by Rachel Schnold, 80–104. Tel Aviv: Beit Hatefutsoth, 2007.
Gerasimova, Inna. "Novaya istoriya starogo pamyatnika." *Mishpokha* 22 (2008): 90–97.
Gershenson, Olga. *The Phantom Holocaust: Soviet Cinema and Jewish Catastrophe*. New Brunswick: Rutgers University Press, 2013.

Gershenson, Olga. "Dobrovol'naya amneziya: Kholokost v sovremennom rossiiskom kino." In *Kholokost: 70 let spustya: materialy mezhdunarodnogo foruma i 9-i mezhdunarodnoi konferentsii "Uroki Kholokosta i sovremennaya Rossiya,"* edited by Il'ya Al'tman, Igor' Kotler and Jürgen Zarusky, 183–191. Moscow: Tsentr "Kholokost," 2015.

Gershenson, Olga. "The Jewish Museum and Tolerance Center in Moscow: Judaism for the masses." *East European Jewish Affairs* 45, no. 2–3 (2015): 158–173.

Gerstenfeld, Manfred. "The Multiple Distortions of Holocaust Memor"y. *Jewish Political Studies Review* 19, no. 3/4 (2007): 35–55.

Gilbert, Shirli. *Music in the Holocaust: Confronting Life in the Nazi Ghettos and Camps*. New York: Oxford University Press, 2005.

Gitelman, Zvi. "The Soviet Union." In *The World Reacts to the Holocaust*, edited by David S. Wyman, 295–324. Baltimore: Johns Hopkins University Press, 1996.

Gitelman, Zvi. "Politics and Historiography of the Holocaust in the Soviet Union." In *Bitter Legacy. Confronting the Holocaust in the USSR*, edited by Zvi Y. Gitelman, 14–42. Bloomington and Indianapolis: Indiana University Press, 1997.

Glantz, David M. "The Struggle for the Caucasus." *The Journal of Slavic Military Studies* 22, no. 4 (2008): 588–711.

Gnilovskoi, Vladimir, ed. *Landshafty i ekonomicheskaya geografiya Severnogo Kavkaza: sbornik statei*. Stavropol: Stavropol'skii gosudarstvennyi universitet, 1977.

Golbert, Rebecca. Holocaust Memorialization in Ukraine. *Polin, Studies in Polish Jewry* 20 (2008): 222–243.

Goldberg, Amos. *Holocaust Diaries as "Life Stories."* Göttingen: Wallstein, 2008.

Goldberg, Amos. "The 'Jewish Narrative' in the Yad Vashem Global Holocaust Museum." *Journal of Genocide Research* 14, no. 2 (2012): 187–213.

Gontmakher, Mikhail. *Evrei na donskoi zemle: istoriya, fakty, biografiya*. Rostov-on-Don: Rostizdat, 2007.

Grabowski, Jan. "The Role of 'Bystanders' in the Implementation of the 'Final Solution' in Occupied Poland." *Yad Vashem Studies* 42, no. 1 (2015): 113–131.

Grechko, Andrei. *Bitva za Kavkaz*. Moscow: Ministerstvo oborony SSSR, 1971.

Green, Warren P. "The Fate of the Crimean Jewish Communities: Ashkenazim, Krimchaks and Karaites." *Jewish Social Studies* 46 (1984): 169–76.

Green, Warren P. "The Nazi Racial Policy towards the Karaites." *Soviet Jewish Affairs* 8 (1978): 36–44.

Grimwood, Marita. *Holocaust Literature of the Second Generation*. New York: Palgrave Macmillan, 2007.

Gronemeyer, Nicole. "Dispositiv. Apparat. Zu Theorien visueller Medien." *Medienwissenschaft. Rezensionen/Reviews* 1 (Hamburg: Universität Hamburg, 1998): 9–21.

Gross, Jan T. *Neighbors: The Destruction of the Jewish Community in Jedwabne, Poland*. Princeton: Princeton University Press, 2011.

Gurkin, Vladimir, and Kruglov, Aleksandr. "Oborona Kavkaza: 1942 god." *Voenno-istoricheskii zhurnal* 10 (1992): 11–18.

Gusenkova, Tamara, ed. *"Rasskazhu vam o voine . . . " Vtoraya mirovaya i Velikaya Otechestvennaya voiny v uchebnikakh i soznanii shkol'nikov slavyanskikh stran*. Moscow: Rossiiskii institut strategicheskikh issledovanii, 2012.

Halbwachs, Maurice. *On Collective Memory*, edited and translated by Lewis A. Coser. Chicago: University of Chicago Press, 1992.

Halbwachs, Maurice. *Das Gedächtnis und seine sozialen Bedingungen [Les cadres sociaux de la Mémoire]*. Frankfurt am Main: Suhrkamp, 2012.

Hamilton, Paula, and Shopes, Linda, ed. *Oral History and Public Memories*. Philadelphia: Temple University Press, 2008.

Hans, Jan. "Das Medien-Dispositiv." *Tiefenschärfe* (Winter Semester 2001/2002): 22–28.

Hansen-Glucklich, Jennifer. *Holocaust Memory Reframed: Museums and the Challenges of Representation*. New Brunswick: Rutgers University Press, 2014.

Harel, Dorit. *Facts and Feelings: Dilemmas in Designing the Yad Vashem Holocaust History Museum*. Jerusalem: Yad Vashem Publications, 2010.

Hartman, Geoffrey H. *The Longest Shadow: In the Aftermath of the Holocaust*. New York: Palgrave Macmillan, 1996.

Harviainen, Tapani. "The Karaites in Eastern Europe and the Crimea: An Overview." In *Karaite Judaism: A Guide to its History and Literary Sources*, edited by Meira Polliack, 633–656. Leiden: Brill, 2003.

Hatuev, Rashid. *Evrei v Karachae*. Chervessk: Karachaevskii NII, 2005.

Hausmann, Guido. "Die unfriedliche Zeit: Politische Totenkult im 20. Jahrhundert." In *Gefallenengedenken im globalen Vergleich: nationale Tradition, politische Legitimation und Individualisierung der Erinnerung*, edited by Manfred Hettling, 413–439. München: Oldenbourg Verlag, 2013.

Hayward, Joel. "Hitler's Quest for Oil: The Impact of Economic Considerations on Military Strategy, 1941–1942." *The Journal of Strategic Studies* 18, no. 4, (1995): 64–135.

Hayward, Joel. "Too Little, Too Late: An Analysis of Hitler's Failure in August 1942 to Damage Soviet Oil Production." *The Journal of Military History* 64 (2000): 769–794.

Heigl, Peter. *Konzentrationslager Flossenbürg in Geschichte und Gegenwart: Bilder und Dokumente gegen das zweite Vergessen*. Regensburg: Buchverlag der Mittelbayerischen Zeitung, 1989.

Hellbeck, Jochen. *Die Stalingrad-Protokolle. Sowjetische Augenzeugen berichten aus der Schlacht*. Frankfurt am Main: Fischer Verlag GmbH, 2012.

Hellbeck, Jochen, ed. *Stalingrad. The City That Defeated the Third Reich*. New York: Public Affairs, 2015.

Hickethier, Knut. "Kommunikationsgeschichte: Geschichte der Mediendispositive Ein Beitrag zur Rundfrage 'Neue Positionen zur Kommunikationsgeschichte'." *Medien und Zeit* 2 (1992): 26–28.

Hickethier, Knut. "Der politische Blick im Dispositiv Fernsehen. Der Unterhandlungswert der Politik in der medialen Republik." In *Die Politik der Öffentlichkeit – Die Öffentlichkeit der Politik. Politische Medialisierung in der Geschichte der Bundesrepublik*, edited by Bernd Weidbrod, 79–96. Göttingen: Wallstein, 2003.

Hickethier, Knut. *Einführung in die Medienwissenschaft*. Stuttgart: Metzler Verlag, 2003.

Hicks, Jeremy. *First Films of the Holocaust: Soviet Cinema and the Genocide of the Jews, 1938–1946*. Pittsburgh: University of Pittsburg Press, 2012.

Hicks, Jeremy. "Otrazhenie Kholokosta v sovetskikh dokumental'nykh fil'makh voennogo vremeni i ego vliyanie na pamyat' o zhertvakh voiny." In *Kholokost: 70 let spustya: materialy mezhdunarodnogo foruma i 9-i mezhdunarodnoi konferentsii "Uroki Kholokosta i sovremennaya Rossiya,"* edited by Il'ya Al'tman, Igor' Kotler and Jürgen Zarusky, 178–182. Moscow: Tsentr "Kholokost," 2015.

Hilberg, Raul. *Perpetrators, Victims, Bystanders: The Jewish Catastrophe 1933–1945*. London: Secker and Warburg, 1995.

Hillbrenner, Anke. "Invention of a Vanished World: Photography of Traditional Jewish Life in the Russian Pale of Settlement." *Jahrbücher für Geschichte Osteuropas* 57, no. 2 (2009): 173–188.

Himka, John-Paul, and Michlic, Joanna Beata, ed. *Bringing the Dark Past to Light: The Reception of the Holocaust in Postcommunist Europe*. Lincoln-London: University of Nebraska Press, 2013.

Hoffmann, Joachim. *Kaukasien, 1942–1943: Das deutsche Heer und die Orientvölken der Sowjetunion*. Freiburg: Rombach Verlag, 1991.

Holliday, Laurel, ed. *Children in the Holocaust and World War II: Their Secret Diaries*. New York: Pocket Books, 1995.

Hooper-Greenhill, Eilean. *Museums and the Interpretation of Visual Culture*. London: Routledge, 2000.

Hrynevych, Vladyslav, and Magocsi, Paul Robert, ed. *Babyn Yar: History and Memory*. Kyiv: Duch i Litera, 2016.

Hynes, Samuel. "Personal Narratives and Commemoration." In *War and Remembrance in the 20th Century*, edited by Jay Winter and Emmanuel Siran, 205–220. London: Cambridge University press, 1999.

Ivanova, Elena. *Vyzyvaya ogon' na sebya: polozhenie evreev pri "novom poryadke" gitlerovskikh okkupantov v 1941–1943 godakh*. Moscow: Tsentr "Kholokost," Rostov-on-Don: Feniks, 2011.

Jäger, Siegfried, and Jäger, "Margret. Das Dispositiv des institutionellen Rassismus. Eine diskurstheoretische Annäherung." In *Konjunkturen des Rassismus*, ed. by Alex Demirovic and Manuela Bojadziev, 212–224. Münster: Westliches Dampfboot, 2002.

Juni, Samuel. "Identity Disorders of Second Generation Holocaust Survivors." *Journal of Loss and Trauma* 21, no. 3 (2016): 203–212.

Kahanovich, Moshe. *Yidisher onteyl in der partizaner-bayegung fun Sovet-Rusland* [in Yiddish]. Rome: Oysg. fun der Tsentraler historisher komishe baym Partizaner farband P. H.H. in Italye, 1948.

Kahn, Miriam. "Heterotopic Dissonance in the Museum Representation of Pacific Island Cultures." *American Anthropologist* 97, no. 2 (1995): 324–338.

Kamenchuk, Irina, and Listvina, Evgeniya. "Kul'tura pamyati kak uslovie formirovaniya tolerantnosti: fenomen Kholokosta i problemy predodavaniya v rossiiskikh shkolakh." *Rossiisko-amerikanskii forum obrazovaniya: elektronnyi zhurnal* 4, no. 3 (2012), http://www.rus-ameeduforum.com/content/en/?task=art&article=1000933&iid=13.

Kamenchuk, Irina. "Fenomen Kholokosta kak faktor vospitaniya tolerantnosti: rossiiskii i mirovoi opyt." In *Proceedings of the 3rd International Academic Conference "Applied and Fundamental Studies,"* 408–413. St. Louis, Missouri: Science and Innovation Center, 2013.

Kardash, Anatolii. "Katastrofa evreev glazami neevreev (svidetel'stva ochevidtsev okkupatsii na territorii byvshego SSSR)." In *Voina na unichtozhenie: natsistskaya politika genotsida na territorii vostochnoi Evropy*, edited by Aleksandr Dyukov and Olesya Orlenko, 448–461. Moscow: Fond "Istoricheskaya Pamyat' ," 2010.

Karner, Stefan." Zum Umgang mit der historischen Wahrheit in der Sowjetunion: Die 'Außerordentliche Staatliche Kommission' 1942 bis 1951." In *Kärntner Landesgeschichte und Archivwissenschaft: Festschrift für Alfred Ogris zum 60. Geburtstag*, edited by Wilhelm Wadl, 509–523. Klagenfurt, 2001.

Karyshev, Mikhail, ed. *Neotvratimoe vozmezdie: po materialam sudebnykh protsessov nad izmennikami rodiny, fashistskimi palachami i agentami imperialisticheskikh razvedok.* Moscow: Voenizdat, 1984.
Kaznacheev, Andrei. *Puti kazach'e-krest'yasnskogo pereselencheskogo dvizheniya na Severnyi Kavkaz.* Moscow: Molodaya gvardiya, 2005.
Keller, Reiner. "Diskurse und Dispositive analysieren. Die Wissenssoziologische Diskursanalyse als Beitrag zu einer wissensanalytischen Profilierung der Diskursforschung." *Forum Qualitative Sozialforschung* 8, no. 2 (2007), http://www.qualitative-research.net/index.php/fqs/article/view/243/538.
Kelly, Laurence. *Lermontov: Tragedy in the Caucasus.* London: Tauris Park Paperbacks, 2003.
Kerman, Judith B., ed. *The Fantastic in Holocaust Literature and Film: Critical Perspectives.* Jefferson: McFarland, 2015.
Kerne, Aaron. *Film and the Holocaust: New Perspectives on Dramas, Documentaries, and Experimental Films.* New York: Continuum, 2011.
Khachaturova, Tatyana, ed. *Deti Kubani v Velikoi Otechestvennoi voine.* Krasnodar: Traditsiya, 2008.
Khanakhu, Ruslan, ed. *Mir kul'tury Adygov: problemy evolyutsii i tselostnosti.* Maykop: Adygea, 2004.
Khapaeva, Dina. *Goticheskoe obshchestvo. Mifologiya koshmara.* Moscow: Novoe literaturnoe obozrenie, 2008.
Kirei, Nikolai. "Evrei Krasnodarskogo kraya." *Byulleten': antropologiya, men'shinstva, mul'tikul'turalizm* 2 (2000): 134–144.
Kirschenbaum, Lisa A. *The Legacy of the Siege of Leningrad, 1941–1995: Myth, Memories, and Monuments.* Cambridge: Cambridge University Press, 2006.
Kizilov, Mikhail. *The Sons of Scripture: The Karaites in Poland and Lithuania in the Twentieth Century.* Berlin-Warsaw: Walter de Gruyter, 2015.
Klein, Kerwin Lee. "On the Emergence of Memory in Historical Discourse." *Representations* 69 (Winter 2000): 127–159.
Kleist, Peter. *Zwischen Hitler und Stalin: 1939–1945.* Bonn: Athenäum-Verlag, 1950.
Klimova, Iryna. "Babyn Yar in sculpture and painting". In *Babyn Yar: History and Memory*, edited by Vladyslav Hrynevych and Paul Robert Magocsi, 259–274. Kyiv: Duch i Litera, 2016.
Klokova, Galina. "Rossiiskie uchebniki istorii o Velikoi Otechestvennoi voine." *Prepodavanie istorii i obshchestvoznaniya v shkole* 1 (2001): 38–43.
Kloubert, Tetyana. "Holocaust Education in postsozialistischen Ländern am Beispiel von Russland, Polen und Ukraine." In *Holocaust Education in the 21st Century*, edited by Eva Matthes and Elizabeth Meilhammer, 214–229. Bad Heilbrunn: Verlag Julius Klinkhardt, 2015.
Koenen, Gerd, and Hielscher, Karla. *Die schwarze Front: der neue Antisemitismus in der Sowjetunion.* Reinbek: Rowohlt, 1991.
Kolesnik, Stanislav, ed. *Sovetskii Soyuz.* Vol. 3, Abramov, Petr, et al., *Rossiiskaya Federatsiya, Evropeiskii yugo-vostok, Povolzh'e, Severnyi Kavkaz.* Moscow: Mysl', 1968.
Konradova, Natalya, and Ryleeva, Anna. "Helden und Opfer Denkmäler in Russland und Deutschland." *Osteuropa* 4–6 (2005): 347–366.
Koposov, Nikolai. *Pamyat' strogogo rezhima: istoriya i politika Rossii.* Moscow: Novoe literaturnoe obozrenie, 2011.

Korobeinikov, Aleksandr, ed. *Ocherki istorii Stavropol'skogo kraya*. Vol. 1. *S drevneishikh vremen do 1917 g*. Stavropol: Knizhnoe izdatel'stvo, 1984.
Koselleck, Reinhart. *Zur politischen Ikonologie des gewaltsamen Todes: ein Deutsch-Französischer Vergleich*. Basel: Schwabe, 1998.
Koselleck, Reinhart. "Formen und Traditionen des Negativen Gedächtnisses." In *Verbrechen Erinnern. Die Auseinandersetzung mit Holocaust und Völkermord*, edited by Volkhart Knigge and Norbert Frei, 21–32. Bonn: BpB, 2005.
Kostyrchenko, Gennadii, ed. *Evreiskii antifashistskii komitet v SSSR, 1941–1948: dokumentirovannaya istoriya*. Moscow: Mezhdunarodnye otnosheniya, 1996.
Kostyrchenko, Gennadii, ed. *Gosudarstvennyi antisemitizm v SSSR: Ot nachala do kul'minatsii, 1938–1953*. Moscow: Materik, 2005.
Kostyrchenko, Gennadii. "Doktrina 'starshego brata' i formirovanie gosudarstvennogo antisemitizma v SSSR v svete ideologicheskoi i etnopoliticheskoi transformatsii Stalinskogo rezhima v 1930-e gg." In *Sovetskie natsii i natsional'naya politika v 1920–1950-e gg.: materialy VI mezhdunarodnoi nauchnoi konferentsii: Kiev, 10–12 oktyabrya 2013 g.*, 41–52. Moscow: Politicheskaya entsiklopediya, 2014.
Kostyrchenko, Gennadii. *Tainaya politika Stalina: vlast' i antisemitizm*. Moscow: Mezhdunarodnye otnosheniya, 2015.
Kovalev, Boris. *Kollaboratsionizm v Rossii v 1941–1945 gg.: Tipy i formy*. Velikii Novgorod: Novgorodskii gosudarstvennyi universitet, 2009.
Kovalev, Boris. "Memorializatsiya zhertv Kholokosta na territorii Novgorodchiny." In *Kholokost: 70 let spustya: materialy mezhdunarodnogo foruma i 9-i mezhdunarodnoi konferentsii "Uroki Kholokosta i sovremennaya Rossiya,"* edited by Il'ya Al'tman, Igor' Kotler and Jürgen Zarusky, 162–168. Moscow: Tsentr "Kholokost," 2015.
Kovriguina, Assia. "Le témoignage en URSS: Qui est l'auteur?" *Fabula / Les colloques, témoigner sur la Shoah en URSS*, http://www.fabula.org/colloques/document2856.php.
Krausnick, Helmut. "Die Einsatzgruppen vom Anschluß Österreichs bis zum Feldzug gegen die Sowjetunion: Entwicklung und Verhältnis zur Wehrmacht." In *Die Truppe des Weltanschauungskrieges: Die Einsatzgruppen der Sicherheitspolizei und des SD 1938–1942*, edited by Helmut Krausnick and Hans-Heinrich Wilhelm, 195–204. Stuttgart: Deutsche Verlags-Anstalt, 1981.
Krikunov, Petr. *Kazaki: Mezhdu Stalinym i Gitlerom: krestovyi pokhod protiv bol'shevizma*. Moscow, Yauza: Eksmo, 2006.
Krinko, Evgenii. *Zhizn' za liniei fronta: Kuban' v okkupatsii (1942–1943)*. Maykop: Adygeyskii gosudarstvennyi universitet, 2000.
Kruglov, Aleksandr. *Tragediya Bab'ego Yara v nemetskikh dokumentakh*. Dnepropetrovsk: Tsentr "Tkuma", ChP "Lira LTD," 2011.
Kudelko, Sergei, et al., ed. *Gorod i voina: Khar'kov v gody Velikoi Otechestvennoi voiny*. Saint-Petersburg: Aleteia, 2013.
Kukhiwtsak, Pyotr. "Mediating Trauma: How do we Read the Holocaust Memoirs?" In *Tradition, Translation, Trauma: The Classic and the Modern*, edited by Jan Parker and Timothy Mathews, 283–298. Oxford: Oxford University Press, 2011.
Kukulin, Il'ya. "Sovetskaya poeziya o Vtoroi mirovoi voine: ritorika skrytoi amal'gamy." In *SSSR vo Vtoroi mirovoi voine: okkupatsiya: Kholokost: Stalinizm*, edited by Oleg Budnitskii and Lyudmila Novikova, 328–351. Moscow: ROSSPEN, 2014.
Kukulin, Il'ya. *Mashiny zashumevshego vremeni: kak sovetskii montazh stal metodom ne ofitsial'noi kul'tury*. Moscow: Novoe literaturnoe obozrenie, 2015.

Kukulin, Il'ya. "Russian Literature on the Shoah: New Approaches and Contexts." *Kritika: Explorations in Russian and Eurasian History* 18, no. 1 (Winter 2017): 165–175.

Kuleshov, Sergei, et al., ed. *Natsional'naya politika Rossii: istoriya i sovremennost'*. Moscow: "Russkii Mir," 1997.

Kumanev, Georgii. "Evakuatsiya naseleniya SSSR: Dostignutye rezul'taty i poteri." In *Lyudskie poteri SSSR v period Vtoroi mirovoi voiny: materialy konferentsii, 14–15 marta 1995 goda*, 137–146. St. Petersburg: Russko-Baltiyskii informatsionnyi tsentr "Blitz", 1995.

Kumanev, Georgii. "Voina i evakuatsiya v SSSR: 1941–1942 gody." *Novaya i noveishaya istoriya* 6, 2006: 7–27.

Kupovetskii, Mark. "Dinamika chislennosti i rasselenie karaimov i krymchakov za poslednie dvesti let." In *Geografiya i kul'tura etnograficheskikh grupp tatar v SSSR*, edited by Igor' Krupnik, 76–93. Moscow: Geograficheskoe obshchestvo SSSR, 1983.

Kurnosov, Aleksei. "O memuarakh uchastnikov partizanskogo dvizheniya (1941–1945 gg.)." In *Istochnikovedenie istorii sovetskogo obshchestva*, edited by Nikolai Ivnitskii. Vol. 1, 289–319. Moscow: Nauka, 1964.

Kurnosov, Aleksei. "Priemy vnutrennei kritiki memuarov: vospominaniya uchastnikov partizanskogo dvizheniya v period Velikoi Otechestvennoi voiny kak istoricheskii istochnik." In *Istochnikovedenie: teoreticheskie i metodologicheskie problemy: sbornik statei*, edited by Sigurd Shmidt, 478–505. Moscow: Nauka, 1969.

Kurnosov, Aleksei. Vospominaniya – interv'yu v fonde Komissii po istorii Velikoi Otechestvennoi voiny Akademii nauk SSSR (Organizatsiya i metodika sobiraniya). *Arheograficheskii ezhegodnik za 1973 god.* (Moscow: Nauka, 1974): 118–133.

Kutler, Laurence. "Holocaust Diaries and Memoirs." In *Holocaust Literature: A Handbook of Critical, Historical, and Literary Writings*, edited by Saul S. Friedman, 521–532. Westport: Greenwood Press, 1993.

Kutsenko, Igor'. *Pobediteli i pobezhdennye. Kubanskoe kazachestvo: istoriya i sud'by*. Krasnodar: Diapazon-V, 2010.

Kuz'min, Sergei. "Ikh zakapyvali zhivymi." In *Natsistskikh prestupnikov – k otvetu!*, edited by Dmitrii Pogorzhel'skii, 44–55. Moscow: Politizdat, 1983.

Landsberg, Alison. *Prosthetic Memory: The Transformation of American Remembrance in the Age of Mass Culture*. New York: Columbia University Press, 2004.

Lang, Berel, ed. *Writing and the Holocaust*. New York: Holmes and Meier, 1988.

Langer, Lawrence L. *The Holocaust and the Literary Imagination*. New Haven: Yale University Press, 1975.

Latour, Bruno. *Reassembling the Social: An Introduction to Actor-Network Theory*. Oxford: Oxford University Press, 2005.

Laub, Dori. "Bearing Witness or the Vicissitudes of Listening." In *Testimony: Crises of Witnessing in Literature, Psychoanalysis, and History*, edited by Shoshana Felman and Dori Laub, 57–74. New York, 1992.

Lebedeva, Natalia. *Podgotovka Nyurnbergskogo protsessa*. Moscow: Nauka, 1975.

Lehr, Johanna, and Bensimon, Patrice. *Case Studies: Intersection of Archives and Fieldwork, Motol (Pinsk Region, Belarus)*, https://www.yahadinunum.org/case-studies-intersection-of-archives-and-fieldwork-motol-pinsk-region-belarus/.

Levshin, Boris. "Deyatel'nost' Komissii po istorii Velikoi Otechestvennoi voiny, 1941–1945 gg." In *Istoriya i istoriki: istoriograficheskii ezhegodnik*, 1974, edited by Melissa Nechkina, 312–317. Moscow: Akademiya Nauk SSSR, 1976.

Leydesdorff, Selma. *Sasha Pecherskii: Holocaust Hero, Sobibor Resistance Leader, and Hostage of History*. New York and London: Routledge, 2017.

Likhanov, Al'bert, et al. *Pionery-geroi*. Moscow: Malysh, 1980.

Linenthal, Edward T. *Preserving Memory: The Struggle to Create America's Holocaust Museum*. New York: Columbia University Press, 2001.

Linets, Sergei. "Evakuatsiya s territorii Severnogo Kavkaza naseleniya letom 1942 goda: otsenka rezul'tatov." *Nauchnaya mysl' Kavkaza* 6 (2003): 89–96.

Linets, Sergei. *Gorod vo mgle . . . (Pyatigorsk v period nemetsko-fashistskoi okkupatsii. Avgust 1942 g. – yanvar' 1943 g.)*. Pyatigorsk: Pyatigorskii gosudarstvennyi lingvisticheskii universitet, 2010.

Lokshin, Aleksandr. "Istoriya rossiiskikh evreev v sovremennykh shkol'nykh uchebnikakh Rossiiskoi Federatsii." In *Evroaziatskii evreiskii ezhegodnik – 5768 (2007–2008)*, edited by Vyacheslav Likhachev, 177–202. Moscow: Pallada, 2008.

Lokshin, Aleksandr. *Istoriya rossiiskikh evreev v shkol'nykh uchebnikakh RF*. Issledovaniya po prikladnoi i neotlozhnoi etnologii. Vol. 209. Moscow: Institut etnologii i antropologii RAN, 2009.

Lord, Beth. Foucault's Museum: Difference, Representation, and Genealogy. *Museum and Society*, 4 no. 1 (2006): 11–24.

Löwenthal, Richard. The Judeo-Tats in the Caucasus. *Historica Judaica* XIV (1952): 61–82.

Lower, Wendy. *Nazi Empire-Building and the Holocaust in Ukraine*. Chapel Hill: University of North Carolina Press, 2005.

Luks, Leonid, ed. *Der Spätstalinismus und die "jüdische Frage": zur antisemitischen Wendung des Kommunismus*. Köln: Böhlau, 1998.

Makhotina, Ekaterina. "Symbole der Macht, Orte der Trauer: Die Entwicklung der Rituellen und symbolischen Ausgestaltung von Ehrenmalen des Zweiten Weltkriegs in Russland." In *Medien zwischen Fiction-Making und Realitätsanspruch: Konstruktionen historischer Erinnerungen*, edited by Monika Heinemann, 279–306. München: Oldenbourg Verlag, 2011.

Makhotina, Ekaterina. "Between 'Suffered' Memory and 'Learned' Memory: The Holocaust and Jewish History in Lithuanian Museums and Memorials after 1990." *Yad Vashem Studies* 44, no. 2 (2016): 207–246.

Malysheva, Elena. *V bor'be za pobedu: sotsial'nye otnosheniya i ekonomicheskoe sotrudnichestvo rabochikh i krest'yan Severnogo Kavkaza v gody voiny 1941–1945*. Maykop: Adygeiskoe knizhnoe izdatel'stvo, 1992.

Malysheva, Elena. *Ispytanie: sotsium i vlast': problemy vzaimodeistviya v gody Velikoi Otechestvennoi voiny 1941–1945*. Maykop: Adygeya, 2000.

Mankoff, Jeff. "Babi Yar and the Struggle for Memory, 1944–2004." *Ab Imperio* 2 (2004): 393–415.

Marchukov, Andrei. *Geroi-pokryshkintsy o sebe i svoem komandire: pravda iz proshlogo. 1941–1945*. Moscow: Tsentrpoligraf, 2014.

Marcuse, Harold. "Holocaust Memorials: The Emergence of a Genre." *American Historical Review* 115, no. 1 (2010): 53–89.

McGlothlin, Erin. *Second-Generation Holocaust Literature: Legacies of Survival and Perpetration*. Rochester, NY: Camden House, 2006.

McGlothlin, Erin. "Narrative Perspective and the Holocaust Perpetrator: Edgar Hilsenrath's The Nazi and the Barber and Jonathan Littell's The Kindly Ones." *The Bloomsbury Companion to Holocaust Literature*, edited by Jenni Adams, 159–178. London: Bloomsbury, 2014.

Meir, Natan M. "The Place of Polish Jewry in a Russian Jewish Museum: the Case of the Jewish Museum and Tolerance Center in Moscow." *Studia Judaica* 32, no. 2 (2013): 101–114.
Melikhova, Julia. "Opyt poiskovo-kraevedcheskogo dvizheniya sovetskogo perioda i perspektivy razvitiya voenno-istoricheskogo turizma v Kurskoi oblasti." *Uchenye zapiski rossiiskogo gosudarstvennogo sotsial'nogo universiteta* 4 (2011): 71–75.
Michelkevičius, Vytautas. *The Lithuanian SSR Society of Art Photography (1969–1989): An Image Production Network*. Vilnius: Vilnius Academy of Arts Press, 2011.
Miller, Alyson. "Stylised Configurations of Trauma: Faking Identity in Holocaust Memoirs." *Arcadia: International Journal of Literary Culture* 49, no. 2 (2014): 229–253.
Miroshnikova, Lyudmila. "Muzei v evreiskoi shkole kak metodicheskii regional'nyi tsentr sokhraneniya pamyati o Kholokoste." In *Pamyat' o Kholokoste: problemy memorializatsii: materialy 6-i mezhdunarodnoi konferentsii "Uroki Kholokosta i sovremennaya Rossiya,"* edited by Il'ya Al'tman, 156–160. Moscow: Tsentr i Fond "Kholokost," 2012.
Miszta, Barbara. *Theories of Social Remembering*. Maidenhead: Open University Press, 1993.
Movshovich, Evgenii. "Kholokost v Rostove." *Yakhad (Rostov-na-Donu)* 3 (1999): 1–7.
Movshovich, Evgenii. *Ocherki istorii evreev na Donu*. Rostov-on-Don: ZAO "Kniga," 2011.
Müller, Beate. "Translating Trauma: David Boder's 1946 Interviews with Holocaust Survivors." *Translation and Literature* 23, no. 2 (2014): 257–271.
Müller, Daniel. "Besetzte Kaukasus-Gebiet." In *Handbuch zum Widerstand gegen Nationalsozialismus und Faschismus in Europa 1933/39 bis 1945*, edited by Gerd R. Ueberschär, 235–243. Berlin: Walter de Gruyter GmbH, 2011.
Müller, Hannelore. *Religionswissenschaftliche Minoritätenforschung: zur religionshistorischen Dynamik der Karäer im Osten Europas*. Wiesbaden: Harrassowitz, 2010.
Müller, Rolf-Dieter. "Gebirgsjäger am Elbrus: Der Kaukasus als Ziel nationalsozialistischer Eroberungspolitik." In *Wegweiser zur Geschichte Kaukasus*, edited by Bernhard Chiari, 55–66. Paderborn: Ferdinand Schöningh, 2008.
Murav, Harriet. *Music from a Speeding Train: Jewish Literature in Post-Revolution Russia*. Stanford: Stanford University Press, 2011.
Nechkina, Melitsa. *Dekabristy*. Moscow: Nauka, 1982.
Nelson, Robert S., and Olin, Margaret, ed. *Monuments and Memory, Made and Unmade*. Chicago: University of Chicago Press, 2003.
Neumann, Arndt. "Das Internet-Dispositiv." *Tiefenschärfe* 2 (2002): 10–12.
Nora, Pierre. "Between Memory and History: 'Les Lieux de Mémoire'." *Representations* 26 (1989): 7–24.
Nora, Pierre. *Realms of Memory: Rethinking the French Past*. Vol. 1, *Conflicts and Divisions*, edited by Lawrence D. Kritzman. New York: Columbia University Press, 1996.
Norkina, Ekaterina. "The Origins of Anti-Jewish Policy in the Cossack Regions of the Russian Empire, Late Nineteenth and Early Twentieth Century." *East European Jewish Affairs* 43, no. 1 (2003): 62–76.
Nowicka, Magdalena. "Zur Diskurs- und Dispositivanalyse des kollektiven Gedächtnisses als Antwort auf einen öffentlichen Krisenzustand. Zwischen Habermas und Foucault." *Forum: Qualitative Sozialforschung* 15, no. 2 (2014). http://nbn-resolving.de/urn:nbn:de:0114-fqs1401228.
Olick, Jeffrey K. "'Collective Memory': A Memoir and Prospect." *Memory Studies* 1, no. 1 (2008): 23–29.
Olick, Jeffrey K. "Collective Memory: The Two Cultures." *American Sociological Association* 17, no. 3 (2009): 333–348.

Olick, Jeffrey K, and Joyce, Robbins. "Social Memory Studies: From 'Collective Memory' to the Historical Sociology of Mnemonic Practices." *Annual Review of Sociology* 24 (1998): 105–140.
Olick, Jeffrey K, Vinitzky-Seroussi, Vered, and Levy, Daniel ed. *The Collective Memory Reader*. New York: Oxford University Press, 2011.
Otershtein, Tankha. "Peredaite eto detyam vashim . . ." *Vekhi Taganroga* 4 (2000): 37–40
Othmer, Karl, ed. *Medien – Bildung – Dispositive: Beiträge zu einer interdisziplinären Medienbildungsforschung*. Wiesbaden: Springer VS, 2015.
Paulsson, Gunnar S. "Bystanders' to the Holocaust: A Re-Evaluation." *Holocaust and Genocide Studies* 18, no. 1 (2004): 110–112.
Pavlova, Tatyana. "Dokumenty chrezvychainoi gosudarstvennoi komissii po ustanovleniyu i rassledovaniyu zlodeyanii nemetsko-fashistskikh zakhvatchikov o Kholokoste." In *Nyurenbergskii protsess: uroki istorii*, edited by Natalia Lebedeva, 61–71. Moscow: Institut Vseobshchei Istorii, 2006.
Perks, Robert, and Thomson, Alistair, ed. *The Oral History Reader*. London and New York: Routledge, 1998.
Persin, Anatolii. *Kraevedenie i shkol'nye muzei: Istoriya, teoriya, praktika*. Moscow: Federal'nyi tsentr detsko-yunosheskogo turizma i kraevedeniya, 2006.
Petermann, Sandra. *Rituale machen Räume: Zum kollektiven Gedenken der Schlacht von Verdun und der Landung in der Normandie*. Bielefeld: Transcript, 2007.
Petrova, Ekaterina. "Osobennosti prepodavaniya temy 'Kholokost' v starshikh klassakh." *Prepodavanie istorii v shkole* 6 (2011): 51–54.
Pieper, Katrin. *Die Musealisierung des Holocaust: das Jüdische Museum Berlin und das U.S. Holocaust Memorial Museum in Washington D.C.* Köln: Böhlau Verlag, 2006.
Pobol', Nikolai, and Polyan, Pavel, ed. *Stalinskie deportatsii, 1928–1953*. Moscow: Materik, 2005.
Pohl, Dieter. *Die Herrschaft der Wehrmacht: deutsche Militärbesatzung und einheimische Bevölkerung in der Sowjetunion; 1941–1944*. München: Oldenburg, 2008.
Polustertye sledy: rossiiskie shkol'niki o evreiskikh sud'bakh: iz rabot uchastnikov vserossiiskogo konkursa starsheklassnikov "Chelovek v istorii. Rossiya – XX Vek." Moscow: "Memorial," 2006.
Polyan, Pavel. *Mezhdu Aushvitsem i Bab'im Yarom: razmyshleniya i issledovaniya*. Moscow: ROSSPEN, 2010.
Popkin, Jeremy D. "Holocaust Memories, Historians' Memoirs: First-person Narrative and the Memory of the Holocaust." *History and Memory* 15, no. 1 (2003): 49–84.
Popov, Andrei. *Dekabristy-literatory na Kavkaze*. Stavropol: Knizhnoe izdatel'stvo, 1963.
Popov, Oleg, and Cherkasov, Aleksandr, ed. *Rossiya – Chechnya: tsep' oshibok i prestuplenii*. Moscow: "Zven'ya," 1998.
Portelli, Alessandro. *The Death of Luigi Trastulli and Other Stories: Form and Meaning in Oral History*. Albany: State University of New York Press, 1991.
Potemkina, Marina. "Evakuatsiya i natsional'nye otnosheniya v sovetskom tylu v gody Velikoi Otechestvennoi voiny." *Otechestvennaya istoriya* 3 (2002): 148–156.
Poulsen, Niels Bo. "Rozsliduvaniya voennikh zlochiniv 'po-sovets'ki': kritichnii analiz materialiv nadzvichainoi derzhavnoi komisii." *Golokost i suchasnist'* 1, no. 5 (2009): 27–45.
Pribylov, Georgii. *Mikhizeevskaya tragediya ili "Kubanskaya Khatyn'."* The village of Mostovskoi, 1993.

Privler, Maks, Kremyanskaya, Anna, and Kremyanskii, Pavel, ed. *Evreiskie deti v bor'be s natsizmom: vsemirnyi dokumental'nyi sbornik*. Vol. 1. Tel-Aviv: Biblioteka Matveya Chernogo, 2001.

Pronshtein, Aleksandr. "Rol' kazachestva i krest'yanstva v zaselenii i khozyaystvennom osvoenii Dona i stepnogo predkavkaz'ya v XVIII – pervoi polovine XIX veka." *Izvestiya Severo-Kavkazskogo nauchnogo tsentra vysshei shkoly: obshchestvennye nauki* 1 (1982): 55–59.

Pupyshev, Nikolai. "Iz vospominanii o A.S. Shcherbakove." *Druzhba narodov* 2 (1985): 162–174.

Pykhalov, Igor'. *Za chto Stalin vyselyal narody? Stalinskie deportatsii – prestupnyi proizvol ili spravedlivoe vozmezdie?* Moskow: Yauza-Press, 2008.

Rafailov, Aleksandr. *Zazhivo pogrebyennye*. Kislovodsk: MIL, 2004.

Rebrova, Irina. "'Evreiskii vopros' na Kubani v ofitsial'nykh dokumentakh voennogo vremeni: Analiz ideologicheskikh, smyslovykh i vremennykh shtampov." In *Natsistskaya politika genotsida na okkupirovannykh territoriyakh SSSR*, edited by Evgenii Rozenblat, 127–132. Brest: Brestskii gosudarstvennyi universitet, 2014.

Rebrova, Irina. "Evakuatsiya evreev na Severnyi Kavkaz: motivatsiya i puti sledovaniya (po dannym ustnykh interv'yu s perezhivshimi Kholokost)." In *Trudy po evreiskoi istorii i kul'tury: materialy XXII mezhdunarodnoi ezhegodnoi konferentsii po iudaike*, edited by Viktoriya Mochalova. Vol. 52, 108–122. Moscow: Sefer, 2016.

Rebrova, Irina. "Traumatische Kindheit: Holocaust und Überlebenspraktiken jüdischer Kinder in den besetzten Gebieten des Nordkaukasus im Zweiten Weltkrieg." In *Kindheiten im Zweiten Weltkrieg*, edited by Francesca Weil, Andre Postert, and Alfons Kenkmann, 393–410. Halle: Mitteldeutscher Verlag, 2018.

Rieth, Adolf. *Den Opfern der Gewalt: KZ-Opfermale der Europäischen Völker*. Tübingen: Wasmuth, 1968.

Ritchie, Donald A., ed. *The Oxford Handbook of Oral History*. Oxford: Oxford University Press, 2011.

Rogoff, Irit. "From Ruins to Debris: the Feminization of Fascism in German-history Museums." In *Museum – Culture: Histories, Discourses, Spectacles*, edited by Daniel J. Sherman, 223–249. London: Routledge, 1994.

Rosen, Alan, ed. *Literature of the Holocaust*. Cambridge: Cambridge University Press, 2013.

Rothberg, Michael. *Multidirectional Memory: Remembering the Holocaust in the Age of Decolonization*. Stanford: Stanford University Press, 2009.

Rothberg, Michael. "Introduction: Between Memory and Memory: From Lieux de Mémoire to Noeuds de Mémoire." *Yale French Studies* 118/119 (2010): 3–12.

Rowe-McCulloch, Maris. "Poison on the Lips of Children: Rumors and Reality in Discussions of the Holocaust in Rostov-on-Don (USSR) and Beyond." *The Journal of Holocaust Research*, 33 no. 2 (2019): 157–174.

Rozett, Robert. "Published Memoirs of Holocaust Survivors." In *Remembering for the Future: The Holocaust in an Age of Genocide*, edited by John K. Roth and Elisabeth Maxwell. Vol. 3. *Remembering for the Future*, 167–171. Basingstoke, Hampshire: Palgrave, 2001.

Rozhdestvenskaya, Elena. "Reprezentatsiya kul'turnoi travmy: muzeefikatsiya Kholokosta." *Logos* 27, no. 5 (2017): 103–130.

Rozhdestvenskaya, Elena, Semenova, Viktoria, Tartakovskaya, Irina, and Kosela, Krzysztof, ed. *Collective Memories in War*. London-New York: Routledge, 2016.

Rubenstein, Philip and Taylor, Warren. "Teaching about the Holocaust in the National Curriculum." *British Journal of Holocaust Education* 1, no. 1 (1992): 47–54.
Rubtsov, Yurii. *Mekhlis: Ten' vozhdya*. Moscow: Veche, 2011.
Ryabov, Oleg. "'Rodina-mat'": istoriya obraza." *Zhenshchina v rossiiskom obshchestve*, 3 (2006): 33–46.
Ryvkin, Mikhail, and Frenkel', Aleksandr. Pamyatniki i pamyat'. *VEK* 7, no. 8 (1989): 23–25.
Scheide, Carmen. "Kollektive und individuelle Erinnerungsmuster und der Großen Vaterländischen Krieg (1941–1945)." In *Stalinistische Subjekte: Individuum and System in der Sowjetunion und der Komintern, 1929–1953*, edited by Brigitte Studer und Heiko Haumann, 435–453. Zürich: Chronos, 2006.
Scherbakowa, Irina. *Zerrissene Erinnerung: der Umgang mit Stalinismus und Zweitem Weltkrieg im heutigen Russland*. Göttingen: Wallstein, 2010.
Schneider, Werner. "Dispositive . . . – überall (und nirgendwo)? Anmerkungen zur Theorie und methodischen Praxis der Dispositivforschung." In *Medien – Bildung – Dispositive: Beiträge zu einer interdisziplinären Medienbildungsforschung*, edited by Karl Othmer, 21–40. Wiesbaden: Springer VS, 2015.
Schroeter, Klaus R. "Pflege als Dispositiv: zur Ambivalenz von Macht, Hilfe und Kontrolle im Pflegediskurs." In *Soziologie der Pflege. Grundlagen, Wissensbestände und Perspektiven*, edited by Klaus R. Schroeter and Thomas Rosenthal, 385–404. Weinheim: Jeventa, 2005.
Schumann, Wolfgang, ed. *Deutschland im Zweiten Weltkrieg*. Vol. 2, *Vom Überfall auf die Sowjetunion bis zur sowjetischen Gegenoffensive bei Stalingrad (Juni 1941 bis November 1942)*. Berlin: Akademie-Verlag, 1975.
Selyunin, Vladimir. *Promyshlennost' i transport yuga Rossii v voine 1941–1945 gg*. Rostov-on-Don: Knizhnoe izdatel'stvo, 1997.
Semiryaga, Mikhail. *Kollaboratsionizm: priroda, tipologiya i proyavleniya v gody Vtoroi mirovoi voiny*. Moscow: ROSSPEN, 2000.
Seravin, Aleksandr and Sopov, Igor'. *Ekspress analiz prepodavaniya istorii Rossii i regiona v sub"ektah Severo-kavkazskogo federal'nogo okruga*. Vladikavkaz, 2014.
Shapira, Dan. "Beginnings of the Karaite Communities of the Crimea Prior to the Sixteenth Century." In *Karaite Judaism: A Guide to its History and Literary Sources*, edited by Meira Polliack, 709–728. Leiden: Brill, 2003.
Shevyakov, Aleksei. "Gitlerovskii genotsid na territoriyakh SSSR." *Sotsiologicheskie issledovaniya* 12 (1991): 3–11.
Shiyanov, Yury, ed. *Arz – imya sobstvennoe*. Budennovsk: Budenovskaya Tipografiya, 2004.
Shneer, David. *Through the Soviet Jewish Eyes: Photography, War, and the Holocaust*. New Brunswick: Rutgers University Press, 2012.
Shpagin, Sergei. "Problemy memorializatsii zhertv Kholokosta: sootnoshenie istorii i postistorii." In *Pamyat' o Kholokoste: problemy memorializatsii: materialy 6-i mezhdunarodnoi konferentsii "Uroki Kholokosta i sovremennaya Rossiya,"* edited by Il'ya Al'tman, 22–25. Moscow: Tsentr i Fond "Kholokost," 2012.
Shrayer, Maxim D. *I Saw It: Ilya Selvinskii and the Legacy of Bearing Witness to the Shoah*. Boston: Academic Studies Press, 2013.
Sim, Stuart. "Postmodernism and Philosophy." In *The Routledge Companion to Postmodernism*, edited by Stuart Sim, 3–14. London, New York: Routledge, 2001.
Simkin, Lev. *Poltora chasa vozmezdiya*. Moscow: Zebra E, 2013.
Sinitsyn, Andrei. "Chrezvychainye organy sovetskogo gosudarstva v gody Velikoi Otechestvennoi voiny." *Voprosy istorii* 2 (1955): 32–43.

Slezkine, Yuri. *The Jewish Century*. Princeton: Princeton University Press, 2004.
Snyder, Timothy. *Black Earth: The Holocaust as History and Warning*. London: The Bodley Head, 2015.
Sodaro, Amy. *Exhibiting Atrocity: Presentation of the Past in Memorial Museums*. New Brunswick: New School University, 2011.
Sorokina, Marina. "'Svideteli Nyurnberga': ot ankety k biografii." In *Pravo na imya: biografiya kak paradigma istoricheskogo protsessa: vtorye chteniya pamyati V. Iofe 16–18 aprelya 2004*, edited by Irina Flige, 50–63. Saint Petersburg: "Memorial," 2005.
Sorokina, Marina. "People and Procedures: Toward a History of the Investigation of Nazi Crimes in the USSR." *Kritika: Explorations in Russian and Eurasian History* 6, no. 4 (2005): 1–35.
Sorokina, Marina. "Lyudi i protsedury. K istorii rassledovaniya natsistskikh prestuplenii v SSSR." In *Voenno-istoricheskii almanakh Viktora Suvorova*, edited by Viktor Suvorov and Dmitrii Hmelnitskii. Vol. 2, 71–126. Moscow: Dobraya Kniga, 2013.
Spangenberg, Peter M. "'Weltempfang' im Mediendispositiv der 60er Jahre." In *Medienkultur der 60er Jahre. Diskursgeschichte der Medien nach 1945*, edited by Irmela Schneider, Thorsten Hahn, and Christina Bartz. Vol. 2, *149–158*. Wiesbaden: Westdeutscher Verlag, 2003.
Sredin, Gennadii, ed. *Politorgany sovetskikh vooruzhennykh sil: istoriko-teoreticheskii ocherk*. Moscow: Vrenizdat, 1984.
Staff, Else. *Justiz im Dritten Reich: eine Dokumentation*. Frankfurt am Main: Fischer, 1964.
Stark, Jared. "Broken Records: Holocaust Diaries, Memoirs, and Memorial Books." In *Teaching the Representation of the Holocaust*, edited by Marianne Hirsch and Irene Kacandes, 191–204. New York: Modern Language Association of America, 2004.
Steinmetz, Rüdiger. *Das digitale Dispositif Cinéma: Untersuchungen zur Veränderung des Kinos*. Leipzig: Leipziger Universität Verlag, 2011.
Stepanenko, Sergei. "Sudebnye protsessy nad voennymi prestupnikami i ikh posobnikami kak akt realizatsii norm mezhdunarodnogo gumanitarnogo prava (na materialah Krasnodarskikh sudebnykh protsessov 1943–1974 godov)." *Istoricheskie, filosofskie, politicheskie i juridicheskie nauki, kul'turologiya i iskusstvovedenie. Voprosy teorii i praktiki* 1, no. 5 (Tambov, 2010): 161–165.
Stewart, Victoria. "Holocaust Diaries: Writing from the Abyss." *Forum for Modern Language Studies* 41, no. 4 (2005): 418–426.
Stolov, Valerii. "Evreyskaya istoriya v rossiiskoi shkole." *Novaya evreyskaya shkola* 1 (1998): 163–182.
Struk, Janina. *Photographing the Holocaust: Interpretations of the Evidence*. London-New York: I.B. Tauris, 2004.
Sturken, Marita. "The Remembering of Forgetting: Recovered Memory and the Question of Experience." *Social Text* 57 (1999): 103–125.
Sukhorukov, Vasilii. *Istoricheskoe opisanie zemli voiska Donskogo*. Rostov-on-Don: Novyi biznes, 2001.
Tcherkasski, Aleksandra. "Vneshnie i vnutrennie vliyaniya na razvitie sovetskogo memorial'nogo landshafta o Vtoroi mirovoi voine." *Bylye gody* 25, no. 3 (2012): 82–89.
Tcherkasski, Aleksandra. "Mesto Kholokosta v sovetskom memorial'nom landshafte." In *Materialy XX ezhegodnoi mezhdunarodnoi mezhdistsiplinarnoi konferentsii po iudaike*, ited by Viktoriya Mochalova. Vol. IV, 85–103. Moscow: Tsentr "Sefer," 2013.

Tcherkasski, Aleksandra, and Terushkin, Leonid. "Strategii sovetskoi evreiskoi obshchestvennosti po uvekovechivaniyu pamyati na primere ustanovki pamyatnikov evreyam-zhertvam fashizma." *Bylye gody* 29 (2013): 69–74.
Terushkin, Leonid. "Pis'ma i dnevniki kak istochnik po istorii Vtoroi mirovoi voiny i Kholokosta (po materialam arkhiva nauchno-prosvetitel'nogo tsentra "Kholokost")." In *Kholokost: novye issledovaniya i materialy: materialy XVIII mezhdunarodnoi ezhegodnoi konferentsii po iudaike*, edited by Victoria Mochalova. Vol. IV, 94–104. Moscow: Sefer, 2011.
Terushkin, Leonid. "'I am the Only Witness Who Survived': Fragments from the Family Correspondence." In *Materials of Conference in Riga 2015–2016*, edited by Menahems Barkahans, 523–535. Riga: Society "Shamir," 2016.
Terushkin, Leonid. "Pis'ma rodstvennikov, sosedei i druzei v evakuatsiyu – neizvestnye svidetel'stva o Kholokoste." In *Istoriya. Pamyat'. Lyudi: materialy VIII mezhdunarodnoi nauchno-prakticheskoi konferentsii*, edited by Maria Makarova, 158–172. Almaty: Assotsiatsiya evreiskikh natsional'nykh organizatsii "Mitsva," 2017.
Teske, Hermann. *General Ernst Köstring: Der militärische Mittler zwischen dem Deutschen Reich und der Sowjetunion 1921–1941*. Frankfurt am Main: Mittler und Sohn, 1965.
Teitelbaum, Yurii. "Memorializatsiya Kholokosta v Krasnodarskom krae: opyt i problemy." In *Pamyat' o Kholokoste: problemy memorializatsii: materialy 6-i mezhdunarodnoi konferentsii "Uroki Kholokosta i sovremennaya Rossiya,"* edited by Il'ya Al'tman, 53–59. Moscow: Tsentr i Fond "Kholokost," 2012.
Tikhonov, Vitalii. "Materialy komissii po istorii Velikoi Otechestvennoi voiny 1941–1945 gg. Akademika I.I. Mintsa kak istoricheskii istochnik." In *Rossiiskoe gosudarstvo: istoki, sovremennost', perspektivy*, edited by Aleksandr Bessudnov. Vol. 2, 32–40. Lipetsk: Lipetskii gosudarstvennyi pedagogicheskii universitet, 2012.
Toker, Leona. "On the Eve of the Moratorium: The Representation of the Holocaust in Ilya Ehrenburg's Novel The Storm." In Grinberg, Marat, et al. *Representation of the Holocaust in the Soviet Literature and Film*. Search and Research 19, 37–58. Jerusalem: Yad Vashem, the International Institute for Holocaust Research, 2013.
Tonkin, Elizabeth. *Narrating our Past: The Social Construction of Oral History*. Cambridge: Cambridge University Press, 1994.
Trubina, Elena. "Uchast' vspominat': vektory issledovanii pamyati." In *Vlast' vremeni: sotsial'nye granitsy pamyati: sbornik statei*, edited by Valentina Yarskaya and Elena Yarskaya-Smirnova, 25–42. Moscow: OOO "Variant," 2011.
Tsvetkov, Valentin. *Na zemle predkov: stranitsy istorii ZAOPZ "Volya" i stanitsy Chelbasskoi*. Maykop: "Poligraf-Yug," 2012.
Tumanov, Valerii. *Shkol'nyi muzei*. Moscow: Tsentr detsko-yunosheskogo turizma i kraevedeniya, 2003.
Umansky, Andrej. "Metodika issledovanii, provodimykh tsentrom 'Yahad-In Unum' i ikh rezul'taty na primere polevykh rabot v gorode Motol', Belorussiya." In *Voina na unichtozhenie: natsistskaya politika genotsida na territorii vostochnoi Evropy*, edited by Aleksandr Dyukov and Olesya Orlenko, 462–470. Moscow: Fond "Istoricheskaya pamyat'," 2010.
Umansky, Andrej. "Geschichtsschreiber wider Willen? Einblick in den Quellen der 'Außerordentlichen Staatlichen Kommission' und der 'Zentralen Stelle.'" In *Bewusstes Erinnern und bewusstes Vergessen: Der juristische Umgang mit der Vergangenheit in den Ländern Mittel- und Osteuropas*, edited by Angelika Nußberger and Caroline von Gall, 347–374. Tübingen: Mohr Siebeck, 2011.

Ursu, Dmitrii. "Metodologicheskie problemy ustnoi istorii." In *Istochnikovedenie otechestvennoi istorii: sbornik statei*, edited by Vladimir Kuchkin (Moscow: Nauka, 1989), 3–32.
Vaksberg, Arkadii. *Iz ada v rai i obratno: evreiskii vopros po Leninu, Stalinu i Solzhenitsynu*. Moscow: "Olimp," 2003.
Varivoda, Natalia. Zaselenie Severnogo Kavkaza slavyanskim naseleniem v XVIII veke. *Istoricheskii vestnik. Nalchik* 3 (2006): 176–197.
Vasil'ev, Il'ya, and Svanidze, Nikolai. *Sobibor. Vozvrashchenie podviga Aleksandra Pecherskogo*. Moscow: Eksmo, 2018.
Vedder, Ulrike. *Das Testament als literarisches Dispositiv: kulturelle Praktiken des Erbes in der Literatur des 19. Jahrhunderts*. München: Fink, 2011.
Venkov, Andrei. *Donskoe kazachestvo v Grazhdanskoi voine (1918–1920)*. Rostov-on-Don: Izdatel'stvo Rostvskogo universiteta, 1992.
Vetter, Matthias. *Antisemiten und Bolschewiki: zum Verhältnis von Sowjetsystem und Judenfeindschaft 1917–1939*. Berlin: Metropol-Verlag, 1995.
Vilenskii, Semen, Gorbovitskii, Grigorii, and Terushkin, Leonid, ed. *Sobibor. Vosstanie v lagere smerti*. Moscow: Vozvrashchenie, 2010.
Voloshin, Viktor, and Ratnik, Valentina. *Vchera byla voina: Taganrog v gody nemetsko-fashistkoi okkupatsii (oktyabr' 1941 – avgust 1943 gg.)*. Taganrog: OOO "Lukomor'e," 2008.
Vyazemskii, Evgenii. "Pedagogicheskie podkhody k izucheniyu kursa 'Istoriya Kholokosta i sovremennost.'" *Methodist* 10 (2011): 12–17.
Vyazemskii, Evgenii. Seminar dlya rossiiskikh prepodavatelei po problemam Kholokosta. *Sovremennoe dopolnitel'noe professional'noe pedagogicheskoe obrazovanie* 2 (2015): 76–81.
Walker, Janet. *Trauma Cinema: Documenting Incest and the Holocaust*. Berkeley: University of California Press, 2005.
Wengler, Joannah Caborn, Hoffarth, Britta, and Kumiega, Lukasz, ed. *Verortungen des Dispositiv-Begriffs: analytische Einsätze zu Raum, Bildung, Politik*. Wiesbaden: Springer VS, 2013.
Wertsch, James V. "Collective Memory." In *Memory in Mind and Culture*, edited by Pascal Boyer and James V. Wertsch, 117–137. Cambridge: Cambridge University Press, 2009.
Winkel, Georg. "'Dispositif turn' und Foucaultsche Politikanalyse." In *Verortungen des Dispositiv-Begriffs: analytische Einsätze zu Raum, Bildung, Politik*, edited by Joannah Caborn Wengler, Britta Hoffarth, and Lukasz Kumiega, 167–198. Wiesbaden: Springer VS, 2013.
Winkler, Christina. "Rostov-on-Don 1942: A Little-Known Chapter of the Holocaust." *Holocaust and Genocide Studies* 30, no. 1 (2016): 105–130.
Winter, Jay. *Sites of Memory, Sites of Mourning: The Great War in European Cultural History*. Cambridge: Cambridge University Press, 1995.
Winter, Jay. "Remembrance and Redemption. A Social Interpretation of War Memorials." *Harvard Design Magazine* 9 (1999): 1–6.
Winter, Jay. "The Generation of Memory: Reflections on the 'Memory Boom' in Contemporary Historical Studies." *Archives and Social Studies: A Journal of Interdisciplinary Research* 1 (2007): 363–397.
Wittgenstein, Ludwig. *Philosophical Investigations*. Oxford: Blackwel, 1997.

Yerushalmi, Yosef Heyim. *Zakhor: Jewish History and Jewish Memory (The Samuel and Althea Stroum Lectures in Jewish Studies)*. Seattle: University of Washington Press. 1996.
Young, James E. *Writing and Rewriting the Holocaust: Narrative and the Consequences of Interpretation*. Bloomington: Indiana University Press, 1990.
Young, James E. *The Texture of Memory: Holocaust Memorials and Meaning*. New Haven: Yale University Press, 1993.
Young, James E. "Holocaust Museums in Germany, Poland, Israel, and the United States." In *Contemporary Responses to the Holocaust*, edited by Konrad Kwiet and Jürgen Matthäus, 249–274. Wesport: Praeger Publishers, 2004.
Young, James E. *Constructing Memory: Architectural Narratives of Holocaust Museums*. Bern: International Academic Publishers, 2013.
Yudkina, Anna. "'Pamyatnik bez pamyati': pervyi vechnyi ogon' v SSSR." *Neprikosnovennyi zapas*, 3 (2015): 112–134.
Zelizer, Barbie. *Remembering to Forget: Holocaust Memory through the Camera's Eye*. Chicago: University of Chicago Press, 1998.
Zeltser, Arkadi. "Differing Views among Red Army Personnel about the Nazi Mass Murder of Jews." *Kritika: Explorations in Russian and Eurasian History* 15, no. 3 (2014): 563–590.
Zeltser, Arkadi. "Pamyatniki Kholokostu v SSSR: na puti k sovremennoi pamyati." In *Kholokost: 70 let spustya: materialy mezhdunarodnogo foruma i 9-i mezhdunarodnoi konferentsii "Uroki Kholokosta i sovremennaya Rossiya,"* edited by Il'ya Al'tman, Igor' Kotler and Jürgen Zarusky, 141–145. Moscow: Tsentr "Kholokost," 2015.
Zeltser, Arkadi. *Unwelcome Memory: Holocaust Monuments in the Soviet Union*. Jerusalem: Yad Vashem studies, 2018.
Zheltikov, Valentin. *Ekonomicheskaya geografiya*. Rostov-on-Don: Feniks, 2001.
Zhuravlev, Evgenii. "Okkupatsionnaya politika fashistskoi Germanii na yuge Rossii (1941–1943 gg.): tseli, soderzhanie, prichiny krakha." *Nauchnaya mysl' Kavkaza* 1 (2001): 36–43.
Zhuravlev, Evgenii. "Voenno-politicheskii kollaboratsionizm na yuge Rossii v gody nemetsko-fashistkoi okkupatsii (1941–1943 gg.)." *Vestnik RUDN, seriya istoriya Rossii* 4 (2009): 20–34.
Zhuravlev, Sergei, ed. *Vklad istorikov v sokhranenie istoricheskoi pamyati o Velikoi Otechestvennoi voine: na materialakh komissii po istorii Velikoi Otechestvennoi voiny AN SSSR, 1941–1945 gg*. Moscow: Institut rossiiskoi akademii nauk, 2015.
Zubkov, Anatolii. *Kolokola pamyati: o tekh, kto pal za mir, za zhizn', za schast'e nashe*. Rostov-on-Don: Rostovskoe knizhnoe izdatel'stvo, 1985.
Zverev, Sergei, and Storozhev, Nikolai. "Vospitanie nenavisti k vragu v sovetskoi voinskoi kul'ture 30–40-kh gg. XX veka." *Vestnik Sankt-Peterburgskogo gosudarstvennogo instituta kul'tury* 3 (2014): 21–28.

Tables

Table 1: Administrative division and Jewish presence in the North Caucasus[1] before World War II

Administrative-territorial unit	Date of foundation[2]	Area ('000 km2)[3]	Total population[4]	Jewish population[5]	Jews as % of total population
Rostov Oblast	September 13, 1937	100.7	2,892,580	3,3024	1.14 %
Krasnodar Krai, including:	September 13, 1937	80.6	3,172,674	7,351	0.23 %
Adyghe AO	July 27, 1922	3.9	241,799	302	0.12 %
Stavropol Krai,[6] including:	March 13, 1937	102.4	1,950,887	7,791	0.39 %
Karachay AO	April 24, 1926	10.8	150,303	159	0.1 %
Cherkess AO	April 30, 1928	3.3	92,898	98	0.1 %
Kabardino-Balkarian ASSR	December 5, 1936	12.3	359,219	3,414	0.95 %
North Ossetian ASSR	December 5, 1936	6.2	329,205	1,714	0.52 %
Population of the RSFSR			109,397,463	956,599	5.62 %[7]

---

1 Excluding Dagestan ASSR, which was not part of the North Caucasus Krai, and Chechen-Ingush ASSR. These republics were not occupied by the Wehrmacht during World War II.
2 *SSSR. Administrativno-territorial'noe delenie soyuznykh respublik na 1 maya 1940 goda* (Moscow: Izdatel'stvo vedomostei verkhovnogo soveta RSFSR, 1940), 6–10.
3 Ibid.
4 According to the 1939 population census. *Russian State Archive of the Economy* [*Rossiiskii gosudarstvennyi arkhiv ekonomiki*], hereafter *RGAE*, f. 1562, op. 336, d. 966–1001; "Development table," in *GRAE*, f. 15A. National composition of the population in the republics, krais and oblasts in the USSR.
5 Ibid.
6 Stavropol Krai was called Ordzhonikidze Krai from 1937 to 1943.
7 Total Jewish population in the North Caucasus as a percentage of the total Jewish population of RSFSR. Estimated by the author.

Table 2: The North Caucasus under Nazi Rule[8]

Administrative-territorial unit	Period of occupation	Administrative center	Period of occupation of the administrative center
Rostov Oblast	October 17, 1941[9]– August 30, 1943[10]	Rostov-on-Don	November 21–29, 1941 (first occupation); July 24, 1942 – February 14, 1943 (second occupation)
Krasnodar Krai, including:	July 29, 1942 – October 9, 1943[11]	Krasnodar	August 9, 1942 – February 12, 1943
Adyghe AO	August 10, 1942 – February 18, 1943	Maykop	August 10, 1942 – January 29, 1943
Stavropol Krai including:	August 2, 1942 – January 23, 1943	Voroshilovsk[12]	August 3, 1942 – January 21, 1943
Karachay AO	August 14, 1942 – January 18, 1943	Mikoyan-Shakhar[13]	August 14, 1942 – January 18, 1943
Cherkess AO	August 11, 1942 – January 17, 1943	Cherkessk	August 11, 1942 – January 17, 1943
Kabardino-Balkarian ASSR	August 12, 1942 – January 11, 1943	Nalchik	October 28, 1942 – January 3, 1943
North Ossetian ASSR	End of November 1942 – January 1, 1943[14]	Ordzhonikidze	Not occupied

---

8 *Informatsionnyi byulleten' arkhivnogo otdela administratsii Krasnodarskogo kraya*, No. 1 (1995): 71–73; *TsDNIRO*, f. R-9, op. 4, d. 503, l. 1–87; Chernogorov Aleksandr, ed., *Krai nash Stavropol'e: ocherki istorii* (Stavropol: Shat-Gora, 1999), 323–358; Boiko, *Stavropol'e v Velikoi Otechestvennoi*, 467–476.
9 Taganrog, Anastasievskii and Fedorovskii districts were under Nazi rule for almost two years, from October 1941 to August 1943; the rest of Rostov Oblast was occupied from July 10, 1942.
10 Sholokhov district (formerly Veshenskii district) of Rostov Oblast was not occupied.
11 Lazarevskii, Tuapse, Adler and Gelendzhik Districts of Krasnodar Krai were not occupied.
12 In 1943 it was renamed Stavropol.
13 Since 1957 it is called Karachaevsk.
14 Only the northern and western parts of the Republic were occupied.

Table 3: Number of Holocaust victims in the North Caucasus[15]

Administrative-territorial unit	Minimum	Maximum
Rostov Oblast	22,000	26,800
Krasnodar Krai[16]	16,320[17]	22,000
Stavropol Krai[18]	19,700	35,000[19]
Kabardino-Balkarian ASSR	30	500[20]
North Ossetian ASSR	20[21]	3,300[22]
Total in the Region	58,070	87,600
Total in the RSFSR	119,210	140,350
Holocaust victims in the North Caucasus as a % of total victims in the RSFSR	48.7 %	62.4 %

---

15 Based on the data of Al'tman. Al'tman, *Zhertvy nenavisti*, 286. For numbers for each district of Krasnodar and Stavropol Krais and Kabardino-Balkarian ASSR, see: Feferman, *The Holocaust in the Crimea*, 512–522; Umansky, "L'extermination des Juifs."
16 Including Adyghe AO with 407 Holocaust victims. Umansky, "L'extermination des Juifs," 288, 356, 361.
17 Ibid, 369. Umansky provides the German data as well; 7,000 victims according to the indictment of Kurt Christmann of the prosecutor's office of Munich I district court of September 29, 1971. "Indictment StA München I 22 Js 202/61 vs. Dr. Christmann," in *BAL*, B 162/1196, Bl. 20.
18 Including Karachay AO and Cherkess AO with 1,400 Holocaust victims in total. Umansky, "L'extermination des Juifs,"562.
19 Polyan, *Mezhdu Aushvitsem*, 136.
20 Umansky, "L'extermination des Juifs," 592.
21 Ibid, 592.
22 The data by Al'tman is incorrect, since the mass killing of Mountain Jews in the villages of Bogdanovka and Menzhinskoe took place in the territory of Stavropol Krai.

Table 4: Summary data on the work of the regional commissions of the ChGK in the North Caucasus specifying the number of human losses for each region[23]

Administrative-territorial unit	Period of ChGK activity	Number of ChGK members	Area covered	Number of reports on human losses	Number of Nazi victims
Rostov Oblast	December 1942 – January 6, 1945	More than 14,500 members	63 districts, 12 cities	1,810[24]	182,061 "peaceful Soviet citizens"[25]
Krasnodar Krai	April 29, 1943 – May 15, 1944	10,300 Commissions, more than 32,000 members	73 districts, 8 cities	461[26]	61,540 "tortured or killed peaceful Soviet citizens and POWs"[27]
Stavropol Krai	December 1942 – June 1943	No data	41 districts, 7 cities	110	31,645 "peaceful Soviet citizens" and 277 POWs[28]

---

23 Author's calculations on the basis of the memoranda of the regional commissions of the ChGK "on accounting for damage and atrocities caused by the German fascist invaders." "ChGK reports for Kabardino-Balkarian ASSR, North Ossetian ASSR, Stavropol and Krasnodar Krais, and Rostov Oblast," in *GARF*, f. R-7021, op. 7 d. 109; op. 12 d. 70; op. 16, d. 586; op. 17, d. 311; op. 40, d. 841.
24 All together 95,602 records were sent to Moscow from Rostov Oblast. "ChGK report for Rostov Oblast," in *GARF*, f. R-7021, op. 40, d. 841, l. 39.
25 Including "46,289 killed people, 42,509 tortured, 5 poisoned, 139 hanged, 12 burnt, 210 raped, 7,737 killed by bombing, 915 beaten, 125 gone missing, 84,030 became forced labour." "ChGK report for Rostov Oblast," in *GARF*, f. R-7021, op. 40, d. 841, l. 42. The ethnicity and social status of the Nazi victims is not mentioned in the memorandum.
26 All together 39,025 records were sent to Moscow from Krasnodar Krai. "ChGK report for Krasnodar Krai," in *GARF*, f. R-7021, op. 16, d. 586, l. 30–31.
27 Including "48,560 peaceful Soviet citizens and 6,570 POWs." Also, 48,464 (or 31,771 according to other data) Soviet citizens became forced labour. "ChGK report for Krasnodar Krai," in *GARF*, f. R-7021, op. 16, d. 586, l. 21, 31.
28 Also, 990 people became forced labour. "ChGK report for Stavropol Krai," in *GARF*, f. R-7021, op. 17, d. 311, l. 4.

Table 4 (continued)

Administrative-territorial unit	Period of ChGK activity	Number of ChGK members	Area covered	Number of reports on human losses	Number of Nazi victims
Kabardino-Balkarian ASSR	June – December 1943	No data	15 districts, 1 city	111	4,241[29]
North Ossetian ASSR	February – December 1943	9 leaders	8 districts	52	128 "peaceful Soviet citizens"[30]

---

[29] Including "2,188 peaceful citizens and 2,053 POWs." "ChGK report for Kabardino-Balkarian ASSR," in *GARF*, f. R-7021, op. 7, d. 109, l. 15.

[30] Also, 150 people were raped and tortured; 1,323 were killed by bombing, 781 were wounded, and 1,657 became forced labour. "ChGK reports for North Ossetian ASSR," in *GARF*, f. R-7021, op. 12, d. 70, l. 2.

Table 5: Definition of Nazi victim groups in Soviet and Russian history textbooks.

Textbook	Soviet people / "peaceful Soviet citizens"	Communists and Komsomol members	Partisans / underground leaders	Soviet POWs	Sex / Age (women, children and old people)	Local citizens / civilians / peaceful citizens	Russians / Russian folk	Jews	"Gypsies" / handicapped people / other "inferior peoples"
Soviet period									
Pankratova, ed., 11th grade (1952)			*						
Kim, ed., 10th grade (1980)	*	*	*	*	*				
Transitional period (the 1990s)									
Danilov and Kosulina, 9th grade (1997)	*		*				* (1)		
Shestakov et al., 11th grade (1995)	*								
Ostrovskii and Utkin, 11th grade (1995)				*	*	*	*	* (1)	
Levandovskii and Shchetinov, 11th grade (1997)						*	* (1)	* (1)	

	Modern Russian period (since the 2000s)			
Danilov and Kosulina, 9th grade (2013)	*	*	*	*	*
Zagladin, ed, 9th grade (2013)		*	*		
Shestakov et al., 9th grade (2011)	*	*	*	* (1)	*
Volobuev et al., 11th grade (2013)	*		*		*
Zagladin, ed, 11th grade (2007)		*	*		*
Levandovskii et al., 11th grade (2013)		*	*	* (1)	* (1)
Shestakov et al., 11th grade (2012)	*	*			

(continued)

Table 5 (continued)

Textbook	Soviet people / "peaceful Soviet citizens"	Communists and Komsomol members	Partisans / underground leaders	Soviet POWs	Sex / Age (women, children and old people)	Local citizens / civilians / peaceful citizens	Russians / Russian folk	Jews	"Gypsies" / handicapped people / other "inferior peoples"
Regional history textbooks (since the 1990s) (2)									
Bekaldiev, Istoriya Kabardino-Balkarii, 8–9th grades (2013)	*	*	*	*	*				
Krugov, Stavropol'skii Krai, 10–11th grades (2006)	*			*		*		*	*
Zaytsev, ed. Kubanovedenie, 9th grade (2012)	*			*	*	*			
Achmiz, ed. Istoriya Adygei, 9th grade (2002)			*	*	*	*			
Kislitsyn, Istoriya Donskogo Kraya, 9th grade (2004)				*	*	*		*	

(1) – According to Generalplan Ost, that is the planned but not the actual number of victims is given in the textbook.
(2) – I studied only those regional history textbooks, in which the section about the Nazi occupation of the region is presented.

Table 6: Statistical data on oral history interviews studied[31]

Narrator's year of birth	USCSF interviews		Author's interviews	
	Male	Female	Male	Female
Before 1920	–	12	–	1
1920–1929	14	17	5	5
1930–1935	12	17	4	7
1936–1939	2	4	2	4
After 1940	1	1	–	–
Total	29	51	11	17
	80		28	

---

31 Author's calculations. Only interviews with Holocaust survivors dealing with the survival in the North Caucasus are counted.

List of Abbreviations

Archives, Research Centers and International Organisations

BAL	Bundesarchiv Ludwigsburg, Germany
BYHMC	Babi Yar Holocaust Memorial Center in Kyiv, Ukraine
EVZ	German foundation "Remembrance, Responsibility, and Future"
Felitsyn Museum	Krasnodar State Historical and Archaeological Museum-Reserve named after Evgenii Felitsyn [*Krasnodarskii gosudarstvennyi istoriko-arkheologicheskii muzei-zapovednik imeni Evgeniya Felitsyna*] in Krasnodar, Russia
FJCR	Federation of Jewish Communities in Russia [*Federatsiya evreiskikh obshchin Rossii*]
GAKK	State Archive of Krasnodar Krai [*Gosudarstvennyi arkhiv Krasnodarskogo kraya*] in Krasnodar, Russia
GANISK	State Archive of Contemporary History of Stavropol Krai [*Gosudarstvennyi arkhiv noveishei istorii Stavropol'skogo kraya*], in Stavropol, Russia
GARF	State Archive of the Russian Federation [*Gosudarstvennyi arkhiv Rossiiskoi Federatsii*] in Moscow, Russia
GARO	State Archive of Rostov Oblast [*Gosudarstvennyi arkhiv Rostovskoi oblasti*] in Rostov-on-Don, Russia
GASK	State Archive of Stavropol Krai [*Gosudarstvennyi arkhiv Stavropol'skogo kraya*] in Stavropol, Russia
Holocaust Center (and Foundation)	Russian research and educational Holocaust Center (1992) and Foundation (1997) in Moscow, Russia
NA of the IRI RAN	Scientific Archive of the Institute of Russian History at the Russian Academy of Science [*Nauchnyi arkhiv Instituta rossiiskoi istorii Rossiiskoi akademii nauk*], in Moscow, Russia
OECD	Organisation for Economic Cooperation and Development
OHD	Oral History Division of Avraham Harman Institute of Contemporary Jewry at the Hebrew University in Jerusalem, Israel
OSCE	Organisation for Security and Cooperation in Europe
RAS	Russian Academy of Sciences [*Rossiiskaya akademiya nauk*] in Moscow, Russia
RGAE	Russian State archive of the Economy [*Rossiiskii gosudarstvennyi arkhiv ekonomiki*] in Moscow, Russia
RGAKFD	Russian State Film and Photo Archive [*Rossiiskii gosudarstvennyi arhiv kinofotodokumentov*] in Krasnogorsk, Russia
RGASPI	Russian State Archive of Socio-Political History [*Rossiiskii gosudarstvennyi arkhiv sotsial'no-politicheskoi istorii*] in Moscow, Russia
RJC	Russian Jewish Congress

SGMZ	Grigorii Prozritelev and Georgii Prave Stavropol State Historical-Cultural and Natural-Landscape Museum-Reserve [*Stavropol'skii gosudarstvennyi istoriko-kul'turnyi i prirodno-landshaftnyi muzei-zapovednik imeni Grigoriya Prozriteleva i Georgiya Prave*] in Stavropol, Russia
SKIRO PK and PRO	Stavropol Krai Institute of Education Development, Professional Development, and Retraining of Teachers [*Stavropol'skii kraevoi institut razvitiya obrazovaniya, povysheniya kvalifikatsii i perepodgotovki rabotnikov obrazovaniya*] in Stavropol, Russia
Society "Memorial"	International Historical and Educational, Human Rights, and Charitable Society "Memorial" [*Mezhdunarodnoe istoriko-prosvetitel'skoe, blagotvoritel'noe i pravozaschitnoe obshchestvo "Memorial"*]
StA München	Staatsarchiv München, Germany
TsAMO	Central Archive of the Russian Ministry of Defence [*Tsentral'nyi arkhiv Ministerstva oborony rossiiskoi federatsii*] in Podolsk, Russia
TsDNIKK	Center for Documentation of Contemporary History of Krasnodar Krai [*Tsentr dokumentatsii noveishei istorii Krasnodarskogo kraya*] in Krasnodar, Russia
TsDNIRO	Center for Documentation of Contemporary History of Rostov Oblast [*Tsentr dokumentatsii noveishei istorii Rostovskoi oblasti*] in Rostov-on-Don, Russia
UNESCO	United Nations Educational, Scientific and Cultural Organisation
USC	University of South California in Los Angeles, USA
USHMM	United States Holocaust Memorial Museum in Washington D.C., USA
USHMMA	United States Holocaust Memorial Museum Archive
VHA USCSF	Visual History Archive of the University of South California Shoah Foundation
VOOPIiK	All-Russian Society for Protection of Historical and Cultural Monuments [*Vserossiiskoe obshchestvo okhrany pamyatnikov istorii i kul'tury*]
YVA	Yad Vashem Archive in Jerusalem, Israel
"Small Academy"	Krasnodar Municipal Institution of Additional Education for Children "Small Academy" [*"Malaya Akademiya"*] in Krasnodar, Russia

Soviet terms

AO	Autonomous Oblast
ASSR	Autonomous Soviet Socialist Republic
Chekist	An officer of the Soviet secret police, the "Cheka" or "VChK," All-Russian Emergency Commission [*Vserossiiskaya chrezvychainaya komissiya*], later applied to officers of the Soviet security services in general.
ChGK	"Extraordinary State Commission for the Establishment and Investigation of the Crimes of the Fascist German Invaders and their Accomplices, and of the Damage They Caused to Citizens, Collective Farms, Public Organisations, State Enterprises, and Institutions of the USSR" [*Chrezvychainaya gosudarstvennaya komissiya*]

List of Abbreviations — 353

CPSU	Communist Party of the Soviet Union [*Kommunisticheskaya partiya Sovetskogo Soyuza*]
FSB	Federal Security Service of the Russian Federation [*Federal'naya sluzhba bezopasnosti*]
GKO	State Defence Committee of the USSR [*Gosudarstvennyi komitet oborony*]
GlavPU KA	The Main Political Administration of the Workers' and Peasants' Red Army [*Glavnoe politicheskoe upravlenie raboche-krest'yanskoi Krasnoi armii*]
JAC	Jewish Anti-Fascist Committee [*Evreiskii antifashistkii komitet*]
KGB	Committee for State Security of the USSR [*Komitet gosudarstvennoi bezopasnosti*]
MGB	Ministry of State Security of the USSR [*Ministerstvo gosudarstvennoi bezopasnosti*]
Mints Commission	The Commission on the History of the Great Patriotic War of the Academy of Sciences of the USSR
NKVD	People's Commissariat for Internal Affairs of the USSR [*Narodnyi komissariat vnutrennikh del*]
OGIZ	Union of the State Book and Magazine Publishers [*Ob"edinenie gosudarstvennykh knizhno-zhurnal'nykh izdatel'stv*]
ORT	Association for the Promotion of Skilled Trades [*Obshchestvo remeslenogo truda*]
OZET	Public Society for Settling Toiling Jews on the Land in the USSR during 1925–1938 [*Obshchestvo zemleustroistva evreiskikh trudyashchikhsya*]
RSFSR	Russian Soviet Federative Socialist Republic
SMERSH	"Death to Spies" [*Smert' shpionam*], an umbrella organisation for three independent counter-intelligence agencies in the Red Army formed in late 1942 or even earlier, but officially announced only on 14 April 1943
SNK	Council of People's Commissars of the USSR [*Soviet narodnykh kommissarov*], or Sovnarkom
Sovinform- buro	Soviet Informational Bureau [*Sovetskoe informatsionnoe byuro*]
TsK	Central Committee [*Tsentralnyi Komitet*]
VKP(b)	All-Union Communist Party Bolsheviks [*Vsesoyuznaya kommunisticheskaya partiya (bol'shevikov)*]

Other Abbreviations

AOK	Armeeoberkomando (Army Command of the Wehrmacht)
CDA	Critical discourse analysis
EK	Einsatzkommando, a sub-group of the Einsatzgruppen – SS death squads of Nazi Germany
OKW	Oberkommando der Wehrmacht (High Command of the Armed Forces)
POW	Prisoner of war
SK	Sonderkommando, a sub-group of the Einsatzgruppen – SS death squads of Nazi Germany

Index

Adler 342
Aleksandriiskaya village 79
Anastasievskii district 46, 342
Armavir 49, 50, 52, 191
Arzgir (village and district) 123, 140, 170, 201, 203, 204, 207, 237, 242, 287–290, 296

Balabanovka 55, 56, 113
Belaya Glina 113
Belorechensk 230
Beltsy (also Bălți) 1
Berlin 186, 187, 204, 205, 234
Beslenei village 125, 171–173, 207, 236, 243
Blagodarnenskii district 123
Bogdanovka village 41, 52, 126, 127, 140, 178
Bratskii village 145, 298

Canada 248
Chelbasskaya village 118, 121
Cherkessk 101, 232, 342

Dnepropetrovsk 75
Dortmund 205

Essentuki 49, 50, 52, 66, 68, 154, 281

Fedorovskii district 46, 342
Feodosia 254

Ganshtakovka farm, also Menzhinskii farm 41, 52, 169
Gelendzhik 342
Gera 205
Glasgow 205
Gulkevichi 113, 277

Hungary 286

Ipatovo 101

Jerusalem 186, 233, 258

Kaliningrad 141, 249
Kanevskaya village 196
Karachay-Chreckess Republic (also Karachay AO and Cherkess AO, Karachay-Cherkess AO) 30, 34, 47, 52, 89, 125, 173, 174, 225, 229, 236, 257, 276, 341–343
Kerch 162
Kharkov 86, 88, 165
Kirov 249
Kishinev 170
Kislovodsk 50, 52, 67, 86, 92–97, 101, 113, 116, 124, 126, 141, 152, 154, 158, 196, 232, 241, 243, 257, 282, 283, 291
Krasnodar 42, 49, 51, 53, 54, 86, 88, 101, 107, 108, 110, 112, 116, 124, 142, 143, 160, 161, 162, 163, 168, 169, 188, 194, 195, 197, 200, 204, 226, 232, 238, 243, 245, 254, 264, 273, 274, 300, 342
Krasnodar Krai 1, 6, 30, 33, 34, 40, 42–50, 59, 71, 77, 86, 101, 113, 116, 118, 124, 125, 156, 164, 195, 196, 200, 224, 226, 229, 230, 234–236, 248, 276, 300, 341–344
Kryvyi Rih 55
Kurskii district 52
Kuzhorskaya village 116–118
Kyiv (also Kiev) 12, 61, 115, 115, 156, 174, 208, 227, 249

Labinskaya village 50, 77
Ladozhskaya village 59, 113, 114, 244
Lazarevskii district 342
Leningrad 43, 44, 61, 79, 121, 125, 171–173, 195, 207, 236, 243, 265
Lodz (also Łódź) 189
Lviv 118

Mariupol 163
Maykop 116, 243, 342
Meshkovskaya village 50
Mikhizeeva Polyana (also Rabochii poselok) 195

Mikoyan-Shakhar (also Karachaevsk) 52, 53, 342
Millerovskii district 140
Mineralnye Vody (city and district) 34, 35, 67, 68, 92, 101–103, 122, 139, 141, 158, 201, 232, 242, 282, 283
Minsk 104, 115, 192
Moscow 45, 82, 172, 189–194, 206, 209, 211, 233, 240, 244, 260, 268, 291, 294
Mostovskoi village 113, 195
Mozdok (city and district) 36, 59

Nalchik 52, 158, 171, 283, 342
Nevel 104
New York 126, 127
North Caucasus 7, 21–24, 27–48, 50, 52–54, 57, 62, 63, 66, 71, 73, 81, 83–85, 89, 92, 96–101, 104, 105, 107, 112, 113, 115, 116, 118, 120–122, 125, 126, 128, 138, 142–145, 145–150, 152, 155, 158, 159, 161, 168, 170–172, 178, 180, 181, 183, 184, 187–189, 191, 193, 196, 199, 201, 205–208, 217, 227–229, 232, 234, 235, 239, 246, 250–285, 286–289, 293, 298–301
Novoderevyankovskaya village 243
Novominskaya village 71, 113, 121, 138
Novopokrovskii district 118, 300
Novorossiysk 34, 36, 40, 42, 47, 49, 59, 67, 112, 119, 155, 163, 246
Novoselitskoe village 101
Novozybkov 141

Ol'khovyi Rog 113, 140
Ordzhonikidze (also Vladikavkaz) 36, 232, 341, 342
Orel 141

Paris 233
Petropavlovskaya village 52, 277
Poland 16, 42, 52, 133, 154, 163, 186, 189, 253, 275
Poltava 169
Pskov oblast 104, 118
Pyatigorsk 52, 56, 59, 60, 67, 101, 166, 226, 270, 283

Republic of Adygea (also Adyghe AO) 30, 34, 37, 47, 117, 229, 341–343
Republic of Belarus (also Belorussian SSR) 43, 52, 90, 101, 104, 113, 147, 163, 179, 188, 189, 191, 228, 250, 251, 259, 269, 275
Republic of Chechnya (also Chechen-Ingush ASSR) 22, 30, 32, 33, 36, 39, 172, 341
Republic of Dagestan 23, 30, 32, 33, 36, 39, 40, 41, 341
Republic of Ingushetia 30
Republic of Kabardino-Balkaria (also Kabardino-Balkarian ASSR) 30, 34, 37, 41, 47, 52, 126, 224, 225, 229, 245, 276, 341–343, 345
Republic of North Ossetia-Alania (also North-Ossetian ASSR) 30, 34, 36, 47, 224, 229, 341–345
Riga 115
Rivne 192
Romania 42, 275
Rostov Oblast 30–34, 36, 39, 40, 42, 45–51, 55, 56, 101, 140, 178, 205, 224, 227, 229, 234–236, 276, 341–344
Rostov-on-Don 39, 40, 42, 46, 49, 54, 58, 79, 83, 101, 107, 108, 110, 121, 122, 128, 129, 131–137, 141–143, 154–156, 158, 175–179, 188, 192, 199, 201, 202, 204, 205, 207, 208, 227, 232, 235, 282, 297, 299, 342
Russian Empire 22, 31, 32, 39, 40, 41, 53, 190, 192, 224, 288
Russian Federation (also RSFSR) 7, 13, 23, 30, 33, 43, 54, 86, 88, 128, 132, 154, 192, 201, 211, 216, 219, 223, 229, 230, 292, 299, 341, 343

Shakhty 48
Sholokhov district 342
Sochi 35
Staromarevka village 101
Stavropol (also Voroshilovsk) 29, 50, 75, 76, 104, 105, 107, 109, 110, 118, 119, 142, 143, 154, 165, 182–184, 204, 205, 232, 255, 287, 342
Stavropol Krai (also Ordzhonikidze Krai) 30, 32, 33, 34, 40, 41, 44–46, 49–52, 59,

67, 70, 86, 101, 116, 121, 123, 125, 139, 142, 156, 167, 181, 182, 224, 226, 227, 229, 234–236, 242, 243, 259, 269, 276, 270, 300, 342–344
Stepnoe village 59, 116, 121

Taganrog 39, 40, 46, 50, 101, 107, 116, 117, 121, 151, 163, 188, 202, 236, 255, 279, 280, 342
Tatsinskii district 55
Teberda 86, 172, 174
Ternopil 118
Tikhoretsk 50
Trunovskii district 123
Tuapse 232, 342
Turkey 35, 42

Udobnaya village 113

Ukraine (also Ukrainian SSR) 12, 34, 43, 55, 68, 87, 90, 98, 101, 113, 121, 147, 179, 188, 191, 208, 228, 259, 269, 275, 277, 286
Ust-Labinsk (city and district) 1–6, 24–25, 101, 113, 145, 197, 198, 237, 244, 275, 298

Vilnius 115, 253
Vitebsk 189
Vtoraya Zmeevka village 100, 134, 274

Warsaw 189, 255

Yeysk 163

Zaporozhye 55, 56
Zheleznovodsk 52, 67
Zhlobin 104
Zunda village 170